ADVANCES IN THERMAL MODELING
OF ELECTRONIC COMPONENTS
AND SYSTEMS

ADVANCES IN THERMAL MODELING
OF ELECTRONIC COMPONENTS AND SYSTEMS

Bar-Cohen, Kraus: Advances in Thermal Modeling of Electronic Components and Systems, Volume 1

ADVANCES IN THERMAL MODELING OF ELECTRONIC COMPONENTS AND SYSTEMS

Volume 1

Avram Bar-Cohen
Corporate Research and Engineering
Control Data Corporation
Minneapolis, Minnesota

Allan D. Kraus
Department of Electrical Engineering
Naval Postgraduate School
Monterey, California

Hewlett-Packard Company
Waltham Division

 HEMISPHERE PUBLISHING CORPORATION
A member of the Taylor & Francis Group

New York   Washington   Philadelphia   London

TK
7870.25
A38
v./

ADVANCES IN THERMAL MODELING OF ELECTRONIC COMPONENTS AND SYSTEMS: Volume 1

Cover design by Debra Eubanks Riffe.

1 2 3 4 5 6 7 8 9 0   B R B R   8 9 8

ISBN 0-89116-689-0
ISSN 0897-7453

CONTENTS

PREFACE

The 1980s have witnessed an explosive growth in the R& D efforts devoted to thermal phenomena in electronic systems. The incessant drive toward higher gate densities and greater functional capability have led to near-order-of-magnitude increases in heat flux and heat density and have spawned numerous studies of thermal control techniques for single chip and milticip packages in industrial, government and university laboratories. During this same period, the scope of thermal R& D has been dramatically increased by the growing recognition of the MTBF limitations imposed by thermally-induced failures in crystal growth and in the fabrication and attachment of chips and chip packages. The need for thermal modeling and thermal control of peripheral equipment (disk drives, tape drives, printers) has added a new dimension to the interaction between the electronics industry and the thermal community.

These developments have set the stage for the introduction of an annual publication offering an annotated bibliography and state-of-the-art reviews of research, analysis and modeling of Thermal Phenomena in Electronic Systems. The first volume focuses primarily on heat transfer within and from chip packages and opens with an overview of available technology and the primary research thrusts in the Thermal Management of Electronic Equipment by Dr. W. Nakayama of the Mechanical Engineering Research Laboratory of Hitachi, Ltd. In Chapter 2, Professors M. M. Yovanovich of the Department of Mechanical Engineering at Waterloo University and V. W. Antonetti of the Department of Mechanical Engineering at Manhattan College explore the Application of Thermal Contact Resistance Theory to Electronic Packages. Chapter 3, by Professors R. J. Moffat of the Mechanical Engineering Department of Stanford and A. Ortega of the Mechanical Engineering Department at the University of Arizona, examine Direct Air-Cooling of Electronic Components and in Chapter 4 the Application of Heat Pipes to Electronic Cooling is reviewed by Professors

P. J. Marto of the Mechanical Engineering Department at the Naval Postgraduate School and G. P. Peterson of the Mechanical Engineering Department at Texas A& M. A review and presentation of theoretical relations for Thermal Stress Failures in Microelectronic Components, authored by Dr. E. Suhir of the AT& T Bell Laboratories at Murray Hill, appears in Chapter 5. The volume closes with an annotated 1986 Bibliography of Heat Transfer in Electronic Equipment prepared by R. E. Simons of IBM at Poughkeepsie. The Bibliography contains nearly 100 titles, grouped in 10 different categories.

It is the editors' hope and expectation that this volume will be especially valuable to:

- the project and/or department manager in the electronics industry, in need of a conceptual understanding of thermal control trends in high-performance systems and thermally-induced failures in electronic packages;

- the electrical and mechanical packaging engineer in the electronics industry in need of an update on recent developments in thermal control and the analysis of thermal stress in electronic packages;

- the thermal consultant in private practice or in a major development laboratory, in need of a critical review of the recent literature and further insight into thermal limitations on the reliability and performance of electronic systems; and to

- academic and/or government laboratory research, in need of a clear understanding of the state-of-the-art in thermal control and thermally-induced failures of electronic packages, prior to commencing a new research effort.

This volume could not have been completed in so short a time without the dedicated effort of the authors and the production staff. Special thanks are due Pat and Jim Allen of Allen Computype, and we are always happy to work with Florence Padgett of Hemisphere.

Avram Bar-Cohen
Allan D. Kraus

Chapter 1

THERMAL MANAGEMENT OF ELECTRONIC EQUIPMENT: A REVIEW OF TECHNOLOGY AND RESEARCH TOPICS

Wataru Nakayama
Mechanical Engineering Research Laboratory,
Hitachi, Ltd,
502 Kandatsu, Tsuchiura-shi, Ibaraki-ken 300, Japan

Increasing miniaturization of microcircuits on chips of increasing size and new schemes of electrical connection, such as flip-chip bonding and surface mounting, are setting more demanding criteria regarding the thermal field within electronic equipment. While the search for a solution to meet a set of prescribed design criteria is becoming more

Adapted from Applied Mechanical Review, Volume 39, Number 12, December 1986.

complex, the body of available data needed to perform such a search is quite small. This article describes the two primary functions to be implemented by electronic heat transfer research: the definition of thermal design criteria and the establishment of a thermal packaging database. Examples of actual designs of packages are drawn from recent publications to illustrate the points of technical importance. The examples are packages of DRAM chips, flat-leaded packages of logic chips, and modules with dismountable heat sinks. These examples are used to address thermal stress problems, the problems of fin design, and thermal interface management, respectively. In the section on natural convection cooling, the effects of various factors on the uncertainties pertaining to heat transfer coefficient are assessed in light of the correlations proposed in the current literature. The section on forced convection cooling deals with the problem of heat transfer from an array of packages in a parallel-plate channel. The next section is devoted to the research topics of nucleate boiling heat transfer enhancement, from the surface of a small component, and microchannel cooling. The final section deals with a review of the art of thermal management of large scale computers (Japanese).

1   INTRODUCTION

Heat Transfer engineering is playing an increasingly important role in the advancement of electronics technology. For large-scale computers, the heat removal from the chips now ranks among the major technical problems that need to be solved to achieve higher data processing speeds. As the electronic devices performing a variety of functions find their way into factories, transportation vehicles, offices, and homes, constituting relatively hostile environments, they will confront a multitude of heat transfer problems that will affect their reliability.

Extensive research is being conducted on new materials and structures to house chips of an increasing number of input/output terminals in a package, to shorten the interconnection distances among various levels of electronic circuits, to improve the electrical characteristics, such as impedance to the transmission of high frequency signals, and to reduce the cost of fabrication. Once employed in actual devices, these new schemes of packaging and interconnection, almost without exception, will require more stringent control of the thermal environment. Furthermore, it is to be noted that the heat transfer problems in electronic equipment are strongly coupled with considerations of electronic performance. In most devices, the geometry of heat conduction paths and coolant paths is dictated by the arrangement of the electronic circuits. Accordingly, there arises an increasing need for the understanding of heat transfer processes within the equipment. This is reflected in recent publications which serve the purpose of providing the designer the methods of analysis, and the relevant data. The reader is referred, for example, to Kraus and Bar-Cohen (1983). In addition, one may observe the rapid growth in the number of research papers, symposia, and short courses devoted to the engineering of thermal control systems for electronic equipment.

A search of the literature, in an attempt to extract any organized collection of applied mechanics problems, is by no means an easy task. Moreover, the majority of heat transfer problems have not as yet been solved to a satisfactory degree due to the interrelationship of the many geometrical and operative parameters. Under these circumstances, this article cannot provide a neat summary of what has been done and what appears to lie ahead in the area of fundamental heat transfer research related to electronic equipment cooling, as would customarily be expected in a review article. Rather, this article examines several categories of electronic packaging and, in each, attempts to define the relevant research topics, provides an evaluation of the applicability of existing heat transfer correlations to thermal design, and suggests future trends in this field. Not all these items are expounded on in every topical area; for example, in the section on chip packaging, the design examples are used to illustrate the research needs, while in the section on advanced cooling schemes, the emphasis is naturally placed on the fundamental aspects of the thermal problem.

The body of literature is very large, and most of the reports are written from the viewpoint of equipment developers. Here, there is little attempt to provide a comprehensive reference list; the reader is referred to the bibliography of Antonetti and Simons (1985). Moreover, no list and exposition of research topics can be comprehensive within a reasonable length. Thus, this article presents the author's personal views on selected subjects to highlight topics that will require the attention of heat transfer researchers.

The original task was to provide a review of Japanese research and technology, the active developments of which are implied by the dramatic growth of Japanese commercial products. This has proved to be a very difficult undertaking. Although it is possible to list various concepts that materialized in Japanese products, these concepts must be remolded into topics suitable for presentation here. The attempt to perform such a remolding encounters enormous difficulty due to the lack of detailed technical data in the open literature. However, the illustrative examples are drawn mainly from the published designs of Japanese works. This, of course, does not infer a lack of knowledge of a number of important research works done in other countries.

With regard to Japanese technology, most of the latest technical information first appears in commercial journals in the Japanese language. It takes a considerable amount of time for their English versions to become available to the public, and there are versions scattered in various conference proceedings and journals published by IEEE, IEPS, and ASME. This, indeed, is quite inconvenient and, inevitably, frequent reference is made to some commercial Japanese journals as the source of information.

Following this introduction, the subsequent sections are devoted to the role of heat transfer engineering in the electronics technology (Section 2), packaging LSI chips (Section 3), natural convection cooling (Section 4), forced convection cooling (Section 5), advanced schemes of cooling (Section 6) and the thermal management of large scale computers (Section 7).

2 THE ROLE OF HEAT TRANSFER ENGINEERING IN THE DESIGN OF ELECTRONIC EQUIPMENT

2.1 Design Process

Figure 1.1 shows the four levels of the structure of an electronic computer: the chip, the package, the printed wiring board (PWB), and the system. The chip is a rectangular slice of single crystal

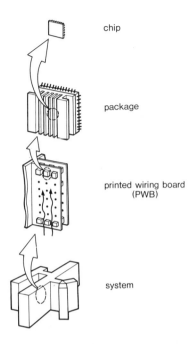

chip

package

printed wiring board
(PWB)

system

Figure 1.1: Structural levels of an electronic computer.

silicon where microscopic patterns of electronic circuits are provided
through a number of thermal, chemical, mass transfer, optical and
mechanical processes. A chip is housed in a package whose primary
function is to seal the chip from the atmosphere. The package contains the electrical leads for the pulsed signals to be transmitted in
and out of the package. A device to enhance heat transfer from the
package to the coolant, such as a finned heat sink, is attached to
a package that possesses a large power dissipation. A package may
contain more than one chip, and the term "module" is sometimes
used for the package which contains more than several chips. Packages are mounted on the PWB where layers of conductor networks
are fabricated to connect the different packages electrically. The system is composed of the PWB's, mutually connected by wiring, the
power supply, the coolant moving device (a fan or pump), and the
peripherals such as memory storage disks and printers.

The goal of thermal design is to limit the temperatures of the
chips to a tolerable range and the management of thermal environment at the four levels of Fig. 1.1 has to be coordinated to achieve

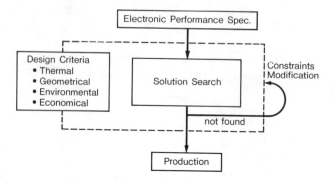

Figure 1.2: Work paths in thermal packaging engineering.

this design goal. While Fig 1.1 shows building blocks or structural levels of a large-scale computer as an illustration, any electronic system may have equivalent structural levels. In many cases, a system is the one which performs a function of mechanical control or process control, and often the thermal environment for the PWB's and other packages is governed by such external heat sources as internal combustion engines.

The task in thermal design is to find a solution which meets the requirements set by the electronic performance specifications and the design criteria (Fig. 1.2). In case a solution is not found within a given time span, some of the criteria have to be re-examined to see if a slight modification can produce a realizable solution. The search for a solution can be pursued in two stages: the planning phase and the design phase. Figure 1.3 illustrates the main steps and the factors in the solution search, where the descriptions related to electronics engineering are encircled by a broken line, those to thermal engineering by double lines, and those to production engineering by a wavy line.

For the purpose of illustration, we shall discuss the case where the specification required is an increase in speed of data processing. Given such a goal, there are basically two ways to achieve its satisfaction: an increased number of transistor junctions on a chip coupled with an increased frequency of the clock pulse and the reduction of connection distance to decrease the transmission time of the signal. Implementation of either or both of these approaches leads to an increase in power dissipation at the chip, board, and system levels.

Given the rates of heat removal at the chip, package, board, and system levels, the examination is made on the basis of the types of cooling systems. These include the kind of coolant, the candidate schemes of enhancement measures for heat transfer at the package and board levels, and the schemes of coolant circulation at the board and system levels. The geometric and operative parameters, such as the size of the cooling devices and the flow rate of the coolant, should fall in certain ranges of permissible values. By referring to the design criteria and the constraints imposed by the bounds on the electrical connection distances, and also by performing approximate heat transfer analyses, the ranges of parameter values can be delineated for each candidate scheme.

At the end of the planning phase, some of the schemes are weeded out, and the specific goals for the thermal design of the selected schemes are tentatively set. These goals may be defined in terms of

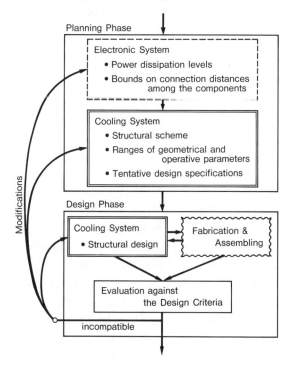

Figure 1.3: Solution search loop.

such parameters as the overall size of the cooling device, the flow rate, the exit temperature of the coolant and the required heat transfer coefficient. The influential factors in the selection of the schemes are: the extent of deviation from the current design, the modifications required on the existing manufacturing facility, and the availability of external suppliers for materials and components. These factors reflect the importance of proven reliability of the design and the cost of manufacturing.

The detailed design of the hardware then follows. At this stage, the interaction with the manufacturing sector is indispensable. Manufacturing of prototype devices, evaluation of their performance, and design modification must be repeated many times as the task of cooling equipment becomes more complex. The ultimate evaluation of the selected cooling strategy is made at the completion of the design phase when the cost of fabrication and the reliability of the cooling system can be predicted with a sufficient degree of accuracy. It may happen that the design criteria are not satisfied by all of the candidate cooling schemes. In this case, further modification of the design or the selection of another cooling strategy must be done. In some cases, one may be required to go back to the level of electronic system planning to determine if a modification of the power dissipation level can lead to a solution which satisfies the prescribed design criteria.

The foregoing is a somewhat streamlined description. In actual design processes, the work steps are often not so clearly defined and a decision regarding the selection of a cooling scheme or a structural design is not made with all the necessary data at hand. It is to be noted from Fig. 1.3 that the search for the proper cooling system is cross-disciplinary in nature. A specification for the optimum thermal design requires a blend of talents within the electronics, heat transfer, materials, and production engineering disciplines.

The thermal design criteria consist of two subgroups: operative and circumstantial (Table 1.1). The operative criteria set the thermal and environmental bounds to be observed during the operation of the equipment. The temperature of a transistor junction, T_j, must be held within a certain range set by the maximum temperature and the range of junction temperature differences among the components (ΔT_j). The purpose here is to secure a required life of the chip and to make the temperature-dependent characteristics of elec-

Table 1.1: Thermal design criteria.

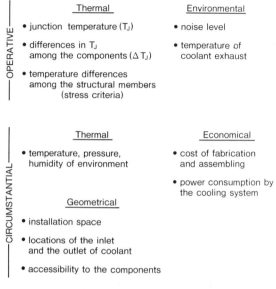

	Thermal	Environmental
OPERATIVE	• junction temperature (T_J) • differences in T_J among the components (ΔT_J) • temperature differences among the structural members (stress criteria)	• noise level • temperature of coolant exhaust

	Thermal	Economical
CIRCUMSTANTIAL	• temperature, pressure, humidity of environment Geometrical • installation space • locations of the inlet and the outlet of coolant • accessibility to the components	• cost of fabrication and assembling • power consumption by the cooling system

tronic circuits uniform throughout the equipment. The temperature differences among the structural members at the package and board levels must also be constrained within a certain range to reduce the level of thermal stress so that the integrity of electrical connections and the reliability of the packages themselves can be maintained.

The concern for personnel working near the electronic equipment is becoming an important factor in the thermal design. The demand on the reduction of noise from air-cooled equipment will be met only by refinements in the thermal design which eliminate redundant air streams. The temperature of air at the exhaust from the equipment as well as the location of the exhaust which are cited as circumstantial factors are often specified, lest personnel be exposed to hot air streams.

The circumstantial criteria set the constraints of thermal, geometrical, and environmental nature. An increasing number of electronic devices will operate in an environment of high temperature and high humidity. System compactness of a device is now becoming important, particularly to space-conscious Japanese customers. The requirements on the accessibility to components for inspection or replacement may exclude some cooling schemes even though they

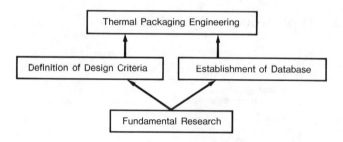

Figure 1.4: Roles of fundamental research.

may be attractive from the viewpoint of heat transfer engineering. The cost of fabrication is often a dominant factor in weeding out proposed cooling strategies. Finally, the power consumption by the cooling system, in addition to the power consumed by the electronic system itself, is becoming a matter of concern for the operators of large-scale computers. This is more critical in Japan than in any other country due to the high cost of electricity. All the factors listed in Table 1.1 act to demand more stringent thermal design in future generations of electronic equipment.

Fundamental research in heat transfer has contributed to the advancement of thermal packaging engineering (Fig. 1.4) in two primary directions: (a) definition of appropriate design criteria and (b) the establishment of a database which makes the search for a solution more rational and efficient.

2.2   Design Criteria Establishment

The upper ceiling on the junction temperature is set from failure rate data taken in the field and from high temperature acceleration testing. A comprehensive reference regarding junction temperature criteria is provided by the US Military Specifications Handbook No. 217D (1982). For a particular design of a chip or a package, the experience in field operation produces a learning factor which enhances the predictability of the failure rate. However, with rapid progress in miniaturization of circuit patterns and enlargement of chip areas, it is desirable to attack the reliability problem on the basis of an understanding of the mechanisms of failure. Table 1.2 shows that the causes of failure can be of an electrical, mechanical, or chemical nature at the chip level; and, as indicated by Wager and Cook (1984) all of these are accelerated by the increase in the junction temperature. Here the junction temperature is a vaguely defined quantity. It usually refers to the average temperature of a chip or the temperature at a location where a temperature-sensing diode is embedded. A real junction is a zone formed by p-type and n-type semiconductors, a layer of silicon dioxide, aluminum conductors, and a passivation layer. The length scale of those subzones is represented by the width of the conductor which is presently of the order of 1-2 μm and will be reduced to sub-μm in the foreseeable future.

A direct influence on the reliability of the junction are the temperatures and the thermal stresses at the interfaces of the subzones, particularly those between the materials of substantially different thermophysical properties. Hence, it is more appropriate to include in the design criteria the peak temperature and the maximum temperature gradient in the junction zone rather than the average temperature of a chip. However, the attempt to find a detailed temperature distribution in a chip raises a considerable difficulty, due to the miniaturized size and the immense number of junctions.

The suppression of thermal stress on the package and board levels is becoming important with the advent of new packaging and mount-

Table 1.2: Heat transfer research needs for the definition of thermal design criteria.

Level	Trends of Technology	Modes of Failure	Research Needs
chip	• miniaturization of circuit patterns • enlargement of chip area	• thermal activation energy ├ electrical—conductor/contact degradation ├ mechanical—fatigue of interconnections └ chemical—corrosion	• analytical and experimental tools to investigate thermal fields
package	• flip–chip mounting • multi–chip packaging ↓ increase in the number of I/O pins ↓ enlargement of package size	• thermal fatigue of solder joints • fracture of bonded interfaces in the package	• thermomechanical properties of packaging materials
board	• surface mounting	• thermal fatigue of solder joints	• thermomechanical properties of composite materials

ing techniques. The parts vulnerable to thermal stress in the new structures are the microscopic solder joints which are devised to shorten the connection distances between the electronic circuits at different levels. Solder balls or posts having a diameter of approximately 100 μm are used to provide the connections between the chip and the substrate (flip-chip mounting). Flatpacks and leadless chip carriers are the type of packages which utilize solder bonding for electrical connections to the PWB without insertion of lead pins into the via holes of the PWB.

The stress caused by mismatching of thermal expansion coefficients among the components and also by the presence of noncompatible temperature distributions within the package and the PWB, is supported mechanically by the lead pins in the case of dual-in-line packages. (See Table 1.3 for a description of the types of packages.) In the new mounting schemes, solder joints are directly subjected to the stress. The increase in the number of electrical leads coming out of the package is making the thermal stress problem even more serious. Very fine leads and solder joints must support the thermal stress which increases with increasing package size and proper precautions must be taken to minimize this effect. The establishment of design criteria for allowable temperature differences on the package and the PWB requires consideration of thermophysical properties and thermal expansion coefficients of structural materials. The problem of practical interest is the prediction of the thermal expansion of a multilayered component with alternating conductor and insulator layers. A ceramic integrated circuit package and a PWB have such structures with a variety of conductor layer patterns.

2.3   Establishment of the Database

The problems of primary importance in air-cooled equipment are listed in Table 1.4. Yet, there are factors that have heretofore not been considered serious in thermal design but are now receiving an increased level of attention. For example, in some applications, Joule heating within a PWB may reach such a level that a more detailed heat transfer analysis may be required so that control of the temperature distribution of the PWB may be considered.

The prediction of the maximum temperature on a natural-convection cooled PWB is a problem of simultaneous heat transfer with

Table 1.3: Classification of packages from the viewpoint of thermal management.

Type	Sketch	Conventional Names
Molded Plastic		DIP, Shrink DIP, Small Outline
		Flatpack, Leaded Chip Carrier
Chip–in–Cavity		Cer DIP
		Leaded Chip Carrier
		Leadless Chip Carrier
		Pin Grid Array
Dismountable Heatsink		heatsink jacket / coolant / thermal connecter / chip / ceramic substrate

lead pins

Table 1.4: Establishment of database - I: Heat transfer problems in air-cooled equipment.

	Level	Problems of flow and heat transfer
Natural Convection	packages on the board	• conjugate (conduction, convection, radiation) heat transfer from the chip to the air ·
	cabinet	• convection driven by multiple heat sources
Forced Convection	packages on the board	• heat transfer from the packages on the wall of a parallel plate channel
	system	• flow in the space of irregular geometry

interacting modes of conduction, convection, and radiation. The difficulty in solving such a problem is amplified where the board spacing is small with respect to the size of the packages. The prediction of the flow pattern in the narrow channel populated by a number of packages must precede the heat transfer analysis. It is at the system level that the prediction of the flow driven by multiple heat sources poses an even more challenging problem. The forced convection heat transfer from the package assembly and PWB's is affected by many parameters. These include the spacing between the PWB's, the geometry and the size of the packages, the longitudinal and lateral pitch of the package placement, the rates of power dissipation, the flow rate of air, and the thermal conductivities of the package materials. The prediction of flow at the system level is generally difficult. This could be partially attributed to the lack of data concerning the flow resistance along flow paths formed by the components of various shapes within the array of PWB's. It could be useful to compare an array of PWB's having finned packages mounted in a high placement density to the core of a compact heat exchanger, although different results are sought. For forced convection cooling of PWB arrays, a detailed temperature distribution among the elements of the array is sought, while for a compact heat exchanger the quantities of interest are the averaged fluid temperatures and the heat transfer rate.

Air is a convenient coolant but incapable of removing the high heat fluxes expected in the future generations of high-speed computers. A few models of mainframe computers have already employed advanced cooling systems. In the conceptual phase of a cooling system, the pattern of electrical connections between different levels of electronic circuits and the density of these connections help set the constraints on the design of heat transfer paths (Table 1.5). The crucial part in the structure of a module is the thermal interface between the chip and the module. Where the flip-chip mounting is employed, a large thermal resistance in the heat path through solder joints requires the removal of heat from the other side of the chip. This is discussed in more detail in Sections 3 and 6. While the housing of the chips in a module (packaged chips) renders a certain ease of maintenance in the field, the module structure adds a thermal resistance which may not be tolerable for chips of very high power dissipation. In anticipation of the advent of high power chips in the

Table 1.5: Establishment of database - II: Primary factors to be considered in the design of advanced thermal packaging.

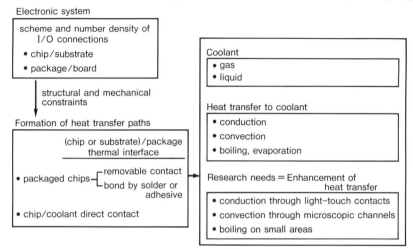

future, cooling of chips in direct contact with the coolant is becoming a research topic of practical importance.

There are a wide variety of feasible cooling schemes even within the constraints imposed by the electronic system. The designer can certainly choose the kind of coolant and the modes of heat transfer to be used. Furthermore, the current research on enhancement of heat transfer can provide the basis for the development of advanced cooling systems.

A discussion of the specific topics of conduction at the interface between the heat dissipating component and the cooling device, convection through microscopic channels, and boiling over small areas will be considered in the appropriate sections that follow.

3 PACKAGING LSI CHIPS

Bar-Cohen (1985) critically reviewed various thermal packaging technologies in use in state-of-the-art electronic systems. In what follows, fundamental problems of packaging will be reviewed with illustrations drawn from actual package designs.

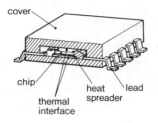

Figure 1.5: Primary components of a package.

3.1　Package Structure

The primary components of a package (module) are, as shown in Fig. 1.5, a chip (chips), electrical leads, heat spreader, and a protective cover. When viewed as a thermal system, a package (module) is composed of a heat source (heat sources), heat spreaders, and thermal interfaces. The heat source is, of course, a chip. The heat spreader is a structural member which serves to diffuse heat from a confined area to a wider area. The package wall and the electrical leads serve as heat spreaders. For high-power chips, active measures are often taken to diffuse heat in the immediate vicinity of a chip by bonding a highly- conductive plate to the chip. Fins attached to the exterior of a package are also heat spreaders.

The package is an assembly of parts bonded together by solder, adhesive, molding compound, and, in certain instances, mechanical parts such as bolts and springs. Heat flows inevitably through the interfaces of different materials. The thermal conductivity of a substance filling the interstices is often smaller than those of the mating materials. Hence, the management of thermal resistance at the interfaces is a key to the achievement of a low internal thermal resistance. A crucial factor affecting the design of the thermal interface is the mismatch of thermal expansion coefficients (TEC) among structural members. Where the TEC mismatching is large, the placement of a third member between the mating members to act as a "buffer" may be required. This causes an increase in the thermal resistance and the fabrication cost.

3.2　Package Classification

Plastic packages or ceramic packages refer to the covering material and the terminology "single-chip" or "multichip" refers to whether one or more chips are contained in the package. The arrangement of electrical leads that come out of the package yields such names as dual-in-line package (DIP), flatpack, chip carrier, and pin-grid-array package (PGA). The adjectives "leaded" and "leadless" indicate whether the package is provided with the electrical leads of metal pins or just the pads which later are to be soldered to the PWB. A somewhat different classification scheme, related to the nature of the thermal problems within the package, is used in this discussion. Table 1.3 shows the correspondence between the proposed classification and the conventional names of package.

3.2.1 Molded plastic package

Because of small power dissipation from a chip, relatively large internal thermal resistance is tolerable in a molded plastic package. Interfaces exist among the chip, the metal pad next to the chip, the electrical leads, and the resin which surrounds the chip and the metal components. The thermal resistance at the interfaces between each of these is negligible in comparison to the resistance across the resin. The advances in packaging technology for this type of package depend on improvements in thermophysical and mechanical properties of the leadframe metal and the resin. Analysis of heat conduction is required not only to determine the internal thermal resistance but also to find the stress distribution caused by the TEC mismatch among the chip, the resin, and the leadframe.

3.2.2 Chip-in-cavity package

In this type of package, the package cover is a box formed by two flat members sealed on four sides. A chip, or an assembly of a chip and a heat spreader, is bonded to one of the flat walls. The multi-chip package (several chips are housed with a ceramic substrate) is included in this category because of a similarity of the heat flow paths to those in a single-chip package. Materials used for the cover of a package are ceramics, resin, or metal.

LSI logic chips and high-speed memory chips are housed in packages of this type with the power dissipation rate per chip attaining levels as high as several watts. The exterior surface of a package exposed or located close to the coolant flow is where most of heat is discharged from the package. Thus, the key to a good thermal design is the strategy of diffusing heat in the vicinity of a chip and then letting it flow with negligible thermal resistance to the primary heat discharge surface.

3.2.3   Multi-chip module with a dismountable heat sink

A module design employed in recent large-scale high-speed computers has a number of chips mounted in proximity to each other on a multilayered ceramic substrate. The power dissipation per chip can reach levels as high as 10 W. Here, a large number of lead pins are required to transmit signals into and out of the module. The pins occupy a substantial area of the backside of the substrate. This requires removing heat from the chip's surface facing away from the substrate. The dense placement of chips precludes the use of heat spreaders.

Chips must be accessible for inspection, repair, and replacement. Hence, the cooling device needs to be dismountable. Important physical factors affecting the design of the module are the bow of a ceramic substrate and the tilt of a chip against the local substrate plane. These geometrical aberrations are small, but pose significant thermal resistances between the chips and the cooling device. The deviation from planarity of the chip surfaces is quite unpredictable. Thus, an attempt to correlate and use thermal resistance data is a delicate matter.

Under these circumstances, the heat sink must be equipped with thermal connectors which can compensate for the geometrical aberrations and connect the chips to the wall of the cooling jacket. The contact pressure on chips must be low enough to avoid damage to the chips. Thus, heat transfer through low-contact-pressure interfaces is a subject of prime importance for this type of module.

It is now the place to bring the aspects concerning the nature of physical processes of heat transfer as related to electronics technology into the discussion. These are: thermal stress on VLSI memory chips in a molded plastic package; internal resistance in chip-in-cavity packages; thermal resistance at the interface between the component and the heat sink; and thermal interface management. There are interrelated problems in thermal, materials, and fabrication engineering that must be solved in order to advance the packaging technology

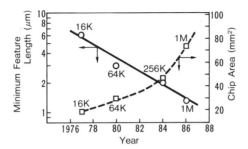

Figure 1.6: **Minimum feature length and chip area of DRAM chips.**

and each type of package described in the foregoing sections possesses a family of problems arising from its own structural characteristics. In what follows, topics are chosen to highlight the nature of problems pertaining to these types of packages.

3.3 Thermal Stress on VLSI Memory Chips in Molded Plastic Packages

Miniaturization of circuit patterns is the most dramatic on MOS (metal-oxide-semiconductor) memory chips. Figure 1.6 shows how the feature length of the circuit has been reduced to increase the memory capacity of a chip. Suzuki and Yoshizumi, (1984) have shown that the size of chips has increased, but only by 50-70% while a four-fold increase of memory capacity has occurred. Figure 1.7, taken from *Electronics Parts and Materials*, (1986), shows an example of packaging geometry and indicates the reduction of resin area around the perimeter of a chip.

For conventional plastic DIPs, where a chip occupies a relatively small area within the package, Mitchell and Berg (1979) point out that leadframes play an important role in spreading heat. Andrews, Mahalingham and Berg (1981) have shown that incorporation of heat spreaders also helps to decrease the internal thermal resistance. As the chip size increases, the effect of such metallic components on the heat transfer becomes less significant and internal thermal resistance approaches a value dictated by the thermal conductivity, the thickness of the resin, and the area of the chip. Suzuki and Yoshizumi (1984) have indicated that, today, almost 90% of the resins used for

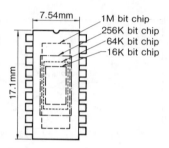

Figure 1.7: Growing size of a DRAM chip molded in a DIP package.

packaging are epoxy with various fillers. The fillers give the resin different thermal and mechanical properties, permeabilities to moisture, and viscosities during molding. The thermal conductivity can differ two- or three-fold within the family of epoxy resins; e.g., 0.0063 W/cm K for resin A and 0.021 W/cm K for resin E. Research is being performed to develop resins having high thermal conductivity and low glass transition temperature, the latter property being important for the reduction of the stress level in the package.

For a large-chip VLSI package, the most pressing concern is the stress imposed on a chip due to the mismatching of the thermal expansion coefficients among the structural members. During fabrication the chip undergoes a period of tensile stress when it is attached to the metal pad by solder. After molding and curing the resin, the stress balance leaves a compression stress on the chip and the pad and a tensile stress on the resin. The general features of the stress distribution can be predicted by the finite element method. However, the prediction of stress at crucial parts in the package is a sophisticated art. Figure 1.8 shows the stress distributions observed by the photoelastic technique by Kotake and Takasu (1980) and Liechti and Theobald (1984). The packages displayed in Fig. 1.8 were not intended for use as VLSI RAM chips. However, Fig. 1.8 serves to illustrate the type of problems that become more acute in VLSI packages. Figure 1.8(a) shows the cross section of the assembly of a silicon chip and a Kovar (Ni-Co-Cu alloy) pad, and indicates the lines of constant-stress. The chip in this example has an area of 5 × 5 mm and is 0.5 mm thick. The magnitude of compression stress was reported to depend on the properties of the resin and the type of curing process. Isochromatic lines in the cross section of the resin are shown in Fig. 1.8(b). The numerical figures are the indices of isochromatic lines near the edges of the chip in Fig. 1.8 and indicate the presence of singular points. The stress situation becomes increasingly severe in VLSI chips with the increase in chip size and the reduction in resin volume. An understanding of the stress distribution near the singular points is becoming essential to the attainment of VLSI packages of high reliability. Singular points are not limited to the immediate neighborhood of the chip, but exist also where the lead pins come out of the resin. A promising analytical technique has been developed for the two-material wedge problem by Hein and Erdogan (1971). However, applications of this method have rarely been reported in the open literature.

In the manufacturers' laboratories, new materials for the resin and the metal leadframe are being studied. For instance, precipitate-hardened copper alloy has been shown by Kuze (1986) to be one of the recent products that possesses high mechanical strength and thermal conductivity with a relatively small thermal expansion coefficient.

As miniaturization proceeds, the structural integrity and the operative reliability of circuits on the chip become increasingly vulnerable to mechanical, thermal, chemical, and electrical strains. Design criteria need to be refined to take into account more of the microscopic

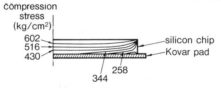

(a) Stress in a cross section of a 5mm×5mm chip
mounted on a Kovar pad by epoxy resin
(the stress distribution after molding)
(Kotake and Takasu (1980))

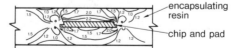

(b) Fractional isochromatic fringe patterns in a
longitudinal mid–section of a package
(Liechti and Theobald (1984))

Figure 1.8: Stress distributions in molded packages studied by photoelastic analysis.

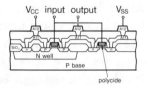

Figure 1.9: Cross section of a CMOS inverter used on the periphery of a memory cell array.

features of the structure on the chip. A new methodology is required to analytically determine the temperature and stress distributions in microcircuits. *Electronic Parts and Materials* (1986) gives a typical example of a microcircuit. It is illustrated in Fig. 1.9. Here the minimum feature length is 1.3 μm. If an active zone of 4 mm × 10 mm × 5 μm on a 1 M-bit DRAM chip is discretized by a 1.3 μm rule, the number of elements amounts to the order of 10^8. Effective measures to deal with the problems inherent in dealing with a large-matrix problem have yet to be developed.

3.4   Internal Thermal Resistance in the Chip-in-cavity Package

Figure 1.10 illustrates the three principal types of finned packages: (a) a cavity-down package; (b) a chip is bonded to the end of a metal stud having circular or rectangular fins; and (c) the cavity-up package where a large percentage of the heat dissipated must be conducted through the cavity's side walls to the finned surface.

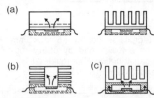

Figure 1.10: Conduction paths in chip-in-cavity packages.

Besides the location of the chip, the type of bonding of the chip to the substrate (die-bonding, flip-chip bonding) is an important factor and has a definite effect on the internal thermal resistance. In die-bonding the surface of the chip opposite to the active circuit area is bonded to the substrate by solder or adhesive. The bonding layer usually poses minor thermal resistance. In flip-chip bonding, large thermal resistance is inevitable across the solder posts. The selection of the structural and bonding schemes is made on the balance among factors such as the heat dissipation rate, the requirement for short electrical connection and the cost of fabrication.

Even when the packaging scheme is set, there is room for the designer to consider various ideas for the reduction of the internal thermal resistance. Kobayashi et al, (1985) and Ura and Asai (1983) provide examples which are shown in Fig. 1.11. In Fig. 1.11(a), a logic LSI chip of 2000-5000 gates is attached to a plate of silicon carbide (SiC) ceramic. Here, the size of the chip is 6.8×6.9 mm, the rate of heat dissipation from the chip is 6 W, and the thermal resistance from the chip to the air stream flowing with a velocity of 5 m/s is 5 K/W. The chip is die-bonded to the SiC substrate by low-temperature solder. The substrate is produced by using berylia as sintering material, and the proper sintering condition gives the ceramic a high thermal conductivity of 2.7 W/cm K and a thermal expansion coefficient (TEC) of $3.7 \times 10^{-6} K^{-1}$. This is close to the TEC for silicon and the low TEC helps to alleviate the thermal stress problem and simplify the package structure. An equally low TEC of $3.5 \times 10^{-6} K^{-1}$ and also a high thermal conductivity of 1.6 W/cm K are reportedly achieved in aluminum nitride (AlN) ceramic (Kurokawa et al, 1985). Figure 1.11(b) shows the application of AlN ceramic as a heat spreader in a canned transistor. In this particular example, a tiny silicon epitaxial transistor of 4.5 W power dissipation is solder-mounted on the spreader measuring 2×2 mm with a thickness of 0.5 mm.

Figures 1.11(c) and 1.11(d) show two examples of flip-chip bonding where more than 200 solder posts connect the chip and substrate. In the case of Fig. 1.11(c) (Kohara et al, 1983), the attempt is made to conduct heat from the back side of the chip. The chip in this example is a 3000 gate logic chip of 8×8 mm. Experiments were conducted using copper plates of various sizes bonded by solder to the back side of the chip. There is a gap of about 30 μm between the copper plate and the inner top surface of the cap, so that the kind of gas used to fill this space in the package will have an influence on the internal thermal resistance. It was shown that the larger the copper plate, the lower the thermal resistance and this is an indication of the importance of heat spreading. With the copper plate having an area of 13 × 13 *mm*, the thermal resistance from the chip to the air stream flowing at a velocity of 4 m/s is 5.4 *K/W* when the package is filled with air, and 3.9 *K/W* when the filler gas is hydrogen.

In the example provided by Kohara et al (1984), and as shown in Fig. 1.11(d), the substrate is a plate of SiC ceramic which is 1.8 *mm* thick and this serves as a heat spreader. The thermal resistance between the chip and the air stream is comparable to that of the configuration of Fig. 1.11(c).

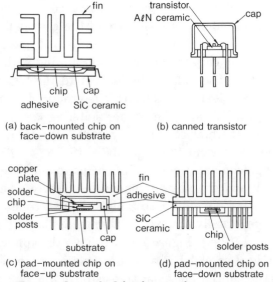

(a) back-mounted chip on
face-down substrate

(b) canned transistor

(c) pad-mounted chip on
face-up substrate

(d) pad-mounted chip on
face-down substrate

Figure 1.11: Examples of chip-in-cavity structure.

One of the factors governing heat conduction in the package is the ratio of package area to chip area. The areas here are defined as those obtained by projection of the package and the chip onto the PWB. The decrease in package "footprint" on the PWB and the trend of increasing chip area both act to reduce the area ratio. Meanwhile, as the number of gates on the logic chip increases, the number of lead pins (I/O pins) must be increased. This results in finer leaded pins providing an increased density on the package. Figure 1.12 shows recent data of package area and I/O pin counts. The solid curves due to Lewis (1984) in the upper figure show the relationship between the package area and the pin counts for flat leaded packages for three pin spacings (16, 20, and 30 mils), and the broken curve is the relationship for pin-grid-array packages with 100 mil lead pitch. The symbols for the actual data pertaining to commercial products are explained in the lower figure, where the ratios of package area to chip area are plotted from the data of Otsuka and Usami (1981) for 30 mil-108 pins, Kohara et al (1983) for PGA, Terasawa, Minami and Rubin (1983) for 20 mil-400 pins, Kobyashi et al (1985) for 20 mil-160 pins and Nikkei Electronics (1985) for 16 mil-180 pins. The largest pin count ever reported in the literature for the flat leaded package is 400, where a 20 mil lead pitch may be the practical limit for such large-pin count packages.

The figures suggest that the flat leaded package may lose its superiority in area economy over the PGA when the pin count increases beyond 400. For the packages with pin counts in a range of 100-200, the chip area increases with increasing number of gates and the package area increases with increasing number of gates, while the package area increases at a lower rate. Small package-to-chip area ratios imply that a heat spreader is no longer effective in reducing the internal thermal resistance. Moreover, a large heat dissipation rate from such logic chips necessitates the adoption of the cavity-down structure where heat is conducted to the base of the fins over a short distance across the substrate and the bonding substance. Major thermal resistance resides on the surface of the fin so that the selection of the fin geometry and the estimate of the heat transfer coefficient on the faces of the fins become tasks of prime importance in the thermal management of flat leaded packages.

One of the pressing tasks in the design and packaging of high-power chips is the management of TEC mismatching of the structural components. The stress distribution in the multilayered structure is a complex problem, and as Yasukawa, Sakamoto and Shida (1983) have observed, the solution requires a computer-aided analysis. For the analysis of temperature distribution in a multilayered substrate, the methods of Kennedy (1960) utilizing Bessel functions, of Ellison (1984) employing two-dimensional Green functions, and Watanabe and Ogiso (1979) based on three-dimensional Green functions have been proposed. Wilson (1979) has suggested that the finite element method (FEM) of analysis must be performed when detailed temperature and stress distributions in the whole package structure are needed. The necessity for such an analysis is increasing as the requirements on package reliability become more demanding. Data for the local heat transfer coefficient on the package surface are indispensable in such detailed analysis. However, these data are scarce. A literature survey has turned out only remotely relevant works; that of Igarashi (1985) concerning the heat transfer coefficient from a square prism, and that of Akino et al (1985) pertaining to the enhancement of heat transfer from a wall due to the presence of a rectangular protuberance.

As mentioned previously, increase in the number of lead pins beyond 400 makes the flat leaded package less attractive, and the multichip module, with a dismountable heat sink, becomes an advantageous form of packaging.

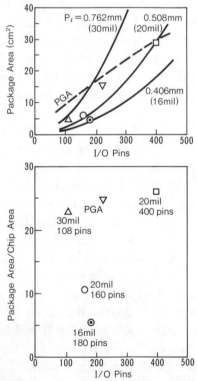

Figure 1.12: **Package area and the number of I/O lead pins.**

3.5 Thermal Resistance at the Interface Between the Component and the Heat Sink

Yovanovich (1978) has provided a concise overview of thermal interface management. In commercial equipment thermal interface management is essentially an art depending in large part on innovative ideas. Figure 1.13 shows several examples, drawn from recent publications. In Fig. 1.13(a) due to Chu, Hwang and Simons (1982), the spherical head of an aluminum piston is pressed lightly onto the surface of a chip by means of a spring. The important factors in this design are the radius of curvature (ρ) and the microroughness (δ) of the piston head, and the kind of gas used to fill the void space. The interfacial resistance of 2.9 K/W is achieved for the chip of 4.57 mm^2 when $\rho = 14$ cm, $\delta = 0.4$ μm, and the filler gas is helium. The example of Fig. 1.13(b), due to Biskeborn, Horvath, and Hultmark (1984) utilizes grease (or so-called thermal paste) to thermally connect the chip's back (top) surface to the module (cap) surface which is cooled by an inpinging air jet. With the 0.1 - 0.35 mm thick thermal paste (1.25 W/m K thermal conductivity), about 45% of the chip's heat passes across this interface with the remainder passing through the solder posts on the active side of the chip. The interfacial resistance of 8 K/W is achieved with the control of the paste thickness to around 0.25 mm.

In the example of Fig. 1.13(c) due to Watari and Marano (1985), a metal stud is pressed onto the face of a chip carrier with thermal compound at the interface. The thermal resistance between the chip and the water coolant is reportedly 5 K/W. In Fig. 1.13(d), originally presented by Nikkei Electronics (1985), flat-leaded chip carriers are mounted on both sides of a PWB and the PWB is sandwiched between cold jackets where bellows serve as thermal connectors. The bellows contain paths for cooling water flow, and have polymer films at the terminal faces to secure contacts with the packages.

Figure 1.13(e), representing the work of Wilson (1982), shows a scheme for cooling from the back side of a substrate. A heat sink having a copper diaphragm and containing water is pressed on the substrate. Figure 1.13(f) due to Nakao et al (1982) indicates an example of thermal interface management for a sealed electronic unit. The sealed unit is designed for use in steel mills, engine rooms and

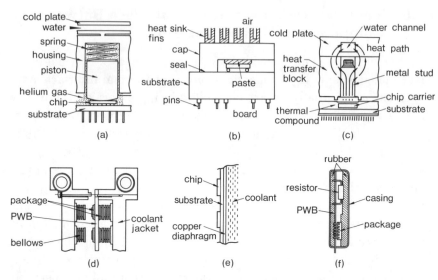

Figure 1.13: Examples of thermal interface management.

other hostile environments. A low resistance thermal path is achieved by the use of rubber having a thermal conductivity of 1.05 W/mK.

3.6 Thermal Interface Management

Table 1.6 lists the facets to be considered in thermal interface management. The factors affecting the interfacial resistance are grouped in reference to where they originate; some are traced back to the pre-assembly phase, some are caused during the assembly of a module, and others develop during the service life of the equipment. In each phase, there are factors pertaining to the surfaces to be thermally connected (termed "parent") and the interstice. The surfaces of a chip, a substrate and a PWB all possess deviations from ideal flatness that may extend over their entire surface. This is called bowing and results from thermal histories imposed upon the components during their production. Bowing is primarily due to TEC mismatching of laminated materials. Examples are in a chip which has a silicon-oxide layer on a silicon base and substrates, as well as PWB's, which have layers of conductors within an insulator matrix. Machining and polishing of the surfaces of a thermal connector, a chip and other intermediary members engender waviness of intermediate length scales and microroughness. A material to fill the interstice has to be selected based upon the design goal for the interfacial thermal resistance and the other factors listed in Table 1.6.

As Table 1.6 implies, the thermal interface management is closely related to the structure and the service environment of the electronic equipment. This makes it hard to find reports of data and analyses in the open literature that focus on the problem of interfacial thermal resistance. Only a limited amount of data have been published. Some of the recent data are plotted in Fig. 1.14 with descriptions of these data given in the associated Table 1.7. It is to be noted that the contact pressure of interest is one to two orders of magnitude lower than those contact pressures quoted in a number of previously published works on thermal contact resistance that relate to other industrial apparatus. Naturally, at such low contact pressures, the interstitial material provides a greater influence on interfacial resistance than it does under circumstances of higher contact pressures. When liquid is used as interstitial material, it must be chemically inactive and possess a low vapor pressure. Silicone oil is one such liquid, and its capillary suction in the microgrooves cut on a silicon plate is utilized to produce a contact pressure (MC-S in Fig. 1.14). Compared to liquid fillers, a gaseous filler produces a much lower interfacial conductance. However, the handling of a gas is much easier during assembling and dismantling of a module. The working

Table 1.6: Factors affecting thermal resistance across an interface.

Phase	Parts	Factors
pre–assembling	parent	bow, wave, microroughness
	interstice	selection of material (gas, liquid, adhesive, solder)
assembling	parent	non–alignment, tilting, elastic–plastic deformation
	interstice	void formation
service	parent	variation of interstitial gap, cleavage formation, due to TEC mismatching of components
	interstice	deterioration of interstitial material

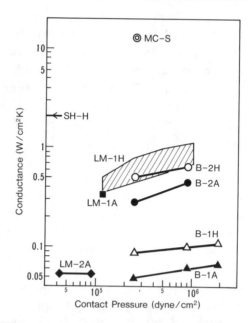

Figure 1.14: Heat transfer conductance at the interface between component surface and light-touch thermal connectors.

principle of the spherical head contact (SH-H) requires little contact pressure other than that necessary to hold the piston head on a chip. The other connectors in Table 1.7 rely on the flexibility of connector members to reduce interfacial resistance with increasing contact pressure. The data in Fig. 1.14 reflect the effect of this flexibility and the affinity between the connector material and the copper surface of the experimental apparatus. The parylene bag encapsulating gallium (LM-H, LM-1A) is transfigured to produce a relatively large contact area. The bundle of fine stainless steel wires (B-1H, B-1A) is not as flexible as the bundle of thicker copper wires (B-2H, B-2A). Those bundle connectors undergo plastic deformations when the contact pressure exceeds certain values, so that the history of force application affects their performance. The combination of a copper membrane and water (LM-2A) ranks as the configuration with the lowest conductance among the test samples. This is probably due to the stiffness of the copper membrane.

The use of adhesive and solder at the interfaces requires heat and chemical treatments at the time of assembly and dismantling a module. Resin adhesives are not as attractive because of their low thermal conductivity (most are around 1 W/mK). Metallic or ceramic particles mixed in the resin increase the thermal conductivity. However, they also increase the apparent viscosity of the resin-particle mixture in the liquid state and, as indicated by Schwink-endorf and Moss (1984), this causes difficulty in producing a thin securely bonded layer of resin in the interstice. The effective heat transfer coefficient for particle-laden resin ranges from 0.03 to $1 W/cm^2 K$. However, solder yields a high coefficient. For example, a coefficient of $31.5 W/cm^2 K$ is obtained when Pb-5% Sn solder (thermal conductivity 0.63 $W/cm\ K$) fills a 0.2 mm thick interstice. Various low-temperature solders are now available and could be candidate interstitial materials in future high-power modules. Heat transfer and solidification of solder in an interstice is an important subject in the definition of the condition of voidless secure bonding.

4   NATURAL CONVECTION COOLING

4.1   Heat Transfer from Naturally Cooled PWB's

Natural convection cooling of PWB's requires little additional hard-ware except, possibly, fins for high-powered devices and, in some

Table 1.7: Description of the data in Fig. 1.14.

Symbol	Thermal Connector	Reference
MC	1.5μm thick silicone oil film sustained by capillary in 1–5μm wide 30μm deep microgrooves cut on a silicon substrate	Tuckerman and Pease (1983)
SH	aluminum spherical head of 14cm radius in contact with silicon	Chu et al. (1982)
B–1	bundle of stainless steel wires, each wire of 4μm in dia. and 2mm long, with the population density of 25465 wires per mm²	
B–2	bundle of copper wires, each wire of 25μm in dia. and 5mm long, with the population density of 509 wires per mm²	Nakayama et al. (1984)
LM–1	gallium encapsulated in a 3μm thick parylene film, pre-assembling height of the connector 0.25–0.75mm	
LM–2	water in a 50μm thick rubber membrane, pre-assembling height of the connector 0.76mm	
Symbols for the interface fluid : H = helium, A = air, S = silicone oil		

cases, partitions to assist in the natural routing of the air. But, in contrast to this structural simplicity, the heat transfer problems rank among the most complex of physical problems. In Fig. 1.15, various factors affecting heat transfer from an array of vertical PWB's are shown. All these factors have comparable effects on the heat transfer process. In order to establish a formula for thermal design, investigation must be performed with a model which is designed to evaluate the effect of just one or, at most, a few of these factors. Models where heat sources (packages) are indistinguishable to each other over the surface of the board are considered as either isoflux plates or isothermal plates. The isoflux model corresponds to the board having negligibly small thermal conductivity in its plane, and when the boards have infinite thermal conductivity in the direction normal to their planes, they constitute a symmetric isoflux channel. On the other hand, when zero thermal conductivity across the board is assumed, an asymmetric isoflux channel, i.e., a pair of walls, one isoflux wall and one insulated, results. The isothermal model corresponds to the board having infinite thermal conductivity in its plane, and the assumption of infinite or zero thermal conductivity across the board leads to a symmetric or asymmetric isothermal channel. It will be shown that a typical actual PWB lies in a range bounded by these extreme models.

For heat transfer from these models, Bar-Cohen and Rohsenow (1984) proposed a set of correlations which encompass a wide range of Rayleigh numbers. Their formulas correlate the date reported by Elenbass (1942), Wirtz and Stutzman (1982), and the results of numerical analysis by Miyatake et al (1972), Miyatake (1973), and Aung (1973). Sparrow and Azevedo (1985) reported another correlation for the case of asymmetric isothermal channels.

The effect of finite conductivity of the wall was studied by Burch, Rhodes, and Acharya (1985). They performed numerical analyses for three cases of wall-to-fluid conductivity ratio: 0.1, 1, and 10. They found that when the conductivity ratio is 10, the local Nusselt number differs from that of an isothermal plate by 12% near the lower end of the plate and 10% at large distances downstream. These cited deviations pertain to a particular channel geometry and Grashof number, but they show a rather small effect of conduction in the conductive walls. Actual PWB's have thermal conductivities

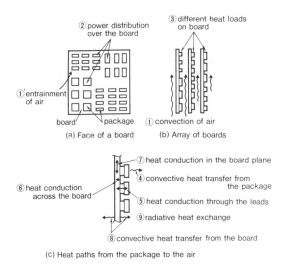

(a) Face of a board   (b) Array of boards

(c) Heat paths from the package to the air

Figure 1.15: Factors affecting natural convection cooled PWB's.

more than ten times that of air. Hence, as long as one assumes uniformly distributed heat generation, heat transfer from the board is likely to be governed by the thermal resistance on the air side.

However, heat conduction in the board is an important factor in actual PWB's where discrete heat sources are mounted. Zinnes (1970) studied the case of discrete heat sources embedded in finitely conductive walls. Jaluria (1984) reported the effect of separation distance between two heat sources embedded in an adiabatic wall. Kishinami and Saito (1984) reported the data for heat transfer from a two-dimensional corrugated wall with heating elements on the crests of the corrugation. These works, which deal with discrete heat sources, indicate the complex interaction among the heat sources, the wall and the fluid.

In addition to the work done with highly simplified models, there have been reports on measurements performed with actual PWB's or close models of PWB's. Unfortunately, reports giving the details of these measurements are scarce. Two works where one finds details to a certain degree are those of Campo, Kerjilian, and Shaukatullah (1982), and Coyne (1984). In those studies, as well as in other similar investigations, the use of empirical factors in a heat transfer correlation to account for the multiple effects of of relevant parameters, was found to be unavoidable.

4.2 Evaluation of the Thermal Parameters of PWB's

As pointed out in Section 2, with the advent of surface mount technology, the prediction of both IC chip temperature and the temperatures of PWB's is becoming important. An increased level of knowledge concerning heat flows is needed and it is expected that this will pertain to both current and future electronic equipment. In view of the diversity of PWB and component arrangements, the development of any design method claiming a certain degree of universality may be a futile task. However, in order to provide a guide for future investigations, some data for actual PWB's and the evaluation of their thermal parameters can be given.

The percentage of heat from the chip that is conducted to the board depends on the kind of metal used for the electrical leads. In the case of a 16-pin DIP, the ratio of thermal resistance through the leads to that on the package surface is of the order of 0.01 for copper alloy leads and 0.1 for 42-alloy (ferrite-nickel alloy) leads. This means that the thermal conductivity of the PWB has a great influence on the heat flow from the package.

A PWB is a lamination of copper and resin insulation and contains a number of electroplated via holes. The thickness of the copper conductors has been set at 35 μm, while the thickness of insulation layer has been reduced over the years. Figure 1.16 due to Yui (1984) shows the trend of decreasing insulation layer thickness. It appears the 0.05 mm is the practical minimum of insulator thickness, and, as indicated by the solid circles in Fig. 1.16, current products have already reached that limit. With this insulator thickness, as many as 16 layers of copper can be incorporated into a 1.6 mm thick board. Due to its laminated structure, a PWB has anisotropic thermal conductivity, the conductivity in the plane (k_x) is higher than the conductivity across the board (k_z) because thermal conductivity is defined in terms of length scales that are far greater than the individual layer thicknesses. The curves of k_x and k_z in Fig. 1.16 define the upper and lower bounds of actual conductivity. The upper bound (k_x) has been computed by considering parallel heat conduction in 35 μm thick copper foils and insulation layers (epoxy

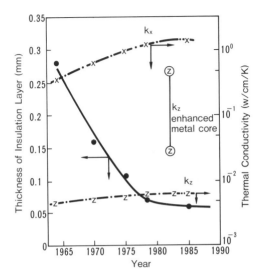

Figure 1.16: Thickness of insulation layer and thermal conductivities of printed wiring board.

resin). The lower bound (k_z) is defined by a series of resistances to heat flow across the alternate layers of copper and insulation. In actual PWB's the presence of insulation in conductor nets increases the resistance to heat flow in the plane, while across the board heat conduction is assisted by electro-plated via holes. Hence, the actual in-plane conductivity and normal conductivity approach one another and would settle somewhere between the bounds of Fig. 1.16. The estimation of the actual thermal conductivity of PWB's requires a novel methodology to cope with the diversity of conductor network patterns.

Heat conduction in the PWB can be enhanced by embedding additional metallic members in the board. PWB's having metal (often copper) layers designed to spread heat in the plane have been marketed under the name of metal-core PWB's. Fukutomi (1983) has shown that in enhanced metal-core PWB's, thermal resistance across the board is reduced by additional electroplated via holes containing copper studs within them. Figure 1.16 shows the range of normal thermal conductivity (k_z) of enhanced metal-core PWB's. At the upper end of the range, the value of k_z is based upon the insertion of 1 mm diam copper studs (49 in an area of 20×20 mm). The value of k_z at the lower end was obtained with the specimen where sixteen electroplated via holes are provided in an area of 20×20 mm. Other designs produced k_z in the indicated range. These measures of enhancing heat conduction are effective at the expense of reduced area for conductor networks.

All of this indicates that the structure of PWB's points to the need to take into account the anisotropy of heat flows in PWB's. The effectiveness of enhancement measures for heat conduction in PWB's depends a great deal on the convection of cooling air. Moreover, the conjugate problems of conductive and convective heat transfer must be solved with the aim of providing a reference point for the evaluation of products that will enter the market in the future.

The interaction of convective flows that develops between neighboring boards is a matter of concern in the thermal design of closely packed equipment. Insofar as the isoflux or isothermal models are concerned, the isolated-plate correlations hold with reasonable accuracy in an unexpectedly wide range of parameters found in actual equipments. To illustrate this, one needs only to compare the correlation of Bar-Cohen and Rohsenow (1984) for symmetric isoflux channels with the correlation for isolated isoflux plates. For example, consider Fig. 1.17. In the upper figure, the heat flux is set at a typical level of 0.01 W/cm^2, and the deviations of the heat transfer coefficient of 10, 20, 30% from the isolated-plate prediction are indicated in an envelope composed of the board height H_* and the board spacing s_*. When boards are set apart by a distance of 1.5 cm, a 10% deviation requires a height of almost 1 m, and at lower board heights, the value of the heat transfer coefficient approaches the isolated-plate limit. The lower figure in Fig. 1.16 shows similar demarcations for the case of a fixed board height of 20 cm. The increase in heat flux increases to Rayleigh number which shifts the system operating to a state closer to the isolated-plate limit. Thus, the current trend of increasing the heat dissipation per PWB increases the validity of the use of isolated-plate correlations for an array of boards with relatively wide separations.

However, this examination of convective heat transfer does not mean to justify the neglect of the effects of neighboring boards in actual thermal design. Radiative exchange between boards and the environment strongly affects the heat transfer from a board. In some cases, the magnitude of the heat flow by radiation compares to the

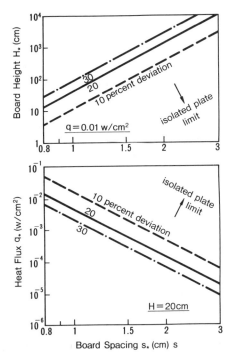

Figure 1.17: Combination of board spacing and board height, or heat flux, giving 10, 20, and 30 percent deviations of the heat transfer coefficient obtained from the isolated-plate correlation.

heat flow by convection. For example, consider the predictive method proposed by Aihara (1967) for the performance of an array of vertical fins. Suppose two PWB's, each 20 *cm* (height) × 20 *cm* (width), are separated by a distance s_*. The space between the two boards is enclosed on one side by a vertical wall, the other sides are open to the environment at 300 *K*. The boards and the base wall are assumed to be at a uniform temperature which is higher than the ambient temperature by ΔT, and all surfaces possess an emmissivity of 0.8. When $s_* = 1$ *cm* and $\Delta T = 32$ *K*, the ratio of radiative heat transfer to net heat transfer rate from a board amounts to 0.09. When s_* is increased to 2 *cm* and $\Delta T = 21$ *K*, this ratio is raised to 0.15. This indicates that there may be cases where the estimation of radiative heat exchange is as important as an attempt to reduce uncertainties in convective heat transfer. The magnitude of uncertainty brought by boundary layer overlapping has been shown in Fig. 1.17 and the

effect of air entrainment at the side of the board as shown in Fig.
1.15 can be expected to result in even more uncertainty.

4.3   Mixed Convection Heat Transfer

Another factor to be noted in naturally cooled electronic equipment is
the effect of updraft generated by the dissipation from lower components on the heat transfer from the upper components. The situation
around the upper components is analogous to combined forced and
free convection. While mixed convection heat transfer has been the
subject of many research works, few studies have dealt specifically
with problems in the geometrical context of electronic equipment.
Where the draft velocity is high, one may use an appropriate correlation for forced convection heat transfer to estimate the temperature of an upper component. A guide to the degree of approximation
in the application of forced convection correlations may be found in
some published works on mixed convection heat transfer such as that
of Raju, Liu, and Law (1984) for a vertical isothermal plate. An examination of these works reveals that the use of forced convection
correlations is not necessarily justified in certain situations found in
electronic equipment.

Even where the use of a forced convection correlation is unavoidable practice, the estimation of fluid velocity is another difficult task.
A growing number of researchers are now investigating the convection pattern in enclosures where multiple heat sources and partial
barriers are present. Two examples of such studies are those of Heya
et al (1984) and Torok (1984). The models in these works have relatively wide internal spaces, while components in the real world tend
to be packed in reduced spaces. Where vent ports are provided on
the frames or enclosures of equipment, an attempt to predict the flow
pattern and the magnitude of the heat transfer coefficient is difficult
because of the lack of information concerning flow resistance at the
screened vent ports. Measurements of the resistance at low flows
across screens has been proposed by Ishizuka et al (1984).

5 FORCED CONVECTION COOLING

5.1 Heat Transfer from Arrays of Packages in a Parallel-plate Channel

Where a forced flow of air is employed to cool packages mounted on PWB's, most of the heat from the chips is transferred to the air from the surfaces of the packages. Thus, an array of heated rectangular bodies on one side of the adiabatic walls of a parallel plate channel serves as a good model for the study of this phenomenon.

Table 1.8 summarizes the relevant parameters and the ranges of their values covered by various investigators. There cannot be said to be any dominant length scale governing the flow and the heat transfer. Hence, any consistent definition of the characteristic length and velocity may be used. Convenience of use in the thermal design must prevail in the definition of parameters used in the presentation of data. From this viewpoint, the Reynolds number defined here is based simply on the approaching air velocity (V_0) and the length of a package (L). The data obtained by means of the naphthalene sublimation technique as used by Sparrow, Niethammer, and Chaboki (1982) and Yanagida, Nakayama, and Nemoto (1984) converted to heat and transfer coefficients on the basis of the heat-mass transfer analogy.

Table 1.8: Studies on arrays of packages cooled by air in forced convection.

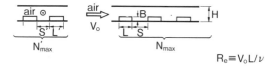

$$R_e \equiv V_0 L / \nu$$

	Investigators	B/L	L'/L	S/L	S'/S	H/B	N_{max}	N'_{max}	R_e
①	Sparrow et al.	0.375	1	0.25	1	2.67	17	3	1248,2309,4368
②	Arvizu & Moffat	1	1	1, 2, 14	1	1~4.6	8	8	1200~6200
③	Yanagida et al.	0.56	3.05	0.59,2.17,5.35	—	5.71	9	1	362
④	Wirtz & Dykshoorn	0.23	1	1	1	1.25~4.62	8	5	1300~13000
⑤	Ashiwake et al.	0.14 ⎰ 0.36	1	0.73	1.63	1.55~4.13	5 10	2	5322~18240

The flow of air in the channel is three-dimensional and is accompanied by flow separation behind the packages and, in some cases, vortex shedding from the packages. Experiments conducted with only one block heated and other blocks unheated produce a heat transfer coefficient which reflects the behavior of flow around the active block. This heat transfer coefficient is considered equivalent to a "local" heat transfer coefficient in an actual array of packages based on the difference between the package temperature and the temperature of the air directly at the upstream end of the package. Figure 1.18 summarizes the data which show the decrease of local heat transfer coefficient as the active block is shifted from the first row to downstream rows. The data were obtained in channels of relatively wide spacings ($H/B \geq 2.67$) and the ratio of heat transfer coefficient on a downstream row (α) to that on the first row (α_1) is plotted against the non-dimensional distance X/L. The distance X is measured from the leading edge of the package in the first row. The reference point is located at the middle of the package so that the curves start from $X/L = 0.5$. Most of the data show a leveling-off of α/α_1 after the second or the third row, implying that the flow becomes fully developed in a relatively short distance behind the first row. Reduced heat transfer on the downstream blocks can possibly be attributed to the formation of stagnant zones in between neighboring packages.

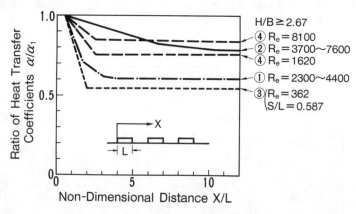

Figure 1.18: Reduction of heat transfer coefficient on downstream packages due to hydrodynamic wakes.

Reduced heat transfer on downstream blocks is, however, not a rule. The data of Arvizu and Moffat (1982) show the enhancement of heat transfer on downstream rows when the channel spacing is close to that of the blocks. In addition, Sparrow, Yanezmoreno and Otis (1984) have shown that the presence of packages of different heights, including the case of missing blocks, generally invites enhancement of the heat transfer on the downstream rows.

A heat transfer coefficient defined in terms of the temperature difference between the ambient temperature is convenient for the designer when the power dissipation of all of the packages on a PWB are equal. The heat transfer coefficient thus defined decreases on downstream rows due to the increase of air temperature. The rise of the mixed mean temperature of air can be computed from the mass flow rate of air in the channel and the heat dissipation from the packages. However, there are indications of imperfect mixing of air from the data obtained with all the blocks heated. The effect of imperfection of mixing can be estimated by comparing the measured heat transfer coefficient to the reference coefficient in the hypothetical case of perfectly mixed flow. The latter is determined by a "local" heat transfer coefficient and a temperature rise of perfectly mixed air. The ratio of the measured heat transfer coefficient to the reference coefficient, denoted here as σ, is equal to the ratio of actual temperature rise of air in the immediate vicinity of a block $(\overline{\Delta T_a})$ to the rise of mixed mean temperature (ΔT_a) : $\sigma = \Delta T_a / \overline{\Delta T_a}$.

An alternate way of determining the temperature scale factor is to conduct experiments with one block heated and others unheated and measure the temperature rises of the unheated blocks downstream of the heated block. The measured temperature rise of an unheated block (adiabatic temperature rise) indicates the measure of dispersion of a thermal wake in a distance between the heated block and the point of measurement.

Figure 1.19 shows a plot of σ against the nondimensional separation distance X_{sep}/L. Here, X_{sep} is the distance from the first heated block to a block located downstream. Each block is given an equal heat load. The curve labeled -4 traces the points (not shown) which are plotted using the equation for adiabatic temperature rise given by Wirtz and Dykshoorn (1984). The finned packages whose data are plotted in Fig. 1.19 have the dimensions described in the inset.

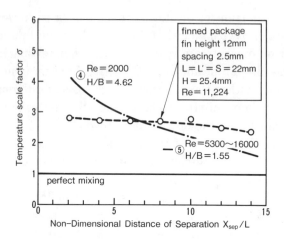

Figure 1.19: Temperature scale factor representing dispersion of a thermal wake.

(See Table 1.8 for the nomenclature.) The model package has a 5.7 mm thick base so that the total height (B) is 17.7 mm, and $H/B =$ 1.44. Four parallel columns of the model packages are provided by Nakayama, Matsushima and Goel (1987) on a channel wall with a lateral separation of 22 mm.

The body of data is too small to really consider any systematic discussion concerning σ. One may observe the magnitude of the thermal wake dispersion and a marked difference between the curve for the array of solid blocks and that for the finned package array. The data of Wirtz and Dykshoorn (1984) appear to show that dispersion of the thermal wake is enhanced in an array of blocks and additional data show that higher Reynolds numbers promote dispersion further. The σ of the finned package array is insensitive to separation distance. A cause for this may be found in that the flow is constrained by the fins and is free to mix only in a zone between the packages, and this process is repeated along a column.

Where the flow is subject to restrictions of small length scales, such as paths between the fins, the thermal wake is less dispersed. The data of Ashiwake (1983) which is curve-5 in Fig. 1.19 also indicates that, in the case of block arrays, a narrow channel spacing restricts the thermal wake dispersion and σ is insensitive to the Reynolds number. This is more explicitly demonstrated by flow visualization experiments. The photographs in Fig. 1.20 were taken with the blocks residing in flowing oil with tiny air bubbles mixed as tracers. In Fig. 1.20(a), little mixing is observed between the streams in the free-flow channels and those over the blocks. The picture on the right-hand side shows the narrowness of the gap between the neighboring plate and the surface of the block. In this study the effect of lateral displacement (δ) of the packages was also investigated. Figures 1.20(b) and (c) show the cases of a partial displacement and a staggered arrangement, respectively. The lateral displacement contributes to a reduced temperature rise of air at the downstream end of a PWB at the expense of an increased pressure drop. However, Ashiwake (1983) has shown that this still offers an advantage in reducing the thermal resistance over an in-line arrangement when compared on the basis of equal pumping power (Ashiwake et al, 1983).

The pressure drop in PWB channels is likely to be influenced most by the blockage ratio of the channel's cross section. The blockage ratio is a percentage of an area occupied by the packages within the cross-sectional area of channel as viewed from the entrance. Due to the lack of sufficient data, as well as the involvement of various geometrical and operative parameters, prediction of the pressure drop in a PWB channel is presently based on tests with a model of a particular design or the experience acquired with configurations of a previous generation. However, in present day large air-cooled computers, finned packages closely packed in a narrow PWB channel are making an entire PWB array look like the core of a compact heat exchanger. With an increased level of geometrical regularity and reduced free-flow areas, the prediction of these pressure drops will become a more accurate art.

Air flow over the finned package is complex due to the presence of free flow areas around the package. Only a fraction of the approaching air enters the space between the fins and the rest is diverted. Despite the complexity of flow, the heat transfer rate from a finned package placed in an open environment can be predicted with reasonable accuracy on the basis of a simple model. Figure 1.21 shows the result of such modeling Ashiwake et al (1983) where the driving potential for flows through the inter-fin paths is given by the dynamic pressure of the approaching flow in front of the finned array.

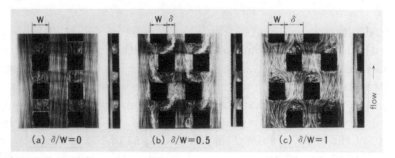

Figure 1.20: Patterns of flow around packages.

The coefficients of flow resistance and heat transfer are borrowed from those of parallel plate channels. It was confirmed from an experiment conducted with a finned package of 20 × 20 mm and with various fin heights that the predicted heat transfer rate agrees with the experimental data when the velocity of the approaching stream is less than several meters per second.

The ordinate of Fig. 1.21 is the Nusselt number based on the package area and the temperature difference between the fin base and the approaching stream. The abscissa is the number of fins on the fixed package area. The fin's thickness is fixed in this example, so that the gap between the fins decreases as the number of fins increases. One of the parameters is the Reynolds number based on the velocity of approaching flow, V_0, and the length of the package, L. Another is the ratio of fin height to the width (also L) of the package. With these parameters fixed, the increase in heat transfer area as the number of fins increases, raises the Nusselt number, but the performance starts to decline where the number of fins becomes too large. This reduced performance is caused by the increased resistance to flow through the inter-fin passages. With higher fin heights, the improvement of heat transfer performance is moderated at smaller fin numbers. In addition, the performance decline comes at a small number of fins when the Reynolds number is low. Most of the fins in use today are made from aluminum by the extrusion process and because of the high thermal conductivity of aluminum, the fin efficiency is above 88% in all the cases of Fig. 1.21.

Any scheme for enhancing heat transfer from fins, such as the serration of fins described by Kishimoto, Sasaki, and Moriya (1984), is beneficial but only when it is employed with an attempt at optimization. There are many parameters that must be considered in an optimization study. First of all, it is rare to find a situation where a finned package is mounted on a PWB and where air flows over it without other nearby obstructions. A number of packages are placed on a PWB, and the back side of a neighboring PWB is in close proximity to the tips of fins. The problem of flow in a parallel plate channel whose cross section is partly occupied by the array of fins was studied by Sparrow, Baliga, and Pantankar (1978) but this work is limited to the case of fully developed flow.

6 ADVANCED SCHEMES OF COOLING

6.1 Direct Cooling of Chips of High Power Dissipation

In an ultimate packaging scheme, chips would be in direct contact with a coolant, or separated from a coolant only by the protection afforded by a material of minimum thickness against chemical and

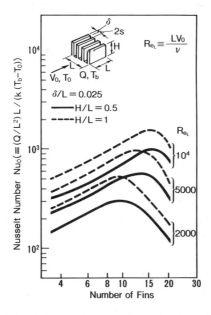

Figure 1.21: Heat transfer performance of finned heat sinks convectively cooled in an open environment.

ate thermal interfaces brings about a drastic reduction of thermal resistance, so that the absence of intermediate structural members enables the designer to not only pack more chips, but also obtain more heat dissipation per unit volume. Despite the attractive features of direct cooling, few commercial systems have ever employed direct cooling methods. (The CRAY-2 supercomputer has its CPU immersed in a flow of liquid coolant. However, the chips are not exposed to the coolant but are in sealed packages.)

In current manufacturing procedures pertaining to electronic equipment, sealed packages containing one or a group of chips are transported from a supplier (either in-house or external) to an assembly line where packages undergo various stages of testing. Individual packaging facilitates the transportation of chips, harnessing and testing on the assembly line and the replacement of failed chips from returned equipment in the field. This convenience has so far prevailed over thermal considerations. In other words, chip heat dissipation apparently has not yet reached a level of consideration that can force a restructuring the current procedure for handling chips. For future generations of mainframe computers, however, direct cooling schemes may emerge as a strong competitor to other less efficient but more common schemes.

As mentioned earlier, the performance of an electronic system is adversely affected by the variation of junction temperatures within the system. This requires uniformity of the heat transfer coefficient and coolant temperature throughout the system. The requirement of uniformity becomes more stringent as the chip heat dissipation rate reaches a level that is high enough to warrant direct cooling. Heat transfer by phase change of a coolant offers a means of securing a substantial uniformity of coolant temperature, and liquid convection in chip mounted channels of very small hydraulic diameter may yield a substantially uniform heat transfer coefficient with an accompanying small temperature rise in the coolant. Phase change of a coolant in microgrooves has the potential to provide both a uniform heat transfer coefficient and a uniform coolant temperature, but the coolant must be selected from among low-boiling point fluids in order to maintain the pressure of the cooling system near atmospheric. In what follows, the research areas of nucleate boiling and microchannel cooling will be reviewed.

what follows, the research areas of nucleate boiling and microchannel cooling will be reviewed.

6.2 Enhancement of Nucleate Boiling Heat Transfer from a Small Surface

A heat transfer surface of ultimate interest here has an area of about 1 cm^2. Nucleate boiling heat transfer on a surface of this size depends critically upon the probability of bubble nucleation. Figure 1.22 shows the correlations of wall superheat versus number density of active nucleation sites. The correlation for water was reported by Nishikawa and Fujita (1977), and that for R-11 by Nakayama et al (1980). The correlations were obtained from experiments where the metal surfaces were provided with various degrees of roughness in an attempt to change the active site density over a wide range. The length of a curve indicates the range of active site density actually observed in the particular experiment. Also shown are the data obtained with a porous surface. This will be described in a later section. All of the data were obtained at atmospheric pressure by decreasing the heat flux from sufficiently high levels. Observation of bubbles with subsequent counting of nucleation sites was possible only at relatively low levels of heat flux. The points to be noted are as follows:

1. On rough surfaces, when a heat flux is maintained at a fixed level, the wall superheat (ΔT) decreased in proportion to the one-sixth power of active site density (N/A). This means that, in order to reduce ΔT by a factor of two, the active site density must be multiplied by 64.

2. The requirement of electrical insulation necessitates the use of a dielectric fluid in direct cooling. However, as the data of boiling R-11 on rough surfaces show, the heat transfer capability of R-11 is low. A direct comparison between the data of water and that of R-11 is feasible only in a limited range of active site density where some data of both fluids overlap. Those comparable data indicate that, when compared on the basis of an equal number of active sites, the heat transfer coefficient of R-11 is one-tenth or less than that of water.

3. These observations suggest that a mere increase in surface roughness is not effective in promoting heat transfer in a di-

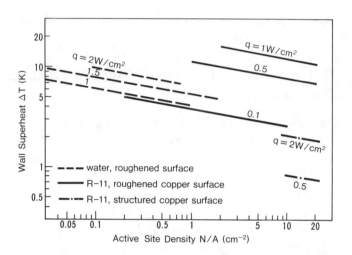

Figure 1.22: Active site density as a function of wall super-heat.

re-entrant type cavities are provided by metallurgical or mechanical means. The data plotted at the lower right corner of Fig. 1.22 were obtained by Nakayama et al (1980) with one such surface - Thermoexcel-E. Its superior performance is evident.

Figure 1.23 shows different types of porous surface structures. They were originally designed to promote heat transfer in shell-and-tube heat exchangers but because of their working principle, they also possess great potential in the cooling of electronic equipment. Geometries of the porous structures formed by metallurgical processes (Fig. 1.23(a)) are often expressed by the size of metal particles before deposition d_p, the thickness of the layer t_p, and the porosity ϵ. Typical values of the parameters, given by O'Neill, Gottzmann, and Terbot (1972) are $d_p < 100\mu m$ and $t_p < 1$ mm. In an example of application to cooling power transistors given by Fujii, Nishiyama, and Yamanaka (1979), copper particles are deposited on the surface of an encapsulant contained with $d_p = 100 - 500$ μm and $t_p \doteq 1$ mm. The porosity in most instances ranges between values of ϵ of 0.2 and 0.6. Oktay (1982) reported an example of actual silicon chips provided with porous deposits in the form of dendrites. The structure shown in Fig. 1.23(b) is described by Nakayama et al (1980). It has continuous tunnels and separated pores and is manufactured by bending the ridges of rugged microfins. The tunnel width of 0.2 mm, the depth 0.4 mm, the pitch 0.55 mm, the pore diameter of 0.1 mm, and the density of pores $260/cm^2$ are a typical set of figures representing the geometry. The sketch in Fig. 1.23(d) shows the cross section of a laboratory model of Thermoexcel-E of Fig. 1.23(b).

The surface (c) in Fig. 1.23 is manufactured by pressing the tips of fins, leaving a continuous slit between the fins. Its typical dimensions, as reported by Stephan and Mitrovic (1981) are a slit width 0.25 mm and a fin pitch 1.35 mm.

The boiling point curves obtained with the surfaces of Fig. 1.23 are compared by Nakayama (1982). An illustration of heat transfer enhancement is given by referring to the data of pool boiling of R-11 at atmospheric pressure. For comparison purposes, a specific level of heat flux (based on the projected area which does not take into account the surface area of the microstructures) is set at $2W/cm^2$, and the ratios of wall superheat on an enhanced surface ΔT_E to that on a smooth copper surface $\Delta T_0 = 13$ K are cited; $\Delta T_E/\Delta T_0 = 0.015$ (a sintered porous layer) as reported by Nishikawa, Ito, and Tanaka (1979), 0.058 (High-Flux), 0.1 (Thermoexcel E and its laboratory model), and 0.3 (Gewa-T). Of course, the performance of each surface depends on certain geometrical and operative parameters, and this comparison is not necessarily based on the optimized surfaces. However, the magnitude of reduction in wall superheat on these surfaces may be noted; even a decrease of wall superheat by one-third means a great advantage in enhancing the temperature-dependent reliability of the transistors via a reduction in junction temperatures. Nakayama et al (1980) and Arshad and Thome (1983) have shown that liquid evaporation inside the cavities plays a major role in heat transfer enhancement, although different modes of liquid motion in the cavities have been shown by Nakayama, Daikoku, and Nakajima, (1982) and Ayub and Bergles, (1985) to be possible.

The high performance of porous surfaces does not necessarily guarantee a great advantage when applied to cooling electronic equipment. Those data reported in most of the cited works belong to the fully-established boiling regime, that is, they were obtained by decreasing the heat flux from a level close to the burn-out or critical point. In actual applications, a heat flux is applied at a fixed level, so that the ability to predict the probability of bubble nucleation at a particular heat flux level is of the utmost importance. Where nucleation of bubbles fails to commence within a sufficiently short time after the application of power, or when nucleation does not commence at all, the chip temperature may increase to an intolerable level.

The problem of temperature overshoot has been studied in the laboratory by increasing the heat provided to the test surface continuously or in steps. Following the nucleation of bubbles the relationship between the heat flux and the wall superheat departs from the one set by natural convection heat transfer (DNC). Some data pertaining to this phenomenon of DNC are plotted in Fig. 1.24. The description of the data and the references are noted in Table 1.9. The points of DNC were read from the boiling curves in the published literature, and it must be remembered that uncertainties are unavoidable. In addition, Oktay (1982) reported the data only in terms of terms of the chip power and the junction temperature. Rather crude assumptions were introduced to replot his data in Fig. 1.24, namely, the chip power dissipates equally from the back side and the active side of a chip, and the saturation temperature of the fluid (FC-86) is $56°C$. The assumption regarding the heat flux was made in view of the probable contribution of the active side where a narrow gap formed by flip-chip bonding would have also served as bubble formation sites. FC-86 and FC-72 are stable chemical compounds of flourine and carbon developed by 3M Company. Both have the chemical composition, C_6F_{14}, and can be regarded as identical

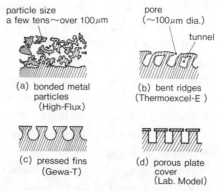

Figure 1.23: Examples of porous surface structure.

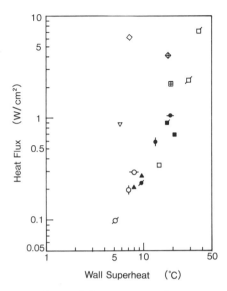

Figure 1.24: Point of departure from natural convection of low boiling-temperature fluids on various surfaces at one atmosphere. See Table 1.9 for the legends.

substances in a heat transfer study, their difference lies only in the impurity content. Other test fluids are R-113 and methanol.

With some reservations in mind regarding the uncertainty involved in plotting the data points, one may draw a tentative conclusion from Fig. 1.24. On smooth and roughened surfaces, a wall superheat could be over 10 K above the saturation temperature. Of equal interest are the data of FC-72 and 86 obtained with various structured surfaces. The temperatures at DNC are found to fall in a range of 5-8 K, while the heat fluxes at DNC are scattered in a wide range. Also, the data of the porous surface tubes reported by Marto and Lepere (1982) show that the DNC in FC-72 occurs at a lower wall superheat than the DNC in R-11.

Some of the characteristics exhibited by the FC-data of the structured surfaces can be explained though mostly qualitatively, while others are considered fortuitous or may be related to unseen physical sources. The level of heat flux at DNC depends on the heat transfer coefficient in the natural convection regime which tends to be high on a surface of small size. The sizes of the test surfaces are: for $\Diamond$, 4.5 × 4.5 mm; for ∇, 20 × 30 mm; for $\bigcirc$, 16.5-18 mm diam

$\times 50\ mm$ in length. Although the trend is qualitatively predictable, the quantitative prediction of natural convection heat transfer from a small surface in a particular set-up is not yet entirely possible.

Table 1.9: Description of the data in Fig. 1.24.

Symbol	Surface Description	Surface material	Fluid	Reference
□ ■	plain tube of 15.8mm o.d.	copper	open symbol FC–72 solid symbol R–113	Marto, Lepere (after overnight cooling)
○ ●	Gewa-T tube of 17.9mm o.d.			
◇ ◆	Thermoexcel-E tube of 16.5mm o.d.			
◌ ◖	High-Flux tube of 18.7mm o.d.			Bergles, Chyu (after boiling and subcooling)
▲	High Flux tube of 25.4mm o.d.	copper	R–113	
▽	Laboratory model of porous surface, 20×30 mm² pore dia.=0.25mm	copper	FC–72	Nakayama et al.
⬓	plain silicon chip	silicon	FC–86	Oktay*
⬔	sandblasted/potassium hydroxide etched chip			
◇	1 mm thick plate having a row of vertical tunnels attached to a chip	copper plate on silicon		
⊕	re-entrant cavity having 32 μm aperture radius	nickel	methanol	Körner, Photiadis
⊞ ■	roughened (0.94 μmm CLA) horizontal plate	stainless steel	⊞ methanol ■ R–113	Joudi, James

*Reported chip power is divided by 2× (chip area (0.2025 cm²)).
From chip temperature, 56 °C is subtracted to estimate wall surheat.

The comparatively small superheat required to activate a bubble in FC-72 was explained by Marto and Lepere (1982) on the basis of a nucleation parameter $(\sigma_s T_s / \rho_v h_{fg})$ where σ_s is the surface tension, T_s is the saturation temperature, ρ_v is the density of vapor and h_{fg} is the latent heat of vaporization. Because FC-72 or FC-86 are considered to be the most favorable candidates for a working fluid in electronic equipment cooling, more investigations are needed to clarify the effects of various relevant parameters on the commencement of nucleate boiling in these fluids.

After commencement of nucleate boiling, the wall superheat (ΔT) is reduced to a certain degree. A review of the data reported in the literature show various modes of transition to fully established boiling. Some data show a large reduction of ΔT triggered by a small increment of heat flux q, while in another extreme, an increase of q yields only a negligibly small decrease in ΔT. The surface structure and the heat capacity of the test surface, coupled with the mode of power application, may have effects on the mode of transition. There are serious difficulties to be overcome in an experimental investigation. It is desirable to use a test element which is of the same material, size, surface structure, and heat capacity as an actual component. However, the determination of heat flow through a surface in question often becomes a complicated task and, in reality, the measurement of surface temperature and power dissipation requires some modification of the chip structure.

In parallel with the investigation into the mechanism of bubble nucleation, active measures to avoid a large temperature overshoot are worthy of consideration. One such measure is to direct an impinging boiling jet onto the surface to be cooled. Ma and Bergles (1983) reported virtual disappearance of the temperature overshoot which had been observed on the same test surface in pool boiling. Such active measures, however, require additional components for the coolant supply. Where the structural simplicity of the pool boiling scheme is favored, the probability of bubble nucleation might be increased by a large virtual surface area provided through an extended surface. A porous stud, studied by Nakayama, Nakajima, and Hirasawa (1984), is an example where the purpose is not limited to the enlargement of surface area but to enhance heat transfer in the established boiling regime. Figure 1.26 shows a sketch of a cylindrical stud having a diameter of 1 cm. The stud is made by bonding several copper plates together. Each copper plate has grooves cut 0.55 mm deep and 0.25 mm wide with the pitch of 0.55 mm on both sides of the plate. The grooves on one side run orthogonal to those on the other, and, at the cross sections of the grooves, they are opened through the plate by holes having a rectangular cross section 0.25 × 0.25 mm. The bonding of plates is made at the crests of the microfins, with the microfins on the bonded faces of the plates orthogonal to each other. The result is a three-dimensional net of passages in the stud, with the possibility of the largest surface area allowable within the constraint imposed by the dimension of the grooves. The passages are open to the environment at the exterior surface of the stud and the bonding was performed by the diffusion bonding process.

The performance of the porous studs is compared with that of a smooth copper stud of a comparable size in Fig. 1.26. The fluid is FC-72, and the pressure is atmospheric. The data were obtained while increasing the heat load. The temperature difference on the abscissa is the difference between the root temperature of the stud

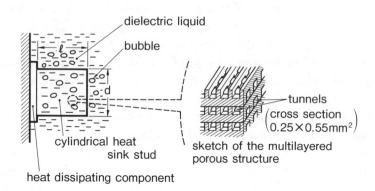

Figure 1.25: Porous stud attached to the face of a heat dissipating component.

and the saturation temperature of FC-72 ($56^{\circ}C$). Compared on the basis of an equal heat load, the temperature difference on the porous stud is 0.2-0.4 of that on the smooth stud. The performance of the porous stud of 5 *mm* in length is better than that of the 1 *cm*-long porous stud in the intermediate range of heat load, while the long stud has a higher peak heat load than the short stud. (The long stud has a 5 *mm* long porous section near its tip; the rest is covered by microfins.) Significant temperature overshoot was not observed in any of these experimental runs. A satisfactory explanation of these experimental results has yet to be found. However, the high cooling performance exhibited by the porous studs shows the potential of this technique.

Figure 1.27 is a photograph showing two-phase convection of FC-72 in microchannels of a transparent model of the porous stud. The model is made of silica glass plates, has two layers of microchannels, vertical channels on the back side and horizontal channels on the front. The channels are connected through holes at their intersections. Rapid pulsation of vapor and liquid columns in the channels was observed. Ejection of bubbles from the outlets on the periphery of the stud (not shown in the photograph), vapor generation in the channels and suction of liquid into the net of channels from the surrounding liquid pool all combine in a dynamic cycle to produce high heat transfer performance.

Most of these experiments were performed with a single test element, while in actual applications chips are arranged in close proximity to each other. Few studies have been reported in the literature on heat transfer from multiple heat sources. Bar-Cohen and Schweitzer (1983) pointed out that at the board (or substrate) level, prediction of two-phase flow in a parallel plate channel is a research topic of practical importance (Bar-Cohen and Schweitzer, 1983). One of the parameters which is of prime importance at the system level is the performance of a condenser. Deterioration of condenser performance, most probably caused by the accumulation of noncondensable gas, raises the pressure and the saturation temperature of coolant. This endangers the structural integrity of the container and causes an increase of junction temperatures. Bravo and Bergles (1976) documented the effect of a noncondensable gas on the system pressure and temperature. Bar-Cohen (1983) has shown that the submerged condenser offers a solution to the problem of noncondensables if the system is allowed to operate within certain thermal bounds set by this mode of condensation. In some applications condensation of

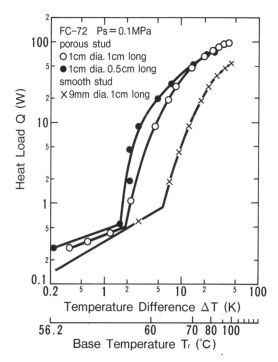

Figure 1.26: **Heat dissipation-temperature curves obtained with heat sink studs.**

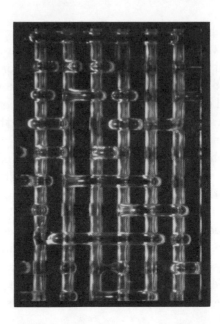

Figure 1.27: Two-phase flow in a net of microchannels.

vapor on a downward facing horizontal surface could be an unavoidable choice. A study of this problem and an attempt to enhance condensation have been recently reported by Yanadori et al (1985).

6.3   Microchannel cooling

Convective heat transfer by liquid in microchannels is a relatively simple process, and its performance can be evaluated, (when the channel geometry is not excessively complex) by referring to the well established relationship pertaining to laminar convective heat transfer and flow resistance. Tuckerman and Pease (1982) fabricated parallel microchannels on a silicon substrate. The dimensions of the microchannels were optimized to provide the highest heat transfer performance for a given pumping power. On a 1×1 cm substrate, where channels, each 50 μm wide and 300 μm high, were provided with the spatial pitch of 100 μm, a thermal resistance of $0.09^{\circ}C/W$ was obtained using water as coolant. Mahalingam (1985) confirmed the superiority of microchannel cooling by providing channels, each 20 μm wide and 1.7 mm high, with the spatial pitch of 30 μm, on a silicon substrate of 5×5 cm. The thermal resistance was reported as $0.03^{\circ}C/W$ at water flow rate of 12 cm^3/s, and $0.02^{\circ}C/W$

at 63 cm^3/s. Sasaki, Kishimoto, and Moriya (1983) reported the data obtained with a 24 × 24 mm silicon slice where four model chips, each 8-mm square, are mounted. Microfins were provided on the back side of the silicon substrate and water was led through the microchannels. The channels, each 900 μm high and having the width (W_c) of 70, 140, or 340 μm, were provided with the spatial pitch of twice the channel width. The data were reported in terms of the permissible heat flux for the chips' maximum temperature rise of 50° C. The heat flux was $150W/cm^2$ when the pressure drop of the water (ΔP) was 9.8×10^3 Pa and $W_c = 140$ μm and 110 W/cm^2 when $\Delta P = 1.9 \times 10^3$ Pa and $W_c = 340$ μm.

As illustrated by these examples, microchannel water cooling yields an extremely low thermal resistance, and a great deal of space saving is possible owing to a very small height of the coolant channels. When the same working principle of laminar convective heat transfer in narrow channels is applied to air cooling, the overall size of a heat sink could be 6.35 × 6.35 mm and 1.27 cm high, and the thermal resistance between the chip and, as shown by Goldberg (1984), air amounts from $4 - 10.7°C/W$ at the air velocity of 12.4 m/s.

Another scheme of gaseous cooling, utilizing Joule-Thomson refrigeration in micropassages, was proposed by Garvey and Little (1981). If R-22 or ammonia is used as a working fluid in ambient temperature refrigeration, it is claimed that the removal of heat at the rate of $10 - 50$ W/cm^2 should be possible using a 2 × 2 × 0.1 cm refrigerator. This could be the highest performance to be achieved by gaseous cooling with small coolant paths. However, actual data proving such a high performance have not been available in the open literature.

Technical difficulties in microchannel water cooling might grow at the system level with increasing system size. Engineering problems of fabricating piping and equalizing flow rates in multiple parallel paths will require attention.

Another mode of microchannel cooling, evaporative heat transfer in capillary grooves, is exactly what one finds in the evaporator section of a heat pipe. Cotter (1984) computed an achievable thermal resistance supposing a counter annular flow in microchannels. As an illustration, methanol operating at 50°C is considered as the working fluid and the triangular pipe having a side of 0.2 mm and

Table 1.10: Summary of reports on the state-of-the-art of Japanese Computers

Fujitsu	M-380	Hiraguri, Ueno, Tsuchimoto	(1981)
	M-780	Yamamoto, Udagawa, Okada	(1986)
	VP-400	Hiraguri, Tabata, Koike	(1985)
Hitachi	M-280	Otsuka, Usami	(1981)
		Usami et al.	(1983)
	M-680	Kobayashi et al.	(1985)
	S-810	Kobayashi et al.	(1984)
NEC	ACOS-1000	Kanai (1981), Saito et al.	
	ACOS-1500	Izutani et al.	(1985)
	SX	Watari, Murano	(1985)

a length of 1 cm is chosen as the channel geometry. If the channels are closely packed in parallel on the surface, a heat removal rate of a few W/cm^2 is possible. If the channels are embedded in a solid as a matrix of parallel pipes occupying 10% by volume, the heat removal rate would amount to a few tens of W/cm^2. These values of heat flux are comparable to the ones experimentally achieved by the porous surface or the porous stud in FC-72 already described.

7 ASSESSMENT OF THE HEAT LOAD IN THE RECENT SERIES OF JAPANESE COMPUTERS

Table 1.10 summarizes the reports on current state of the art of thermal management of large-scale computers. The manufacturer's name and the code name of the machine precede the authors' names. Also referred to in this chapter are the data of packages reported by Kohara et al. (1983, 1984).

Figure 1.28 shows the advances in the integrated circuit technology for high-speed logic chips. The horizontal axis shows the number of logic gates on a chip. The vertical axis of the upper graph corresponds to the time required for the switching operation of a gate, and that of the lower graph the rate of heat dissipation from a chip.

Some data points form pairs as indicated by the distinct symbols. The data of each pair belong to a same machine, the difference in the gate count arising from the function of the gates. The count of gates on a chip for simple operational functions tends to be larger than that of a chip performing complex logic operations. The suffix x in the lower graph indicates that the reported power consumption is the maximum value permitted by an employed circuit integration technology, while the other data represent the levels of power comsumption in the states of normal usage.

The number of gates on a chip has increased owing both to the miniaturization of circuits and to the enlargement of chip area. For example, Usami et al (1983) and Kobayashi et al (1985) reported that the increase in the gate count from 1500 to 5000 was achieved by the reduction of the minimum feature length of circuits from 2 μm to 1.5 μm. The chip area has been increased steadily from 4.5 mm square for a 4-watt chip, 7 mm square for a 6-watt chip, to 9 mm square for a 8.5-watt chip. The high-speed logic chips are equipped with emitter-coupled logic gates or current-mode logic gates, which operate with ceaseless flows of electric current. Although the power consumption on an individual gate decreases with the scaling down of the physical size of the gate, the growth of gate count raises the total heat dissipation from a chip. These gate switching modes will serve as the primary technologies for high- speed chips in the foreseeable future. The increase in the heat dissipation shown in Fig. 1.28 has been achieved in a period of about six years. The extension of this trend shows that a few tens of watts may become a reality in the not-too-distant future. It is becoming more and more evident that the management of the heat dissipated by the chip manifests itself as a controlling factor in the advancement of circuit integration.

Figure 1.29 shows the heat flux at the chip, package (module), and board levels. The horizontal axis shows the area of the component, those of the chip and the package/module are their footprint areas on the PWB. Open symbols are used for the data of the air-cooled machines, and shaded symbols for the data of the liquid-cooled machines. The liquid cooling in those instances is termed more precisely as indirect liquid cooling, where thermal conduction devices

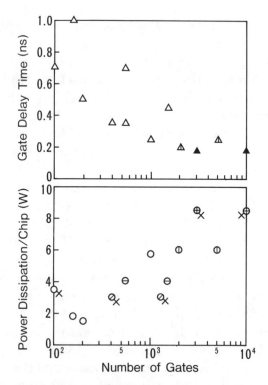

Figure 1.28: Processing speed and power dissipation as a function of the number of gates on logic chips.

are present between the chip and the liquid cooled cold plate (Fig. 1.13(c) and (d)). Double circles at the package/module level are the data of multi-chip modules, which include not only the data of logic modules but also those of high-speed memory modules. From Fig. 1.29 one may make the following observations.

1. The data of heat flux on chips of the liquid-cooled machines are found among those of the air-cooled machines. This indicates that, in the present generation of large-scale computers, the management of heat at the chip level is not the prime motive force for the adoption of liquid cooling.

2. At the package/module level, the heat flux data is spread over a range $0.24 - 2\ W/cm^2$, and the data of component area ranges between 1.6 and 100 cm^2. This wide spread of the package level data results from the diversity of designs.

This also reflects the development of packaging technology since around 1980.

3. It should be noted that the heat flux of the liquid-cooled modules is at the highest level but does not necessarily exceed those of the air-cooled packages by a large margin. This implies that a heat flux of around $2\ W/cm^2$ is still manageable by air-cooling coupled with the use of finned heat sinks.

4. The board level data indicate that this is where the liquid-cooling has become a necessity in order to cope with increased heat load. The limitation of air-cooling first comes from the fact that, due to the small heat capacity of the air, an increasingly high velocity is required for the air stream in order to suppress the temperature rise of the air which, if not controlled, may lead to acoustic noise problems. Liquid, with its large heat capacity, relieves the strained thermal environment at the board level.

5. The liquid-cooled modules are mounted on larger boards than those contained in air-cooled machines. This is reflecting the development of board systems which creates a relatively large

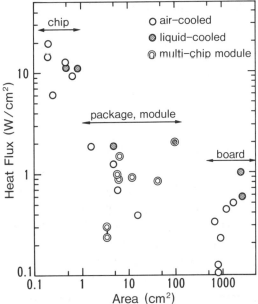

Figure 1.29: Heat flux as a function of component area.

space in front of the board, thereby allowing the installation of rather bulky devices for indirect liquid cooling.

Another cross section of the package-level data is presented in Fig. 1.30, where the vertical axis shows the thermal resistance from the junction to the air, and the horizontal axis the volume of the package. All of the data belong to the air-cooled packages, and again double circles are used to plot the data of the multi-chip modules. Those packages and modules are equipped with finned heat sinks, so that the volume gives a good indication of the size of the heat sinks. In the case of multi-chip modules, the modular thermal resistance is defined by summing up the parallel resistances for heat flows from individual chips to the air. If the component resistance for each of N chips in the module is identical, the modular resistance is equal to $1/N$ times the component resistance. More chips in a module often means an increased placement density of chips accompanied by an increase in the thermal resistance for an individual chip due to the decreased area for heat spreading around the chip. However, this increase in the individual resistance is, in general, small, so that the sum of the parallel resistances results in a smaller modular resistance for those modules containing more chips.

A group of data bounded by the two broken lines in Fig. 1.30 belongs to the packages and modules, where chips are bonded on their back faces to the substrate immediately below the finned heat sink. In such packaging structures, the internal thermal resistance is much smaller than the resistance of the finned heat sink. A certain correlation between the thermal resistance and the modular volume, indicated by the two broken lines, reflects the fact that the heat sinks are designed for different heat loads but with some common norms. The common factors in the heat sink designs are the material of fins, which is aluminum, and the geometry of fins. The thickness of a fin is set around 1 mm, and the gap between the fins is in the range of 2 – 3 mm. With such similarities in the fin geometry the surface area provided by the finned heat sink is almost proportional to the module volume. Where the internal thermal resistance is negligible and the heat transfer coefficient on the fins is assumed fixed, and the thermal resistance decreases in proportion to the surface area, and hence, the volume. The lines bordering the data group of back-mounted face-down modules have a slope of −40 degrees, implying that the internal thermal resistance is small. The bandwidth of the data group is produced partly by the difference in the velocity of the cooling air, which is in the range of 4 − 8 m/s.

The data marked by shaded circles belong to the packages and modules, where the chips' active sides face the the substrate and the electrical connections between the circuits on the chips and the substrate are made through solder balls. A relatively large internal thermal resistance results from this bonding scheme, and attempts have been made to reduce the internal resistance by methods already discussed (see Fig. 1.11(c) and (d). In spite of these attempts, the data of the single chip packages in Fig. 1.30 lie well above the data of the back-mounted face-down package type. The type of I/O leads employed in the Pin Grid Array, requires a larger package base area than the leaded chip carrier where the I/O lead count is relatively small, (Fig. 1.12). This requirement on a larger package area, hence, a larger package volume, is partly attributable to the location of the data points above the data group of the back-mounted scheme which is often employed in combination with the flat leaded I/O scheme. The scheme of pad mounting, however, has a great advantage in reducing the distance for signal exchange between the chip and the substrate, and also in placing chips close to each other in multi-chip modules. The decrease of thermal resistance indicated in Fig. 1.30 for a multi-chip module is due to the same effect of increased surface area on the module as in the case of back-mounted package types. The pad-bonding scheme also has advantages in electronic performance as the number of chips in the module increases. As mentioned earlier, the scheme to suppress temperature rise of coolant at the board level has to be devised in response to the development of PWB system configurations. Figure 1.31 shows the three principal PWB systems used in the current generation of large-scale computers. In Fig. 1.31a, the PWB's are plugged into the mother board. Electrical connections among the PWB's are made through the circuits embedded in the mother board and the cables attached at the free ends of the PWB's. Such PWB arrays are stacked, one over the other, and cooling air is blown from the bottom to the top of the assembly. The increase in the heat load at the board level has been managed by increasing the velocity of the cooling air and the installation of intercoolers for the air between the PWB arrays. Such

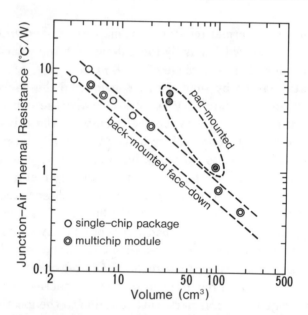

Figure 1.30: Junction-to-air thermal resistance as a function of module volume

measures are no longer effective in dealing with high levels of heat dissipation in the present generation of computers. Kobayashi et al (1985) have shown that a partial solution to the problem of air-cooling has been provided by the development of PWB's having many layers of conductor layers, for instance, as many as twenty layers of copper conductors. The high density PWB's allow the construction of a CPU in one or a few single stacks of a PWB array, where the path length of the air flow is short. This holds the variation of air temperature within the CPU to a minimum.

Short and parallel paths for air flows are also provided in the board arrangement of Fig. 1.31(b), where an array of horizontal PWB's are encased in a frame. Signal communications among the PWB's are made through the conductors in the side boards.

For the board arrangements of Fig. 1.31(a) and (b), the spacing between PWB's is an important factor in making the signal transmission time among the PWB's as short as possible. With the advent of highly integrated modules, however, a CPU can be built with one large board carrying the modules, as illustrated in Fig. 1.31(c). Even where a few such boards are used to form the core of a processor, the spacing between the boards does not lead to a bottleneck in the speed of the data processing. In the large board system, there is a relatively large space in front of the board which allows the installation of cooling devices with less constraints on their volume than in the parallel board array system. The devices for indirect liquid cooling require a relatively large space for their bodies composed of thermal conduction elements and cold plates and for the hoses to supply cooling water to the cold plates. The large board system and the indirect liquid cooling scheme are, therefore, interdependent. Watari and Murano (1985) have described NEC's supercomputer SX which has twelve water-cooled modules whose thermal conduction device is shown in Fig. 1.13(c) on a board of 54.1 × 45.7 cm^2. The Fujitsu M780 has been described by Yamamoto et al (1986) and has a variant of indirect liquid cooling where chip carriers are cooled by the device shown in Fig. 1.13(d). In what is called single-board CPU, 336 chip carriers are mounted on both faces of a board measuring 54 × 48.8 cm^2.

As illustrated by the interrelationship between the PWB system configuration and the cooling system, the development of a new cooling technology and that of a circuit fabrication technology that leads to an increased level of component integration are often complementary. Thermal engineering, therefore, is now an essential discipline required in the early stages of product planning and development.

Owing to the advances in the integration technology, the processors of the computer have become more compact in size, in spite of the requirement of more memory capacity and higher data processing speed. With the progress of integration at the component and system levels, it appears that the following course of development is going to prevail in computer installations.

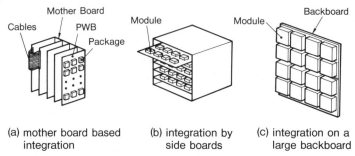

(a) mother board based integration

(b) integration by side boards

(c) integration on a large backboard

Figure 1.31: Schemes of board connection.

1. One of the major concerns of the computer designer is the electric power consumed by the computer system. If a newly developed system can be operated using an amount of electric power less than or equal to that of the existing system while offering an enhanced capability in data processing, it will have a competitive edge in the marketplace because the user will not be required to retrofit its power supply. In this case, the new machine does not yield a marked increase in the total heat load of the computer room. However, components bearing miniaturized structures in the processors as well as in peripherals such as magnetic disc files will require a great deal of attention in order to prevent their thermal environment from drifting out of the designed operative norm.

2. The likely development prompted by the reduction of the physical size of processors is the enhancement of a computing system due to the installation of more processors or different machines in the computer room. This will be accompanied by the growth in the number of peripherals and the total heat load of the computer room will then be multiplied.

3. The heat disposal system will be made compact and energy efficient in response to the above developments of the computer system.

The thermal environment for the computer system is provided by a concerted operation of air-conditioners and water chillers. Figure 1.32 shows the types of heat disposal equipment and their specific functions in pumping heat out of the processors. In Fig. 1.32(a), the air-conditioner supplies chilled air to the underfloor duct, and the air drawn from the duct into the processor carries heat out to the computer room. A computer room is necessarily equipped with stand-alone air-conditioners, because the convenience of installation of processors and peripherals precludes the use of a heat disposal system which carries heat from the electronic equipment directly to the outdoor atmosphere. The room air serves as an intermediate heat transfer medium, and the air-conditioners and the water piping with the cooling tower at the extreme end transport heat from the room to the atmosphere. The temperature of the underfloor air is typically in the range of 16 − 22 $^{\circ}C$, and the relative humidity is in

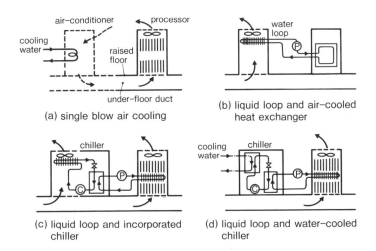

(a) single blow air cooling

(b) liquid loop and air–cooled heat exchanger

(c) liquid loop and incorporated chiller

(d) liquid loop and water–cooled chiller

Figure 1.32: Basic schemes of heat disposal from processors.

the range of 60 – 70%. Most of the refrigeration load for the air-conditioner is sensible heat from the computer. According to an estimate for the heat load in summer time, as given by Kawahara, Kondo, Yoshida (1986), 95% of the refrigeration load is sensible heat, and 85% of that sensible heat comes from the computer.

In Fig. 1.32(b), heat from the water-cooled processor is discharged to the room air by way of a water loop and the air-cooled heat exchanger, as reported by Yamamoto, Udagawa and Ohada (1985), is housed in a frame. Figure 1.32(c) shows a direct-expansion chiller supplying chilled water to the intercooler in the processor. The condenser of the chiller is cooled by air drawn from the underfloor duct. The scheme of Fig. 1.32(d) is made more energy efficient by connecting the condenser of the chiller and the exterior atmosphere by the water loop. The manufacturers of computers provide this water cooling scheme as an option for the customer. In general, the equipment and the schemes shown in Fig. 1.32 coexist in a computer room. Even where liquid cooling is employed for the CPU, the frames of memory storage, I/O and other processors are cooled by air. Thermal systems engineering is becoming more important in the plan and operation of such a composite system.

Air-conditioners and chillers specifically designed for the operation in computer rooms have a refrigeration load in the range of about 25 – 45 kW. Examples of the performance data are shown in Table 1.11. Along with the refrigeration rating (R), the electrical power consumed by the apparatus (E) and the ratio of refrigeration rating to power comsumption (R/E) are shown. The type of condenser is a decisive factor for the energy efficiency of the apparatus; the water-cooled condenser yields higher R/E than the air-cooled condenser. What differentiates the computer room air-conditioning machine from conventional ones is the need to maintain humidity in a prescribed range. In the phase of dehumidification, air is once cooled to condense the moisture and then usually reheated to have its temperature raised to the desired level. Reheating by electric heaters is convenient but power consuming, while regenerative reheating that utilizes heat from the condenser comsumes less power. When the humidity has to be raised, the electric heater to the humidifier is activated. This operation consumes considerable power as exemplified by the figures in the parentheses in Table 1.11. The on-time of the humidifier, however, is usually limited to a short period, such as during the start up for the computer system in dry climate.

The performance data in Table 1.11 reflect the state-of-the-art of technology. More effort is needed to raise the energy efficiency, and reduce the physical size of the apparatus. Various methods of enhancing heat transfer may find applications in the equipment constituting a system of thermal environmental control for computers.

Table 1.11: An example of a set of air conditioner ratings.

With	water–cooled condenser				air–cooled condenser			
With	electric reheaters		regenerative reheaters		electric reheaters		regenerative reheaters	
R Refrigeration rating (KW)	28	42.4	29.8	46	25	37.2	25	37
E Electric power consumption (KW)	11.8 (29.8)	18.3 (45.3)	9.6 (12.6)	14.5 (18.5)	13 (31)	19.5 (46.5)	11.9 (4.9)	17 (21)
R/E	2.42 (0.94)	2.32 (0.94)	3.1 (2.37)	3.18 (2.49)	1.92 (0.81)	1.9 (0.8)	2.1 (1.68)	2.18 (1.76)

( )=when the humidifier is activated to its full capacity

8   CONCLUDING REMARKS

Thermal management of future generations of electronic equipment needs to be based on a fundamental understanding of the physical processes involved. Stringent requirements on temperature distributions at the chip, package, board and system levels can only be met by advances in multifaceted heat transfer engineering. The research topics highlighted in this article are just a small fraction of the large body of thermal problems. There are many other important topics such as packaging gallium arsenide chips, heat transfer in chips and packages which operate in a cryogenic environment and thermal stress in cryogenic devices during the cool-down operation.

Turning attention to the devices operating on the periphery of mainframe computers or on remote terminals, one finds a host of thermal problems. Printers, recorders and optical displays all depend on advanced thermal management to increase the operating speed, expand the capacity or produce a high quality visual output. Turning attention further to the outer environment of electronic equipment, one foresees strains on existing air conditioning systems in offices and factories as heat dissipation from electronic equipment increases. An efficient and compact air conditioning system is a subject of increasing importance in the electronic age.

Major advances in the speed of information processing have been borne out by the silicon bipolar technology whose working principle requires more power to increase the processing speed. Looking at the current trend of increasing power, one wonders how much power consumption is justified for information processing. Although the bipolar technology would not be totally replaced by other technologies in the foreseeable future, the day may come when a far greater mass of data will be processed with much greater speed and drastically lower power consumption than can be witnessed today. It appears too early to speculate on the future of new schemes such as optical computers and biocomputers. It is, however, certain that the ground laid down in the era of bipolar technology will continue to serve as the base of more sophisticated heat transfer engineering in the era of new technologies.

9 REFERENCES

Aihara, T (1967). The method of performance estimation for naturally cooled vertical fin heat exchangers, *Trans JSME* **35**, 77-86 (in Japanese).

Akino, A, Suzuki, K, Sanogawa, K and Okamoto, Y (1983). Visualization of temperature and heat transfer distribution around a protuberance in a parallel-plate-channel by thermosensitive liquid crystal film, *Flow Visualization* **3**, 40-46.

Andrews, J, Mahalingham, L, and Berg, H (1981). Thermal Characteristics of 16 and 40 pin plastic dips, *IEEE Trans CHMT-4*, 455-461.

Antonetti, V W and Simons, R E (1985). Bibliography of heat transfer in electronic equipment, *IEEE Trans CHMT-8*, 289-295.

Arshad, J, and Thome, J B (1983). Enhanced boiling surfaces: heat transfer mechanism, mixture boiling, *ASME/JSME Thermal Eng Joint Conf* **1**, 191-197.

Arvizu, D C, and Moffat, R J (1982). The use of superposition in calculating cooling requirements for circuit board mounted electronic components, *Electron Components Conf* **32**, 133-144.

Ashiwake, N, Nakayama, W, Daikoku, T, and Kobayashi, F (1983). Forced convective heat transfer from LSI packages in an air-cooled wiring card array, *Heat Transfer in Electronic Equipment - 1983, ASME HTD* **28**, 35-42.

Aung, W (1973). Heat transfer in electronic systems with emphasis on asymmetric heating, *Bell Syst Tech J* **52**, 907-925.

Ayub, Z H, and Bergles, A E (1985). Pool boiling from Gewa surfaces in water and R-113, in *Augmentation of Heat Transfer in Energy Systems, ASME HTD* **52**, 57-66.

Bar-Cohen, A (1983). Thermal design of immersion cooling modules for electronic components, *Heat Transfer Eng* **4**, 35-50.

Bar-Cohen, A, (1985). Thermal management of air and liquid- cooled multi-chip modules, *ASME/AICHE Natl Heat Transfer Conf*, Denver, CO.

Bar-Cohen, A, and Rohsenow, W M (1984). Thermally optimum spacing of vertical, natural convection cooled, parallel plates, *J Heat Transfer* **106**, 116-123.

Bar-Cohen, A, and Schweitzer, H (1983). Thermosyphon boiling in vertical channels, *Heat Transfer in Electronic Equipment-1983, ASME HTD* **28**, 13-20.

Bergles, A E, and Chyu, M C (1982). Characteristics of nucleate pool boiling from porous metallic coatings, *J Heat Transfer* **104**, 279-285.

Biskeborn, R G, Horvath, J L, and Hultmark, E B (1984). Integral cap heat sink assembly for the IBM 4381 processor, in *Proc 1984 Int Electronic Packaging Society Conf*, 468-474.

Bravo, H V, and Bergles, A E (1976). Limits of boiling heat transfer in a liquid-filled enclosure, in *Proc 1976 Heat Transfer and Fluid Mechanics Inst.*, Stanford Univ. Press, CA, 114-127.

Burch, T, Rhodes, T, and Acharya, S (1985). Laminar natural convection between finitely conducting vertical plates, *Int J Heat Mass Transfer* **28**, 1173-1186.

Campo W, Kerjilian, G, and Shaukatullah, H (1982). Prediction of component temperatures on circuit cards cooled by natural connection, *IEEE Trans CHMT-5*, 499-501.

Chu, R C, Hwang, U P, and Simons, R E (1982). Conduction cooling for an LSI package: A one-dimensional approach, *IBM J Res Dev* **26**, 45-54.

Cotter, T P (1984). Principles and prospects for micro heat pipes, in *Proc 5th Int Heat Pipe Conf*, Preprints IV, 126-133.

Coyne, J C (1984). An analysis of circuit board temperatures in electronic equipment frames cooled by natural convection, *Fundamentals of Natural Convection/Electronic Equipment Cooling, ASME HTD* **32**, 59-65.

Department of Defense (1982). Reliability prediction of electronic equipment, MIL-HDBK-217D, Washington DC.

Electronic Parts and Materials (1984). **23** (11), 21-101 (Edition for 256 K DRAM, in Japanese).

Electronic Parts and Materials (1986). **25** (1), 21-77 (Edition for 1 M DRAM, in Japanese).

Elenbass, W (1942). Heat dissipation of parallel plates by free convection, *Physica* **9**, 1-28.

Ellison, G N (1984). *Thermal Computations for Electronic Equipment*, Van Nostrand Reinhold, New York.

Fujii, M, Nishiyama, E, and Yamanaka, G (1979). Nucleate boiling heat transfer from micro-porous heating surface, in *Advances in Enhanced Heat Transfer, ASME*, 45-51.

Fukutomi, N (1983). Multiwire PWB for VLSI mounts, *Nikkei Electronics*, 137-158 (in Japanese).

Garvey, S, and Little W A (1981). The potential for integrating microminiature refrigeration into hybrid electronic packages, *Int J Hybrid Microelectronics* 4, 296-298.

Goldberg, N (1984). Narrow channel forced air heat sink, *IEEE Trans CHMT-7*, 154-159.

Hein, V L, and Edrogan, F (1971). Stress singularities in a two-material wedge, *Int J Fracture Mech* 7, 317-330.

Heya, N, Takeuchi, M, and Kimura T (1984). Natural convection heat transfer in enclosures with partial partitions, *Trans JSME* Ser B 50, 724-732 (in Japanese).

Hiraguri, T, Tabata, A, and Koike, Y, (1985). Vector processors - the FACOM VP series, *Fujitsu*, 36 (4), 303-311 (in Japanese).

Hiraguri, T, Ueno, S, and Tsuchimoto, T (1981). Large-scale computer M-380/382, *Nikkei Electronics*, 26 (276), 179-199 (in Japanese).

Igarashi, T (1985). Heat transfer from a square prism to an air stream, *Int J Heat Mass Transfer* 28, 175-181.

Ishizuka, M, Miyazaki, Y, and Sasaki, T (1984). Aerodynamic resistance for perforated plates and wire nettings in natural convection, *ASME* Paper 84-WA/HT-87.

Izutani, Y, Kuwata, M, Bando, K, Narita, Y, Hashimoto, K, Yoshimoto, R, Akagi, M, Nakamura, T, Inoue, M, Ishibashi, M, Watari, T, Maatsuo, H, and Ishikawa, K (1985). Hardware of ACOS system 1500 series, *NEC-Giho* 38 (11), 9-21 (in Japanese).

Jaluria, Y (1984). Interaction of natural convection wakes arising from thermal sources on a vertical surface, *Fundamentals of Natural Convection/Electronic Equipment Cooling, ASME HTD* 32, 67-76.

Joudi, K A, and James, D D (1977). Incipient boiling characteristic at atmospheric and subatmospheric pressures, *J Heat Transfer* 99, 398-403.

Kanai, H (1981). Low energy LSI and packaging for system performance, *IEEE Trans* **CHMT-4** (2), 173-180.

Kawahara, K, Kondo, S and Yoshida, M (1986). Daikin packaged air conditioner for computer room, *Reito (Refrigeration)* **61** (703), 55-62 (in Japanese).

Kennedy, D P (1960). Spreading resistance in cylindrical semiconductor devices, *J Appl Phys* **31**, 1490-1497.

Kishimoto, T, Sasaki, E, and Moriya, E (1984). Gas cooling enhancement technology for integrated circuit chips, *IEEE Trans* **CHMT-7**, 286-293.

Kishinami, K, and Saito, H (1984). Experimental study on natural convection heat transfer from vertical wavy plates having localized heat sources, *Trans JSME* **50**, 2496-2499 (in Japanese).

Kobayashi, F, Takizawa, K, Ogiue, K, Toda, G, and Wajima, M (1984). Packaging technology for the supercomputer HITACHI S-810 array processor, *Proc IEEE Int Conf Computer Design*, 379-382.

Kobayaski, F, Murata, S, Watanabe, H, Kawashima, S, Anzai, A, Murakami, K, and Ikuzaki, K (1985). Hardware technology for M- 680/682H, *Nikkei Electronics*, 268-288 (in Japanese).

Kohara, M, Hatta, M, Genjyo, H, Shibata, H, and Nakata, H (1984). Thermal stress-free package for flip-chip devices, *IEEE Trans* **CHMT-7**, 411-416.

Kohara, M, Nakao, S, Tsutsumi, K, Shibata, H, and Nakata, H (1983). High thermal conduction package technology for flipchip devices, *IEEE Trans* **CHMT-6**, 267-271.

Korner, W, and Photiadis, G (1977). Pool boiling heat transfer and bubble growth on surfaces with artificial cavities for bubble generation, in *Heat Transfer in Boiling*, Hemisphere, New York, 77-84.

Kotake, H, and Takasu, S (1980). Quantitative measurement of stress in silicon by photoelasticity and is application, *J Electrochem Soc: Solid-State Sci and Tech* **127**, 179-184.

Kraus, A D, and Bar-Cohen, A (1983). *Thermal Analysis and Control of Electronic Equipment*, McGraw-Hill, New York.

Kurokawa, Y, Utsumi, K, Takamizawa, H, Kamata, T, and Noguchi, S (1985). AIN substrates with high thermal conductivity, *IEEE Trans* CHMT-8, 247-252.

Kuze, T (1986). Copper leadframe having high thermal conductivity and mechanical strength, *Nikkei Microdevices*, 145-153 (in Japanese).

Lewis, E (1984). The VLSI package: An analytical review, *IEEE Trans* CHMT-7, 197-201.

Liechti, K M, and Theobald, P (1984). The determination of fabrication stresses in microelectronic devices, in *Proc IEEE CHMT 34th Electronic Components Conf*, 203-208.

Ma, C F, and Bergles, A E (1983). Boiling jet impingement cooling of simulated microelectronic chips, *Heat Transfer in Electronic Equipment - 1983, ASME HTD* 28, 5-12.

Mahalingham, M (1985). Thermal management in semiconductor device packaging, *Proc IEEE* 73, 1396-1404.

Marto, P J, and Lepere, V J (1982). Pool boiling heat transfer from enhanced surfaces to dielectric fluids, *J Heat Transfer* 104, 292-299.

Mitchell, C, and Berg, H M (1979). Thermal studies of a plastic dual-in-line package, *IEEE Trans* CHMT-2, 500-511.

Miyatake, O, and Fujii, T (1972). Free convective heat transfer between vertical parallel plates: One plate isothermally heated and the other thermally insulated, *Heat Transfer-Japanese Res* 3, 30-38.

Miyatake, O, Fujii, T, Fujii, M, and Tanaka, H (1973). Natural convective heat transfer between vertical parallel plates: One plate with a uniform heat flux and the other thermally insulated, *Heat Transfer-Japanese Res* 4, 25-33.

Nakao, K, Fujii, M, Nagaku, T, Sakurai, H, and Nai, K (1982). Novel cooling method for sealed card units, *Mitsubishi Electric Tech Report* 56, 68-71.

Nakayama, W (1982). Enhancement of heat transfer, in *Heat Transfer - 1982* 1, Hemisphere, New York, 223-240.

Nakayama, W, Ashiwake, N, Daikoku, T, and Sato, M (1984). Heat transfer conductance and contact pressure of elastic contactors, *ASME Paper 84-HT-87*.

Nakayama, W, Daikoku, T, Kuwahara, H, and Nakajima, T (1980). Dynamic model of enhanced boiling heat transfer on porous surfaces. Part I: Experimental investigation, *J Heat Transfer* **102**, 445-450.

Nakayama, W, Daikoku, T, Kuwahara, H, and Nakajima, T (1980). Dynamic model of enhanced boiling heat transfer on porous surfaces. Part II: Analytical modeling, *J Heat Transfer* **102**, 451-456.

Nakayama, W, Daikoku, T, and Nakajima, T (1982). Effects of pore diameters and system pressure on saturated pool nucleate boiling heat transfer from porous surfaces, *J Heat Transfer* **104**, 286-291.

Nakayama, W, Matsushima, H, and Goel, P (1987). Forced convective heat transfer from arrays of finned packages, *Proc Int Symp Cooling Tech for Electronic Equipment*, Honolulu, 663-678.

Nakayama, W, Nakajima, T, and Hirasawa, S (1984). Heat sink studs having enhanced boiling surfaces for cooling of microelectronic components, *ASME Paper 84-WA/HT-89*.

Nikkei Electronics (1985). Water-cooled single-board CPU; Hardware technology for large-scale computer M-780, 59-62 (in Japanese).

Nishikawa, K, and Fujita, Y (1977). Correlation of nucleate boiling heat transfer based on bubble population density, *Int J Heat Mass Transfer* **20**, 233-245.

Nishikawa, K, Ito, T, and Tanaka, K (1979). Enhanced heat transfers by nucleate boiling on a sintered metal layer, *Heat Transfer-Japanese Res* **8**, 65-81.

Oktay, S (1982). Departure from natural convection (DNC) in low-temperature boiling heat transfer encountered in cooling micro- electronic LSI devices, *Heat Transfer - 1982*, **4**, Hemisphere, New York, 113-118.

O'Neill, P S, Gottzmann, C F, and Terbot, J W (1972). Novel heat exchanger increases cascade cycle efficiency for natural gas liquefaction, *Adv Cryogenic Eng* **17**, 420-437.

Otsuka, K, and Usami, T (1981). Design of a 108 pin VLSI package with low thermal resistance, in *Proc 3rd European Hybrid Microelectronics Conf*, 94-99.

Raju, M S, Liu, X Q, and Law, C K (1984). A formulation of combined forced and free convection past horizontal and vertical surfaces, *Int J Heat Mass Transfer* **27**, 2215-2224.

Saito, M, Furukatsu, N, Hiki, Y, Tokunaga, T, Watanabe, S, and Tomita, K (1981). Technology of high speed data processing and architecture for large-scale computer ACOS 1000, *Nikkei Electronics*, 174-200 (in Japanese).

Sasaki, S, Kishimoto, T, and Moriya, K (1983). Microfin structure and its cooling characteristics, in *Inst Electron Commun, Semiconductor-Mat Div. Natl Conf*, 173 (in Japanese).

Schwinkendorf, W E, and Moss, J (1984). Thermal conductivity and interface resistance of particle-filled resin, *ASME Paper 84-HT-88*.

Sparrow, E M, and Azevedo, L F A (1985). Vertical-channel natural convection spanning between the fully-developed limit and the single-plate boundary-layer limit, *Int J Heat Mass Transfer* **28**, 1847-1857.

Sparrow, E M, Baliga, B R, and Patankar, S V (1978). Forced convection heat transfer from a shrouded fin array with and without tip clearance, *J Heat Transfer* **100**, 572-579.

Sparrow, E M, Niethammer, J E, and Chaboki, A (1982). Heat transfer and pressure drop characteristics of arrays of rectangular modules encountered in electronic equipment, *Int J Heat Mass Transfer* **25**, 961-973.

Sparrow, E M, Yanezmoreno, A A, and Otis, Jr, D R (1984). Convective heat transfer response to height differences in an array of block-like electronic components, *Int J Heat Mass Transfer* **27**, 469-473.

Stephan, K, and Mitrovic, J (1981). Heat transfer in natural convective boiling of refrigerants and refrigerant-oil-mixtures in bundles of T-shaped finned surfaces, *Advances in Enhanced Heat Transfer - 1981, ASME HTD* **18**, 131-146.

Suzuki, Y, and Yoshizumi, A (1984). Resins for packaging material, *Electronic Parts and Materials* **23**, 74-79 (in Japanese).

Terasawa, M, Minami, S, and Rubin, J (1983). Comparison of thin film, thick film, and co-fired high density ceramic multilayer with the combined technology: T & T HDCM, *Int J Hybrid Microelectronics* **6**, 607-615.

Torok, D F (1984). Augmenting experimental methods for flow visualization and thermal performance prediction in electronic packaging using finite elements, *Fundamentals of Natural Convection/Electronic Equipment Cooling, ASME HTD* **32**, 49-57.

Tuckerman, D B, and Pease, R F (1982). Ultrahigh thermal conductance microstructures for cooling integrated circuits, in *Proc IEEE 32nd Electronics Components Conf*, 145-149.

Tuckerman, D B, and Pease, R F (1982). Optimized convective cooling using micromachined structures, *Electrochem Soc Extended Abstr* **125**.

Tuckerman, D B, and Pease, R F (1983). Microcapillary thermal interface technology for VLSI packaging, in *1983 Symposium on VLSI Technology*, 60-61.

Ura, M, and Asai, O (1983). Development of SiC ceramics having high thermal conductivity and electrical resistivity, in *F C Report (Japan Fine Ceramics Assoc)* **1**, 5-13.

Usami, M, Hososaka, S, Anzai, A, Otsuka, K, Masaki, A, Murata, S, Ura, M, and Nakagawa, M (1983). Status and prospect for bipolar ECL gate arrays, *Proc IEEE Int Conf Computer Design*, 272-275.

Wager, A J, and Cook, H C (1984). Modeling the temperature dependence of integrated circuit failures, in *Thermal Management Concepts in Microelectronic Packaging, ISHM* Tech Monograph Ser 6984-003, 1-43.

Watanabe, N, and Ogiso, K (1979). Analysis of temperature rise in multi-layered substrates, *Trans Inst Electro-Commun Eng* **62-C**, 421-428 (in Japanese).

Watari, T, and Murano, H (1985). Packaging technology for the NEC SX supercomputer, *Proc 1985 IEEE Electronic Component Conf*, 192-198.

Wilson, E A (1979). Thermal Conductivity analysis of multi-chip micro-package, *Proc 5th Int Thermal Conductivity Conf*, 49-55.

Wilson, E A (1982). Accommodating LSI in a high performance computer, *Electronic Packaging and Production*, 142-152.

Wirtz, R A, and Stutzman, R J (1982). Experiments on free convec-
    tion between vertical plates with symmetric heating, *J Heat
    Transfer* **104**, 501-507.
Wirtz, R A, and Dykshoorn, P (1984). Heat transfer from arrays
    on flat packs in a channel flow, in *Proc 4th Int Electronic
    Packaging Conf*, 318-326.
Yamamoto, H, Udagawa, Y, and Okada, T (1986). Cooling and
    packaging for FACOM M-780, *Fujitsu* **37** (2), 124-134.
Yanadori, M, Hijikata, K, Mori, Y, and Uchida, M (1985). Funda-
    mental study of laminar film condensation heat transfer on a
    downward horizontal surface, *Int J Heat Mass Transfer* **28**,
    1937-1944.
Yanagida, T, Nakayama, W, and Nemoto, T (1984). Heat trans-
    fer from arrays of rectangular bodies on the wall of a parallel
    plate channel, *Trans JSME Ser B* **50**, 1294-1301 (in Japanese).
Yasukawa, T, Sakamoto, T, and Shida, S (1983). HISETS: A soft-
    ware system for designing semiconductor device packages, in
    Proc IEEE Int Electronic Devices Meeting, Paper 10.7.
Yovanovich, M M (1978). Thermal contact resistance in microelec-
    tronics, in *NEPCON Proc*, 177-188.
Yui, M (1984). Trend of material development in *Electronics Tech*,
    special issue (in Japanese).
Zinnes, A E (1970). The cooling of conduction with laminar natural
    convection from a vertical flat plate with arbitrary surface
    heating, *J Heat Transfer* **92**, 528-535.

Chapter 2

APPLICATION
OF THERMAL
CONTACT
RESISTANCE THEORY
TO ELECTRONIC
PACKAGES

M. Michael Yovanovich
Microelectronics Heat Transfer Laboratory
Department of Mechanical Engineering
University of Waterloo
Waterloo, Ontario

Vincent W. Antonetti
IBM Corporation[1]
Data Systems Division
Poughkeepsie, NY

---

[1]Currently visiting professor, Department of Mechanical Engineering, Manhattan College, Riverdale, New York

1 INTRODUCTION

There are two primary objectives in the thermal design of electronic equipment. The first is to ensure, during worst case operating conditions, that all component temperatures are maintained below specified maximum functional limits. The second objective is to ensure, during nominal operating conditions, that the relatively lower component temperature reliability requirement is satisfied. Successful thermal management requires an accurate determination of the individual thermal resistances inherent in the package, including the often critical thermal contact resistances.

A survey of the state-of-the-art of heat transfer technology as applied to electronic packages was given by Bergles, Chu and Seely (1977). The more current papers by Kraus, Chu and Bar-Cohen (1982) and Simons (1983) describe thermal management techniques in electronic equipment and the texts of Kraus and Bar-Cohen (1983) and Ellison (1984) deal with thermal design considerations in electronic equipment. Although there are no books on contact resistance available, survey articles by Yovanovich (1978), Madhusudana and Fletcher (1981) and Antonetti and Yovanovich (1984) have been written.

The purpose of this paper is to expand on, and to update, the surveys of contact resistance in microelectronics by Yovanovich (1978), and Antonetti and Yovanovich (1984).

2 SPREADING AND CONSTRICTION RESISTANCE

2.1 Steady State Half-Space Solutions

Heat transfer from a semiconductor junction is often treated as a small singly-connected planar heat source in perfect contact with a chip which, as shown by Kraus, Chu and Bar-Cohen (1982), is modeled as a semi-infinite body. Furthermore, it is assumed that the thermal energy generated by the semiconductor is released uniformly over the junction area. As shown in Fig. 2.1, heat flows from the junction through the chip, spreads in some complex three-dimensional manner and leaves the chip through some interface or

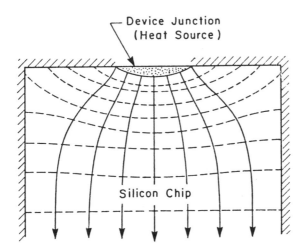

Figure 2.1: Heat flow lines and isotherms when chip is modeled as a half space.

boundary located at a distance which is very large relative to the characteristic dimension of the junction area. This spreading of the heat flow lines gives rise to a spreading resistance. The maximum temperature within the junction area will occur at or near the centroid of the junction area, depending upon whether it is symmetric or nonsymmetric. Yovanovich and Burde (1977) and Yovanovich, Burde and Thompson (1977) have shown that the average junction temperature will always be below the centroidal temperature for all geometries. The spreading (or constriction) resistance is defined as the temperature difference between the average junction temperature and some reference temperature, divided by the total heat flow rate from the junction. It should be noted that an alternate definition of the resistance could be based on either the maximum, or centroidal, temperature. Analytical and numerical solutions have been obtained by Yovanovich (1976), Yovanovich and Burde (1977) and Yovanovich, Burde and Thompson (1977) for numerous symmetric and nonsymmetric, singly-connected, planar geometries subjected to a uniform heat flux. The results of these studies, expressed as a dimensionless spreading resistance defined as $\Psi_\infty = k\sqrt{A_c}R_c$, are shown to be a weak function of the shape of the contact area. The subscript ∞ denotes that the result pertains to a contact area on a semi-infinite body.

For all contact shapes examined, the relationship between the spreading resistance R_c, the thermal conductivity of the conductor (chip) k, and the characteristic dimension of the contact area A_c (junction area) is $k\sqrt{A_c}R_c = 0.467\pm5\%$. In addition, it was observed that the ratio of the average junction temperature to the centroidal temperature was $0.83 \pm 4\%$ for all geometries considered.

An important practical conclusion obtained from these investigations is that a junction of arbitrary shape can be modeled as a circular junction having an equivalent area.

2.2   Effect of Coatings Upon Half-Space Spreading Resistance

The thermal effect of conductive (coating conductivity greater than the substrate conductivity) and resistive coatings (coating conductivity less than the substrate conductivity) upon spreading and constriction resistances of circular contact areas was examined by Negus, Yovanovich and Thompson (1985). Numerical results for the effect of the resistive coatings for a uniform heat flux over a circular contact area are presented in Fig. 2.2 for a range of values of the conductivity ratio $\kappa = k_1/k_2$ where k_1 and k_2 are the coating and substrate conductivities respectively, and the relative thickness of the coating $\beta = \delta/a$ where δ is the coating thickness and a is the radius of the contact area. The dimensionless spreading resistance in this plot is defined to be $k_1 a R_c$. Numerical results for the spreading resistance for an equivalent isothermal contact were correlated for $0.01 \le \beta \le 100$ and $0.01 \le \kappa \le 1$ with a maximum relative error of approximately 2.6% at $\beta = 0.01$ and $\kappa = 0.2$. The correlation gives

$$\Psi = F_1 F_2 + F_3$$

where
$$\beta_1 = \log \beta \text{ and}$$

$$F_1 = 0.12368 - 0.12309\kappa - 0.00085\kappa^2$$

and
$$F_2 = \tanh(2.8479 + 1.3337\beta_1 + 0.06864\beta_1^2)$$

$$F_3 = 0.12325 + 0.1432\kappa - 0.01657\kappa^2. \tag{2.1}$$

This correlation should encompass most anticipated cases of resistive coatings such as oxide and anodized layers. Attempts have been made to correlate the data obtained for conductive coatings, $1 \le \kappa \le 100$, but a single convenient expression such as eq (2.1) has not yet been determined with acceptable accuracy. Two studies were performed by Negus, Yovanovich and Thompson (1985). The first study examined the effect of both heat flux and temperature specified conditions under steady-state conduction, while the second study examined the effect of transient conduction.

2.3 Steady State Flux Tube Solutions

When the spacing between adjacent junctions becomes comparable to the characteristic dimension of the junction area, the spreading resistance will be different from that given by the half-space solution. Kennedy (1960) and Bergles, Chu and Seely (1977) have indicated that, in this case, it is appropriate to treat the junction and chip as a circular, planar heat source attached to one end of an adiabatic circular cylinder (flux tube) of radius equal to one-half the junction spacing. The spreading resistance for this geometric model has been studied by several investigators. Solutions for uniform flux have been given by Roess (1948) and Kennedy (1960). Mikic (1966), Yovanovich (1976) and Negus and Yovanovich (1984) have provided solutions for both uniform flux and isothermal flux distributions. Solutions for other flux distributions and numerical results of the dimensionless spreading resistance (based on the junction radius) as a function of the ratio of the junction radius to the flux tube radius have been provided by Yovanovich (1976).

Negus and Yovanovich (1984) recently obtained the most accurate solution for the dimensionless constriction resistance for an isothermal circular contact on a circular flux tube over a wide range of the relative contact area size. The correlation equation, based upon the definition of dimensionless constriction resistance ($\Psi \equiv 4kaR$)

$$\Psi = 1 - 1.40978\epsilon + 0.34406\epsilon^3 + 0.04305\epsilon^5 + 0.02271\epsilon^7 \qquad (2.2)$$

It is valid over the range $0 \leq \epsilon \leq 0.9$ and has a maximum relative error of 0.02% at $\epsilon = 0.7$, but with round-off gives complete four decimal place agreement with all the correlated data.

Solutions were provided by Negus and Yovanovich (1983) for other combinations of contact areas and flux tubes such as circle/circle

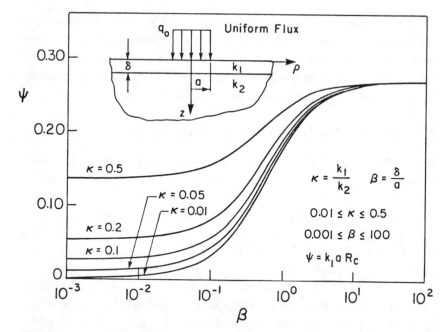

Figure 2.2: Dimensionless constriction resistance for a circular contact on a single layer.

and circle/square for the uniform flux, true isothermal, and equivalent isothermal boundary conditions. Numerical results were correlated using the form of eq (2.2)

$$\Psi = \Psi_\infty + C_1\epsilon + C_3\epsilon^3 + C_5\epsilon^5 + C_7\epsilon^7 \tag{2.3}$$

where $\Psi = 4kaR$ and Ψ_∞ is the half-space solution. The correlation coefficients C_1 through C_7 are given in Table 2.1.

For microelectronic applications, an accurate engineering approximation was given by Negus, Yovanovich and Beck (1987) valid for a circular contact on a circular or square flux tube, or a square contact on a square flux tube and is

$$\Psi = 0.475 - 0.62\epsilon + 0.13\epsilon^3 \tag{2.4}$$

where $\Psi \equiv k\sqrt{A_c}R_c$, $\epsilon = \sqrt{A_c/A_t}$ and the maximum error with respect to the exact solution is less than 2% for $0 \le \epsilon \le .5$ and less than 4% for $0 \le \epsilon \le .7$.

Often, very complex geometries can be handled analytically by superposition of the preceding solutions. In some cases, however, it may be more appropriate, as shown by Kadambi and Abuaf (1983), to use a numerical method to determine the spreading resistance. Furthermore, it should be noted that when dealing with perfect contacts such as the junction-chip interface, the physical properties of the chip do not influence the spreading resistance. The pertinent parameters are the thermal conductivity of the chip, the junction area, and the ratio of the junction area to the projected area of the chip.

2.4 Effect of Coatings Upon Flux Tube Spreading Resistance

Antonetti (1983) obtained a series solution for the *approximate isothermal* circular contact area on a circular flux tube as a function of the relative coating thickness, the conductivity ratio and the relative size of the contact. Antonetti's work should be consulted for a tabulation of the numerical values of the constriction parameter correction factor which is defined as the ratio of the constriction parameter with a coating to that without a coating for the same relative contact area size. Fig. 2.3 shows the effect of the relative coating (layer) thickness, and the substrate-to-layer conductivity $K \equiv k_2/k_1$ when the relative contact area parameter $\epsilon = 0.01$.

Table 2.1: Coefficients for correlations of the constriction parameter Ψ.

Flux Tube Geometry and Contact Boundary Condition	Ψ_∞	C_1	C_3	C_5	C_7
Circle/Circle Uniform Flux	1.08076	-1.41042	.26604	-.00016	.058266
Circle/Circle True Isothermal	1.00000	-1.40978	.34406	.04305	.02271
Circle/Square Uniform Flux	1.08076	-1.24110	.18210	00825	.038916
Circle/Square Equiv. Isothermal Flux	1.00000	-1.24142	.20988	.02715	.02768

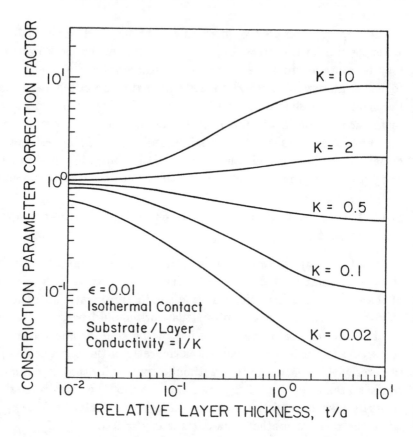

Figure 2.3: Constriction parameter correction factor versus relative layer thickness.

3   CONTACT RESISTANCE AT MECHANICAL JOINTS

3.1   Parameters Influencing Contact, Gap and Joint Conductances

In general, when two surfaces are pressed together the contact is imperfect and the real heat transfer area of the joint is only a small fraction of the apparent area. As shown in Fig. 2.4, actual contact occurs at only a few discrete points. In the absence of an interstitial fluid, the heat flow lines converge and diverge at the microscopic contacting spots.

Steady-state heat transfer across interfaces formed by two contacting solid bodies is usually accompanied by a measurable temperature drop across the joint because there are thermal constriction and spreading resistances to heat flow within a very thin region at the interface. These resistances are in series. The temperature drop ΔT_c at the interface is obtained by extrapolation of the steady-state, one-dimensional temperature distributions from regions *far* from the interface.

The contact conductance or contact coefficient of heat transfer is defined as

$$h = (Q/A_a)/\Delta T_c \tag{2.5}$$

where Q/A_a is the steady-state heat flux based upon the apparent contact area. The thermal contact resistance is defined as the temperature drop at the interface divided by the total heat flow rate. Thus,

$$R = \Delta T_c/Q \tag{2.6}$$

and, therefore, we can write the following relationship between the thermal contact resistance and the contact conductance:

$$R = 1/hA_a \tag{2.7}$$

The thermal resistance concept will be used throughout this chapter because it lends itself to mathematical analysis. Whenever reference is made to thermal conductance, the reciprocal of the product of

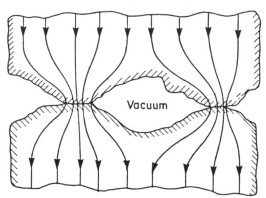

Figure 2.4: Constriction and spreading of heat flow lines at contacting asperities.

the thermal resistance and apparent contact area is implied. Real surfaces are not perfectly smooth (specially prepared surfaces such as those found in bearings can be considered to be almost ideal surfaces) but consist of microscopic peaks and valleys. Whenever two real surfaces are placed in contact, intimate solid to solid contact occurs only at discrete parts of the interface and the real contact area will represent a very small fraction (< 2%) of the nominal contact area.

The real interface is characterized by several important factors:

1) Intimate contact occurs at discrete parts of the nominal interface.

2) The ratio of the real contact area to the nominal contact area is usually much less than 2%.

3) The pressure at the real contact is much greater than the apparent contact pressure. The real contact pressure is related to the flow pressure of the contacting peaks of the asperities.

4) A very thin gap exists in the regions in which there is no intimate contact and it is usually occupied by a third substance.

5) The third substance can be air, other gases, liquid, grease or another metal.

6) The interface is idealized as a line but the actual thickness of the interface ranges from $0.5\,\mu m$ for smooth surfaces to about $60\,\mu m$ for very rough surfaces.

7) Heat transfer across the interface can take place by conduction through the real contact area, by conduction through the substance in the gap or by radiation across the gap if in vacuum. It is possible that all three modes of heat transfer occur simultaneously, but usually they occur in pairs, with solid-solid conduction always present.

The process of heat transfer across an interface is complex because the thermal resistance can depend upon many geometric, thermal and mechanical parameters of which the following are very important:

a) geometry of the contacting solids: surface roughness, asperity slope and waviness

b) thickness of the gap (non-contact region)

c) type of interstitial fluid: gas, liquid, grease, vacuum

d) interstitial gas pressure

e) thermal conductivities of the contacting solids and the interstitial substance

f) hardness or flow pressure of the contacting asperities: plastic deformation of the highest peaks of the softer solid

g) modulus of elasticity of the contacting solids: elastic deformation of the wavy parts of the interface

h) average temperature of the interface: radiation effects as well as property effects.

Because thermal contact resistance is such a complex problem, it is necessary to develop simple thermophysical models which can be analysed and experimentally verified. To achieve these goals the following assumptions have been made in the development of the several contact resistance models which will be discussed later:

a) contacting solids are isotropic: thermal conductivity and physical parameters are constant

b) contacting solids are thick relative to the roughness or the waviness

c) surfaces are clean: no oxide effect

d) contact is static: no vibration effect

e) first loading cycle only: no hysteresis effect

f) relative apparent contact pressure is not too small ($> 10^{-6}$) nor too large ($< 10^{-1}$)

g) radiation is small or negligible

h) heat flux is steady and not too large ($< 10^8 \, W/m^2$)

i) contact is made in a vacuum or the interstitial fluid can be considered to be a continuum if it is not a gas

j) the interstitial fluid perfectly wets both contacting solids.

4 CONFORMING ROUGH SURFACE MODELS

The problem of predicting and measuring contact, gap, and joint thermal conductances has received considerable attention during the past three decades because of the importance of the topic in many heat-transfer systems. Significant progress has been made in our understanding and ability to predict thermal contact conductance. The present state of knowledge has reached a point where simple, explicit correlations have been developed for the contact, gap, and joint conductances.

The purpose of this section is to establish correlations for conforming rough surfaces when interstitial fluids such as gases are present in the gap. The proposed correlations will be compared with recently obtained empirical data to demonstrate the validity of the assumptions used to develop the models.

4.1   Contact Conductance Correlation

The total constriction resistance of the ith contact spot is

$$R_{ci} = \psi_{ci1}/4k_1a_i + \psi_{ci2}/4k_2a_i \tag{2.8}$$

where ψ_{ci1} and ψ_{ci2} are the thermal constriction (or spreading) parameters which depend upon the relative size of the contact spot. Because of geometric and thermal symmetry about the contact plane we can put

$$\psi_{ci1} = \psi_{ci2} = \psi_{ci} \cong [1 - (a_i/b_i)]^{1.5} \tag{2.9}$$

provided $0 < a_i/b_i \leq 0.3$.

If we let $k_s = 2k_1k_2/(k_1 + k_2)$, the harmonic mean thermal conductivity, then eq (2.8) can be written as

$$R_{ci} = \psi_{ci}/2k_s a_i \tag{2.10}$$

The total contact resistance of N contact spots thermally connected in parallel is therefore,

$$\frac{1}{R_c} = \sum_{i=1}^{N} \frac{1}{R_{ci}} = 2k_s \sum_{i=1}^{N} \frac{a_i}{\psi_{ci}} \tag{2.11}$$

The contact conductance can be derived by means of the definition

$$Q_c = h_c A_a \Delta T_c = \Delta T_c / R_c \tag{2.12}$$

Therefore,

$$h_c = \frac{2k_s}{A_a} \sum_{i=1}^{N} \frac{a_i}{\psi_{ci}} \tag{2.13}$$

Noting that $a_i/b_i \leq 0.3$ and $0.85 \leq \psi_{ci} < 1$, the specific constriction parameter ψ_{ci} appropriate to each contact spot can be replaced by the mean value of the constriction parameter ψ_c based upon the total set of contact spots.Therefore, we have

$$\psi_{ci} \cong \psi_c = (1 - \epsilon)^{1.5} \tag{2.14}$$

where $\epsilon = a/b = \sqrt{A_r/A_a}$. The mean value of the constriction parameter depends upon a and b, the contact spot and associated flux tube radii, respectively, determined by the total real and apparent areas.

A detailed geometric analysis of interacting conforming, rough surfaces yields the following important geometric results:

1) Contact conductance parameter

$$\sum_{i=1}^{N} \frac{a_i}{A_a} = \frac{1}{4\sqrt{2\pi}}(m/\sigma)\exp(-x^2) \tag{2.15}$$

2) Relative real contact area

$$\epsilon^2 = A_r/A_a = \frac{1}{2}erfc(x) \tag{2.16}$$

3) Contact spot density

$$n = \frac{1}{16}(m/\sigma)^2\frac{\exp(-2x^2)}{erfc(x)} \tag{2.17}$$

4) Mean contact spot radius

$$a = \sqrt{8/\pi}(\sigma/m)\exp(x^2)erfc(x) \tag{2.18}$$

where $x = Y/\sqrt{2}\sigma$ and Y/σ is called the relative mean plane separation. The surface parameters σ and m are the effective rms surface roughness and the effective absolute surface slope, respectively. They are determined from

$$\sigma = \sqrt{\sigma_1^2 + \sigma_2^2} \tag{2.19}$$

and

$$m = \sqrt{m_1^2 + m_2^2} \tag{2.20}$$

An assumption of plastic deformation of the contacting asperities during the first loading cycle leads to a relationship between the relative real contact area and the relative contact pressure. A force balance on the real and apparent contact areas gives

$$P/H = A_r/A_a = \epsilon^2 = (1/2)erfc(x) \qquad (2.21)$$

This relationship between ϵ and x allows one to compute the other surface parameters. After substitution of the contact conductance parameter, eq (2.15), into the contact conductance expression, eq (2.13), we obtain after multiplying by σ/mk_s the dimensionless contact conductance,

$$(\sigma/m)(h_c/k_s) = \frac{1}{2\sqrt{2\pi}} \frac{\exp(-x^2)}{(1-\epsilon)^{1.5}} \qquad (2.22)$$

with $x = erfc^{-1}(2P/H)$ and $\epsilon = \sqrt{P/H}$. This complex equation was correlated by Yovanovich (1981) to yield the simple equation

$$(\sigma/m)(h_c/k_s) = 1.25(P/H)^{0.95} \qquad (2.23)$$

which agrees with the exact expression to within $\pm 1.5\%$ for $2 \leq Y/\sigma \leq 4.75$.

Example 1

Consider a 5 × 5 mm chip bonded to and centrally located on a 25 × 25 mm alumina ($k = 21$ W/mK) substrate which is 2 mm thick. Assuming the edges and top surface are insulated, determine the thermal resistance from the chip bondline to the bottom surface of the substrate.

The total resistance from chip bondline to substrate bottom surface consists of a spreading resistance and a resistance due to the bulk material. First the spreading resistance. The relative contact size is calculated

$$\epsilon = \sqrt{\frac{A_c}{A_t}} = \sqrt{\frac{5 \times 5}{25 \times 25}} = 0.20$$

The dimensionless constriction (or in this case spreading) resistance from eq (2.4) is

$$\Psi = 0.475 - 0.62\epsilon + 0.13\epsilon^3 = 0.475 - 0.62(0.20) + 0.13(0.20)^3 = 0.352.$$

Then the spreading resistance is

$$R_c = \frac{\Psi}{k\sqrt{A_c}} = \frac{0.352}{21\sqrt{2.5 \times 10^{-5}}} = 3.35^{\circ}C/W.$$

To this the bulk resistance of $0.002/21(6.25 \times 10^{-4}) = 0.15°C/W$ is added. This results in a total resistance from chip bondline to substrate of $3.50°C/W$.

4.2 Gap Conductance

Heat transfer across the interstitial gap formed by the contact of conforming rough surfaces is difficult to analyze because of the complexity of the local geometry which determines whether the local heat transfer can be modelled as continuum, slip or rarefied. To overcome these difficulties it is necessary to model the gap conductance from an overall point of view. To this end it is assumed that the local gap conductance can be modelled as heat conduction between two isothermal parallel plates which are separated by an effective thickness

$$t + M \qquad (2.24)$$

where t is the local gap thickness and M is a gas parameter.

Yovanovich, DeVaal and Hegazy (1982) have shown that the total gap heat flow rate, Q_g, and the overall gap conductance, h_g, are determined by integration over the effective gap area, A_g

$$Q_g = \iint_{A_g} \frac{k_{g,\infty} \Delta T_g dA_g}{t + M} = \iint_{A_g} h_g(x,y) \Delta T_g dA_g \qquad (2.25)$$

Assuming that the variation of ΔT_g is localized around each contact spot, and therefore $\Delta T_g \cong \Delta T_c$ over the major portion of the gap, we have

$$h_g = \frac{1}{A_g} \iint_{A_g} \frac{k_{g,\infty} dA_g}{t + M} \qquad (2.26)$$

where

$$h_g = \frac{1}{A_g} \iint_{A_g} h_g(x,y) dA_g \qquad (2.27)$$

is the mean value of the gap conductance over the effective gap area which is given by

$$A_g = A_a(1 - \epsilon^2) \cong A_a \qquad (2.28)$$

because $\epsilon < 0.3$.

For gaps formed by two conforming rough surfaces having Gaussian height distributions, an expression for the fraction of the projected gap area, dA_g, is given by

$$dA_g = (A_a/\sqrt{2\pi})\exp[-(Y/\sigma - t/\sigma)^2/2]d(t/\sigma) \qquad (2.29)$$

where t is the local gap thickness and σ is the effective rms surface roughness.

Combining Eqs (2.26) and (2.29) leads to the gap conductance and its integral

$$h_g = (k_{g,\infty}/\sigma)I_g \qquad (2.30)$$

where

$$I_g = \{1/\sqrt{2\pi}\} \int_0^\infty \frac{\exp[-(Y/\sigma - t/\sigma)^2/2]}{(t/\sigma + M/\sigma)}d(t/\sigma) \qquad (2.31)$$

The upper limit can be put to a value of 3 for numerical computational purposes because the area under the curve becomes negligible for $u > 2.5$. Hegazy (1985) has shown that the approximation of Yovanovich, DeVaal and Hegazy (1982)

$$I_g(Y) \doteq [Y/\sigma + M/\sigma]^{-1} \qquad (2.32)$$

is accurate provided $(Y/M) < 1$, where Y/σ is the relative mean plane separation given by Yovanovich (1981).

$$\frac{Y}{\sigma} = 1.184\left[-\ln\left(3.132\frac{P}{H}\right)\right]^{0.547} \qquad (2.33)$$

Therefore, the gap conductance can be expressed to a first order approximation as

$$h_g = \frac{k_{g,\infty}}{Y + M} \qquad (2.34)$$

where $k_{g,\infty}$ is the gas conductivity under continuum conditions and the gas parameter is given by

$$M = \alpha\beta\Lambda. \qquad (2.35)$$

The accommodation parameter, α, is defined as

$$\alpha = (2 - \alpha_1)/\alpha_1 + (2 - \alpha_2)/\alpha_2 \qquad (2.35a)$$

where α_1 and α_2 are the accommodation coefficients at the solid-gas interfaces. The fluid property parameter, β, is defined by

$$\beta = (2\gamma)/[(\gamma + 1)Pr] \qquad (2.35b)$$

where γ is the ratio of the specific heats, and Pr is the Prandtl number. The mean free path, Λ, of the gas molecules is given in terms of $\Lambda_{g,\infty}$, the mean free path at some reference temperature and pressure,

$$\Lambda = \Lambda_{g,\infty}(T_g/T_{g,\infty})(P_{g,\infty}/P_g) \qquad (2.35c)$$

The dimensionless gap conductance C_g divided by the gas-to-solid thermal conductivity ratio as a function of the relative contact pressure and the reciprocal of the gas parameter is plotted in Fig. 2.5 for the pertinent range of the independent parameters.

4.3 Dimensionless Joint Conductance

The contact and gap heat transfer rates are approximately independent for most practical contact problems. If the radiative heat transfer rate across the gap is assumed to be negligible, the total or joint conductance, h_j, for conforming rough surfaces is equal to the sum

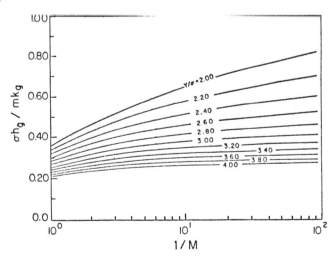

Figure 2.5: **Dimensionless gap conductance versus gas parameter M and relative gap thickness Y/σ.**

of the contact and gap conductance developed above. Therefore,

$$h_j = h_c + h_g \tag{2.36}$$

Multiplying by $(\sigma/m)k_s$ to non-dimensionalize gives

$$(\sigma/m)(h_j/k_s) = (\sigma/m)(h_c/k_s) + (\sigma/m)(h_g/k_s) \tag{2.37}$$

or for convenience

$$C_j = C_c + C_g \tag{2.38}$$

Example 2

Consider a case where heat is to be transferred across a contact formed by an aluminum cold plate and a stainless steel part containing heat producing electronic equipment. Four screws hold the assembly together exerting a total load of 1.38×10^6 N/m^2(200 psi). The projected (or apparent) contact area is 0.00056 m^2. The contacting surfaces are assumed to be flat and the mechanical load is assumed to be uniformly distributed. The specified surface finishes are $1.27\mu m$ (50μ in) for the aluminum, and $0.64\mu m$ (25μ in) for the stainless steel.

Assume the thermophysical properties are as follows. Aluminum: thermal conductivity $k_1 = 170W/mK$, microhardness = $120kg/mm^2$ ($1180 \times 10^6 N/m^2$). Stainless steel: thermal conductivity $k_2 = 16$ W/mK, microhardness = $210kg/mm^2$. Furthermore, assume that the average asperity slope for both metal surfaces is unknown. As a good rule of thumb, for relatively smooth surfaces $\sigma/m = 5$ to $9\mu m$. For this example assume $\sigma/m = 7\mu m$.

First determine the contact conductance (across the solid microcontacts). This begins with a calculation of the harmonic mean thermal conductivity

$$k_s = \frac{2k_1k_2}{(k_1 + k_2)} = \frac{2(170)(16)}{(170 + 16)} = 29.2 \ W/mK.$$

The hardness of the softer material (aluminum) is used to determine the relative contact pressure

$$\frac{P}{H} = \frac{1.38 \times 10^6}{1180 \times 10^6} = 1.17 \times 10^{-3}.$$

For the metallic microcontacts, therefore, the contact conductance from eq (2.23) is

$$h_c = \frac{1.25k_s \left(\frac{P}{H}\right)^{0.95}}{\left(\frac{\sigma}{m}\right)} = \frac{1.25(29.2)(1.17 \times 10^{-3})^{0.95}}{7 \times 10^{-6}} = 8550W/m^2 K.$$

Next, determine the gap conductance (i.e., the heat transfer coefficient across the interstitial air space). These begin with the computation of the effective RMS surface roughness from eq (2.19)

$$\sigma = \sqrt{(1.27)^2 + (0.64)^2} = 1.42\mu m = 1.42 \times 10^{-6}m.$$

Next use eq (2.33) to calculate the distance between the mean planes

$$Y = 1.184\sigma \left[-\ln\left(3.132\frac{P}{H}\right)\right]^{0.547}$$
$$= 1.184(1.42\times 10^{-6})\left[-\ln\left(3.132\ 1.17\times 10^{-3}\right)\right]^{0.547} \stackrel{=}{=} 4.32\times 10^{-6}m$$

In order to determine the gas parameter M, first find α, β, and Λ for air at about one atmosphere and $300\ K$ must be found. From Hegazy (1985), the accomodation coefficient for air against most clean metals is 0.9. Then from eq (2.35a)

$$\alpha = \frac{2 - \alpha_1}{\alpha_1} + \frac{2 - \alpha_2}{\alpha_2} = \frac{2 - 0.9}{0.9} + \frac{2 - 0.9}{0.9} = 2.44.$$

The interstitial fluid parameter from eq (2.35b) is

$$\beta = \frac{2\gamma}{(\gamma + 1)\,Pr} = \frac{2(1.4)}{(1.4 + 1)(0.71)} = 1.65.$$

Using eq (2.35c), the mean free path of the air is

$$\Lambda = 6.44 \times 10^{-8}\left(\frac{300}{288}\right) = 6.67 \times 10^{-8}m.$$

Now the gas parameter M is determined from eq (2.35) as

$$M = \alpha\beta\Lambda = 2.44(1.65)(6.67 \times 10^{-8}) = 0.27 \times 10^{-6}m.$$

From eq (2.34), the contact conductance across the interstitial air gap is

$$h_g = \frac{k_{g,\infty}}{Y + M} = \frac{0.026}{(4.32 + 0.27)10^{-6}} = 5664W/m^2 K.$$

The total or joint conductance per eq (2.36) is the sum of the contact and gap conductances.

$$h_j = h_c + h_g = 8550 + 5664 = 14,214/m^2 K$$

or in terms of thermal resistance

$$R_j = \frac{1}{14,214(5.6 \times 10^{-4})} = 0.126^{\circ}C/W.$$

Notice that the magnitudes of h_c and h_g are about the same. The situation could be improved by copper instead of stainless, or by coating the stainless part with a soft copper. In both instances the harmonic mean thermal conductivity of the contacting materials would be increased, thus increasing the contact conductance. Moreover, the softer copper would also increase the metallic contact area, further increasing the contact conductance.

Finally, it is important to note that the assumption of flat contacting surfaces is rarely achieved in practice. Therefore the joint resistance calculated here should be considered a lower bound, and that the actual joint resistance will be somewhat higher.

4.4   Experimental Verification of Rough Conforming Models

Experimental data have been obtained by Hegazy (1985) which provide ample evidence that the preceding conforming, rough surface contact and gap conductance models are accurate. Data were obtained under vacuum conditions for a variety of metals. Each interface consisted of a relatively smooth, lapped surface and a rough, bead-blasted surface of identical material. The surface parameter (σ/m) was 8.20 μm to 12.4 μm for the smoothest interfaces and it was 38.3 μm to 59.8 μm for the very rough interfaces. The mean interface temperature ranged from $372K$ to $451K$. The bulk hardness of the metals, H_b, ranged from 1010 MPa to 1727 MPa and the contact hardness, H_c, was determined to be 1972 MPa for the very rough Zirconium-4 interface at the highest load and 4113 MPa for the smooth SS304 interface at the lightest load.

The contact hardness, H_c, was determined by means of a mechanical model developed by Yovanovich, Hegazy and Antonetti (1983) for bead-blasted surfaces. The model is based upon a least squares fit of Vickers microhardness measurements obtained at several loads, as shown in Fig. 2.6 for several metals, and a geometric-mechanical model which is based upon a force balance applied to a contact formed by a typical asperity produced by bead-blasting. The effective contact hardness can be determined by means of the following expression based upon a relative contact pressure of $P/H = 10^{-3}$:

$$H_c = c_1(0.95\sigma/m)^{c_2} \qquad (2.39)$$

where c_1 and c_2 are the Vickers microhardness correlation coefficients,

$$H_v = c_1 d_v^{c_2} \qquad (2.40)$$

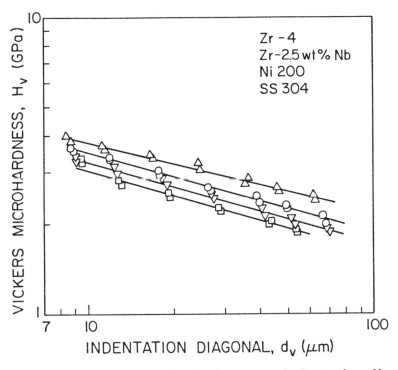

Figure 2.6: Vicker's microhardness versus indentation diagonal for several metals.

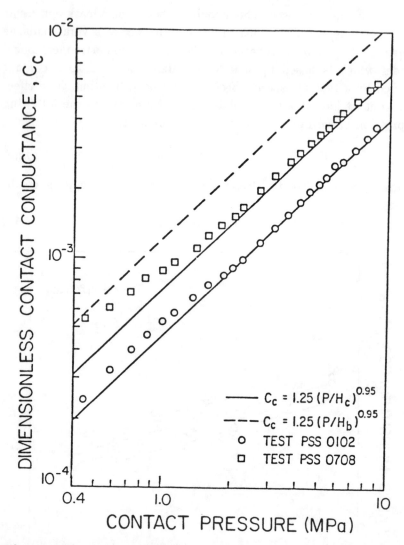

Figure 2.7: Comparison of contact conductance theory and test for stainless steel 304 in vacuum.

and d_v is the measured Vickers diagonal. The microhardness values are GPa and the Vickers diagonal is in μm. The correlation coefficients are given in Table 2.2.

Table 2.2: Hegazy (1985) Vickers microhardness coefficients,
$H_v = c_1 d_v^{c_2}$.

Material	H_b, GPa	c_1	c_2	Max. % Diff.	RMS % Diff.
Zr-4	1.913	5.677	-0.278	3.4	1.7
Zr-2.5wt%Nb	1.727	5.884	-0.267	10.2	2.7
Ni200	1.668	6.304	-0.264	4.8	1.8
SS304	1.472	6.271	-0.229	4.2	1.4

Hegazy also developed an approximate, semi-general microhardness correlation of the metals given in Table 2.2

$$H_c = (12.2 - 3.54 H_b)(\sigma/m)^{-0.26} \qquad (2.41)$$

where H_c, the contact hardness, and H_b, the bulk hardness are in GPa, and the effective surface parameter (σ/m) is in μm. This relationship shows clearly how H_c depends upon H_b and the surface roughness. Thermal contact conductance values were experimentally determined for five different interfaces for the metals listed in Table 2.2. For apparent contact pressures ranging between approximately 0.45 MPa to 890 MPa, the measured contact conductance for all metals is in very good to excellent agreement with the new model predictions. Typical results are presented in Fig. 2.7 for two stainless steel interfaces, the smoothest and the roughest. The circular data points represent the roughest surface tested.

Theoretical values based upon the appropriate values of the contact microhardness are approximately 4100 MPa and 2510 MPa respectively, and the measured bulk hardness was 1470 MPa. The dashed line represents the model predictions based upon the bulk hardness. The differences observed at light contact pressures $P < 2MPa$ are attributed to several factors which are not accounted for in the present model. One of the major factors may be thermal strains induced at the microcontacts due to very large temperature drops and local temperature gradients. Fig. 2.8 shows the very good agreement between the conforming, rough model predictions and the data for three metals for a wide range of the relative contact pressure.

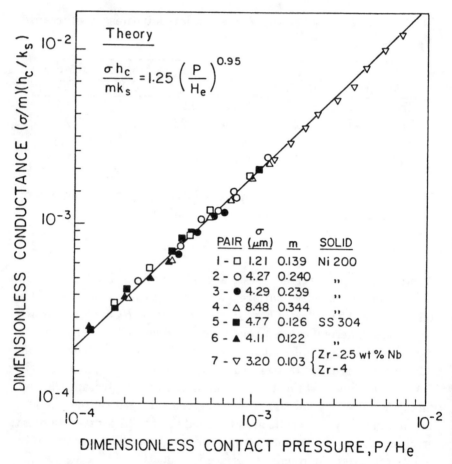

Figure 2.8: **Comparison of contact conductance theory against some test data for clean, bare surfaces in vacuum.**

To verify the joint conductance model with a gas at the interface, Hegazy (1985) measured joint conductances only for nitrogen gas at a pressure of approximately 570 Torr between stainless steel surfaces having $\sigma/m = 36.9\mu m$. The gap parameter Y/M was approximately 29 to 38 over the entire load range. The measured joint conductance results are compared with the model predictions in Fig. 2.9, where it can be seen that the joint results are in good agreement over the entire load range, but the measured values are above the predictions over this load range. At the light load the gap conductance is the dominant mode of heat transfer, and at the high load the contact and gap conductances are comparable. The gap and joint conductance models clearly predict the effect of increasing load, but underpredict the joint conductance by approximately 10 to 15%.

The gap and joint conductance models were also verified by Hegazy (1985) where measurements of contact and joint conductances were made for a stainless steel interface with nitrogen gas at a pressure of approximately 40 Torr. The surface parameter $\sigma/m = 32.3\ \mu m$, the gap parameter Y/M was approximately 2.4. The dimensionless contact, gap and joint conductances are plotted versus the dimensionless contact pressure in Fig. 2.10 where it can be seen that the

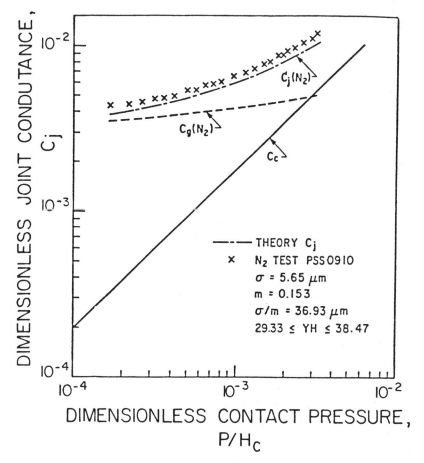

Figure 2.9: Comparison of joint conductance theory and test for stainless steel 304 in a nitrogen environment.

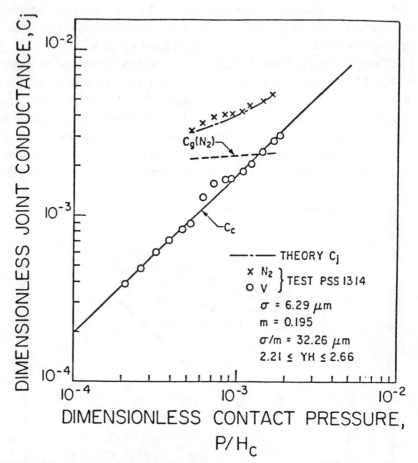

Figure 2.10: Comparison of contact, gap and joint conductances for stainless steel 304 and nitrogen.

joint conductance results are in good agreement over the limited load range, but the contact conductance results and the theory are in good agreement over the entire load range.

5   ENHANCEMENT OF CONTACT CONDUCTANCE

In many electronics packages the thermal conductance across a particular interface must be improved for the thermal design to meet its performance objective. If the joint cannot be made permanent because of servicing or other considerations, the contact heat transfer coefficient can be enhanced, that is utilizing one of the following well-known techniques: application of a thermal grease, insertion of a soft metal foil, or coating one or both of the contacting surfaces with a relatively soft metal.

Thermal greases, however, cannot be employed in many critical electronic assemblies because of the possibility of the grease evaporating and contaminating nearby sensitive components. Foils are attractive from a theoretical point of view but in practice they are not used very often. This is because soft foils tend to wrinkle which can result in an increase rather than a decrease in the contact resistance. Furthermore, soft foils are flimsy, often deflecting under their own weight, which makes them difficult to handle and apply effectively.

Metallic coatings are free of the contamination problems associated with thermal greases and the handling problems associated with soft foils. In addition, when one of the contacting surfaces of a joint is coated, Fried and Kelley (1965), Mal'kov and Dobashin (1969) and Al-Astrabadi, O'Callaghan, Snaith and Probert (1981) have reported as much as an order of magnitude improvement in the contact conductance, depending upon the materials used, the texture of the contacting surfaces, and the metallic layer thickness.

More recently, a new thermo-mechanical model for coated contacts has been developed by Antonetti (1983) and Antonetti and Yovanovich (1985) and shown to predict accurately the thermal contact conductance data obtained from experiments performed on conforming, rough, nickel specimens put in contact with nominally flat, smooth, nickel specimens coated with a silver layer. The purpose of this section is to illustrate the utility of this new model by employing it to analyze a common electronics packaging problem: heat transfer across an aluminum joint operating in a vacuum or in a gaseous environment.

5.1   Theoretical Background

The Antonetti (1983) and Antonetti and Yovanovich (1985) works provided a comprehensive treatment of the theoretical development and experimental verification of the thermo-mechanical model. In

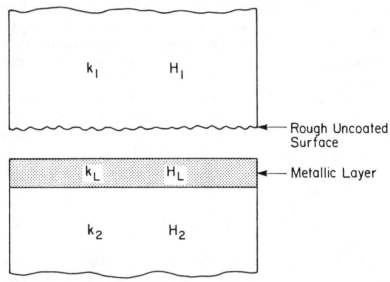

Figure 2.11: Schematic of the coated joint.

the following discussion, therefore, only those portions of the theory needed to apply the model to a thermal design problem will be presented. The general expression for the contact conductance of the coated joint shown in Fig. 2.11 operating in a vacuum is

$$h'_c = h_c \left(\frac{H}{H'}\right)^{0.93} \left[\frac{k_1 + k_2}{C k_1 + k_2}\right] \qquad (2.42)$$

where h_c is the uncoated contact conductance, H is the hardness of the softer of the substrates, H' is the effective hardness of the layer-substrate combination, C is a constriction parameter correction factor (see eq (2.51)), which accounts for the heat spreading in the coated substrate, and k_1, k_2 are the thermal conductivities of the substrates respectively.

As can be seen from eq (2.42), the coated contact conductance is the product of three quantities: the uncoated contact conductance h_c; a mechanical modification factor $(H/H')^{0.93}$; and the thermal modification factor in brackets. The uncoated contact conductance can be determined by means of the conforming, rough surface correlation equation:

$$h_c = 1.25 \left(\frac{m}{\sigma}\right)\left(\frac{2k_1 k_2}{k_1 + k_2}\right)\left(\frac{P}{H}\right)^{0.95} \tag{2.43}$$

where H is the flow pressure (microhardness) of the softer substrate (H_1 or H_2), m is the combined average absolute asperity slope for the two contacting surfaces, and σ is the combined rms surface roughness which can be obtained from profilometer measurements.

For a given joint, the only unknowns in eq (2.42) are the effective microhardness H' and the constriction parameter correction factor C. Thus, the key to solving coated contact problems is the determination of these two quantities.

5.2   Mechanical Problem

Strictly speaking, the effective microhardness must be obtained empirically for the particular layer(coating)-substrate combination under consideration. This requires a series of Vickers microhardness measurements which will result in an effective microhardness plot similar to that shown in Fig. 2.12 (a silver layer on a nickel substrate). If the materials and facility required to generate the required effective microhardness plot are not available, then, to a first approximation, it can be assumed that the general form of the plot for the particular layer-substrate combination under consideration is the same as that shown in Fig. 2.12, and the following equations can be used to estimate the effective microhardness. In the region $t/d < 1.0$:

$$H' = H_2 \left(1 - \frac{t}{d}\right) + 1.81 H_L \left(\frac{t}{d}\right) \tag{2.44}$$

where H_L is the microhardness of the layer and H_2 is the microhardness of the substrate, with both values obtained from a Vickers microhardness test.

Similarly, in the region $1.0 \leq t/d \leq 4.90$

$$H' = 1.81 H_L - 0.21 H_L \left(\frac{t}{d} - 1\right) \tag{2.45}$$

where the relative thickness t/d is determined from

$$\frac{t}{d} = 1.04 \frac{t}{\sigma}\left(\frac{P}{H'}\right)^{-0.097} \tag{2.46}$$

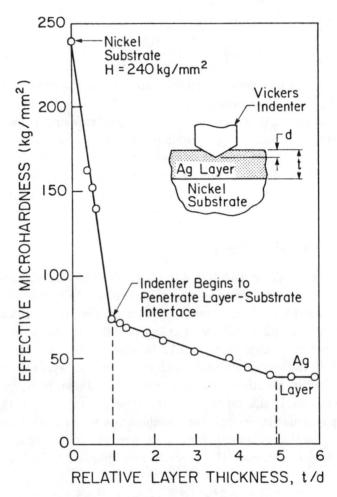

Figure 2.12: Vicker's microhardness measurements for silver layer on nickel 200 substrate.

When $t/d > 4.90$, the effective microhardness is equal to the layer microhardness H_L.

Because t/d depends upon the effective microhardness, and t/d must be known to determine the effective microhardness, an iterative approach is required. Convergence is rapid, however, due to the fact that t/d is a weak function of the effective microhardness. It is recommended that the arithmetic average of the layer and substrate microhardness values be used as the initial guess for the iterative procedure.

Moreover, it is important to realize that very often the substrate surfaces may have been work-hardened, particularly if finished by a lapping process. In this case, it is incorrect to use the bulk hardness of the substrate for H_2. Rather the reader is referred to the work of Yovanovich, Hegazy and DeVaal (1982) for the method to be followed to determine a proper value of the effective substrate microhardness.

5.3 Thermal Problem

The thermal portion of the analysis involves finding the solutions of Laplace's equation within the layer and the substrate subject to the perfect contact boundary condition at their common interface. If the dimensionless constriction (spreading) parameter is defined as

$$\Psi(\epsilon', \phi_n) = 4k_2 a' R_c' \qquad (2.47)$$

then, after determining the constriction resistance R_c from the thermal analysis, the constriction parameter with a layer present at the contact is shown by Antonetti (1983) to be

$$\Psi(\epsilon', \phi_n) = \frac{16}{\pi \epsilon'} \sum_{n=1}^{\infty} \frac{J_1^2(\delta_n' \epsilon')}{(\delta_n')^3 J_0^2(\delta_n)} \phi_n \gamma_n \rho_n \qquad (2.48)$$

Equation (2.48) is the expression for the dimensionless constriction parameter for an uncoated contact (with a uniform flux heat flux condition at the contact area and the resistance based upon the area average temperature) multiplied by three modification factors.

The first of these ϕ_n, accounts for the influence of the layer; the second γ_n, accounts for the contact temperature basis used to determine the constriction resistance; and the third ρ_n, accounts for the contact spot heat flux distribution. For abutting surfaces, it is usual to assume that the contact spots are isothermal. The modification factors in this case are $\gamma_n = 1.0$,

$$\phi_n = K \left[\frac{(1+K) + (1-K)e^{-2\delta_n' \epsilon' \tau'}}{(1+K) - (1-K)e^{-2\delta_n' \epsilon' \tau'}} \right] \qquad (2.49)$$

and

$$\rho_n = \frac{\sin(\delta_n' \epsilon')}{2J_1(\delta_n' \epsilon')} \qquad (2.50)$$

The parameter K in the above equation is the ratio of the substrate to layer conductivities. The constriction parameter correction factor is now defined as the ratio of the constriction parameter with a layer to that without a layer, for the same value of the relative contact spot radius:

$$C = \frac{\Psi(\epsilon', \phi_n)}{\Psi(\epsilon')} \qquad (2.51)$$

The constriction parameter correction factor is tabulated in Antonetti (1983) and in abridged form by Antonetti and Yovanovich (1985).

5.4   Alternate Contact Conductance Analysis

It should be noted that an expression for the contact conductance of a coated joint has been derived by Antonetti (1983)

$$h'_c = \frac{2a'k'}{\Psi(\epsilon')} \left(\frac{N'}{A_a}\right) \qquad (2.52)$$

where the effective thermal conductivity k' of the coated joint is defined as

$$k' = \frac{2k_1 k_2}{C_2 k_1 + k_2} \qquad (2.53)$$

and the constriction parameter can be determined from

$$\Psi(\epsilon') = (1 - \epsilon')^{1.5} = (1 - \sqrt{P/H'})^{1.5} \qquad (2.54)$$

The average contact spot radius can be determined from

$$a' = 0.77 \left(\frac{\sigma}{m}\right) \left(\frac{P}{H'}\right)^{0.097} \qquad (2.55)$$

By means of a force balance at the joint, the total number of contact spots per unit apparent area can be estimated by

$$\frac{N'}{A_a} = \frac{1}{\pi(a')^2} \left(\frac{P}{H'}\right) \qquad (2.56)$$

It is interesting to note that, although quite different algebraically, eq (2.52) gives numerical values of the contact conductance identical to those of eq (2.42). The advantage of using eq (2.52) is that it permits one to appreciate how the various parameters contribute to the coated contact conductance change as the coating thickness is varied.

The preceding theory is valid for conforming, rough surfaces. There is at present no theory which can handle out-of-flat surfaces, although in most practical situations the contact conductance values predicted by eqs (2.42) and (2.52) can be used to estimate the upper bound of the contact conductance.

5.5   Experimental Verification of the Theory

The thermo-mechanical models outlined above were experimentally verified in by Antonetti (1983) and Antonetti and Yovanovich (1985) wherein Nickel 200 specimens were tested. One of the specimens was coated with a thin layer of pure silver and the other was bead-blasted to varying degree of surface roughness. The uncoated surfaces were first tested to verify the conforming, rough surface model (eq (2.23)). These test results are shown in Figs. 2.13 and 2.14 as the lower bounds on the joint conductance. The theory and experiment are in very good agreement over the entire load range for the two cases reported here. The effects of load and surface roughness are clearly seen in these plots.

The effect of silver layers of different thicknesses are also reported in Figs. 2.13 and 2.14. The increase in the joint conductance is significant even when the layer is only $1\,\mu m$ thick. The joint conductance is increased by approximately a factor of 10 when the layer is approximately $6\,\mu m$ thick for the surface whose roughness is approximately $1.28\,\mu m$, and approximately $40\,\mu m$ for the surface whose roughness is approximately $4.27\,\mu m$. From the plots in Figs. 2.13 and 2.14 it can be seen that these layer thicknesses can be thought of as *infinitely* thick layers, and the corresponding joint conductances can be said to be the upper bounds on the joint conductance.

The agreement between the experiment and the theory are very good over the entire load range and all layer thicknesses. By the introduction of effective thermal conductivity, k', and effective microhardness, H', to nondimensionalize the joint conductance and the contact pressure, Antonetti (1983) was able to show in a general way the very good agreement between the theoretical and measured values of the joint conductance, as seen in Fig. 2.15.

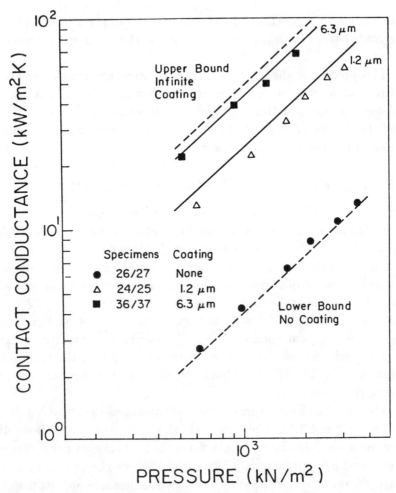

Figure 2.13: Contact conductance versus contact pressure for silver layer on nickel substrate ($\sigma = 1.28\,\mu m$).

5.6   Application to Microelectronics

The theory outlined in the previous section will now be applied to a common problem in electronics packaging: heat transfer across an aluminum joint. Assume the joint in question has an apparent contact area of $6.41 \times 10^{-4}\,m^2$ (approximately $1\,in^2$), and the range of contacting surface finishes being considered are as given in Table 2.3. What is desired is a parametric study showing the variation in joint contact conductance as a function of metallic coating type and thickness, surface finish, and the pressure on the joint.

The thermophysical properties of the coatings to be considered are tin and silver as well as the aluminum substrate material are presented in Table 2.4.

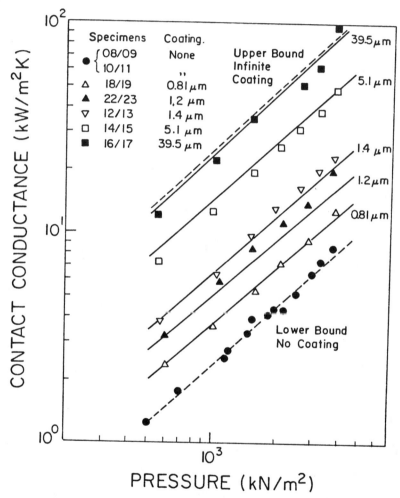

Figure 2.14: Contact conductance versus contact pressure for silver layer on nickel substrate ($\sigma = 4.27\,\mu m$).

Fig. 2.16 shows the effect of the layer on the contact conductance. As seen from this figure, except for a very thin layer (about one micrometer), the performance curves are arranged according to layer microhardness. Lead with the lowest microhardness has the highest contact conductance, and silver with the highest microhardness has

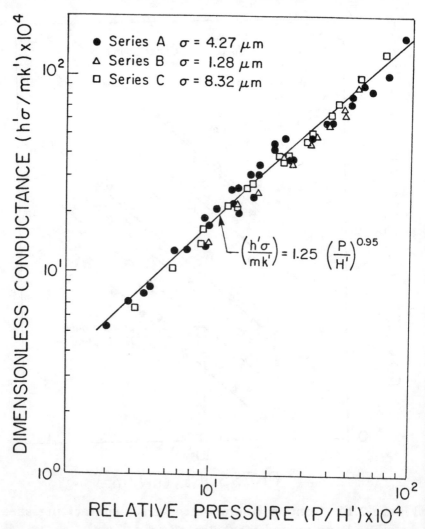

Figure 2.15: Dimensionless contact conductance versus relative contact pressure for silver layer on nickel substrate.

Table 2.3: Assumed surface characteristics of joints.

Uncoated Side of Contact

(σ) rms Roughness μm	(m) Avg Absolute Asperity Slope *radians*
1.0	0.10
4.0	0.20
8.0	0.30

Coated Side of Contact

0.20	0.02

Table 2.4: Assumed properties of materials.

	k (W/mK)	H (kg/mm^2)
Lead	32.4	3.0
Tin	58.4	8.5
Silver	406.0	40.0
Aluminum	190.0	85.0

the lowest contact conductance. The thermal conductivity of the coating appears to play a minor role. The unusual shape of the curves is attributable to the fact that the assumed effective hardness curve shown in Fig. 2.16 has three distinct zones. Moreover, because the microhardness of silver is much closer to aluminum than are the microhardnesses of lead and tin, the transition from one region to the next is not abrupt in the silver on aluminum effective microhardness curve, and this is reflected in the smoother contact conductance plot for the silver layer shown in Fig. 2.16. Fig. 2.17

was generated by drawing a curve through three points computed for each layer material using the surface characteristics found in Table 2.3. From Fig. 2.17, it can be seen that the smoother the finish of the uncoated surface, the higher the joint conductance. Fig. 2.18 shows the effect of joint pressure on the joint conductance. It should be noted that in the model which has been used, the load is assumed to be uniformly applied over the apparent contact area. The alternate analysis technique was used to generate Table 2.5. Here the change in the parameters influencing the contact conductance can be appreciated as the tin coating thickness increases. Table 2.5 clearly shows that as the layer thickness is increased, the primary reason for the corresponding increase in the contact conductance is the dramatic increase in the total number of microcontacts.

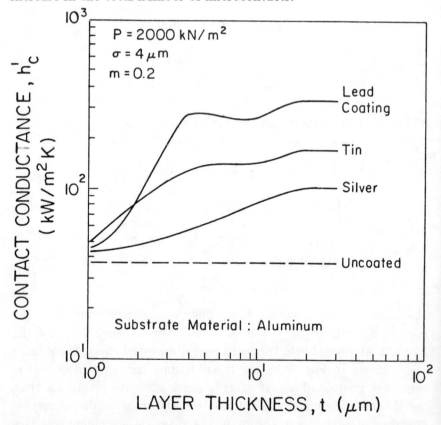

Figure 2.16: Effect of layer thickness upon joint conductance.

Table 2.5: Summary of calculated theoretical parameters for a tin layer on an aluminum substrate (in a vacuum) ($\sigma = 4\mu m$, $m = 0.20$)

P	t	t/d	H'	$\Psi(\epsilon')$	C	k'	a'	N'	h'_c
2000	0.0	0.00	85.0	0.927	1.000	190.0	8.6	6700	36600
2000	1.0	0.45	53.8	0.909	1.355	161.4	8.9	9700	47900
2000	2.0	0.84	26.8	0.872	1.627	144.7	9.6	17000	84300
2000	4.0	1.58	14.4	0.827	2.046	124.7	10.2	28100	134000
2000	8.0	3.09	11.7	0.809	2.566	106.6	10.4	33200	141000
2000	16.0	6.66	8.5	0.777	3.015	94.6	10.7	42900	174000
2000	∞	—	8.5	0.777	3.253	89.3	10.7	42900	165000

P	=	Pressure (kN/m^2)
t	=	Layer Thickness (μm)
t/d	=	Relative Layer Thickness
H'	=	Effective Hardness (kg/mm^2)
$\Psi(\epsilon')$	=	Constriction Parameter
C	=	Constriction Parameter Correction Factor
k'	=	Effective Thermal Conductivity (W/mK)
a'	=	Mean Contact Spot Radius (μm)
N'	=	Total Number of Contact Spots
h'_c	=	Thermal Contact Conductance ($W/m^2 K$)

5.7 Ranking Coating Performance

Yovanovich (1973), in studying the effect of soft foils on joint conductance, proposed that the performance of different foil materials be ranked according to the parameter (k/H), using the properties of the foil material. He showed empirically that the higher the value of this parameter, the greater was the improvement in the contact conductance over a bare joint. Following this thought, in the Antonetti and Yovanovich (1985) work, it was proposed, although not proven experimentally, that the performance of coated joints be ranked by the parameter $k'/(H')^{0.93}$. Table 2.6 shows the variation in this parameter as the layer thickness is increased. Table 2.6 suggests as well, that even if the effective hardness of the layer-substrate combinations being considered are not known, the relative performance of coating candidates can be estimated by assuming an infinitely thick coating (where the effective hardness equals the layer hardness).

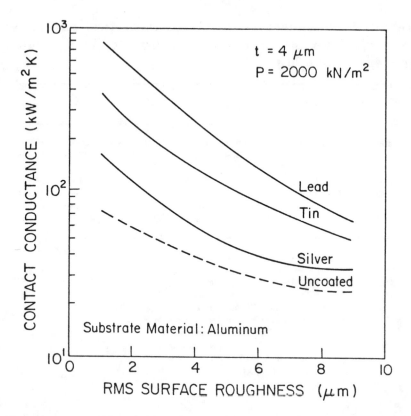

Figure 2.17: Effect of surface finish upon joint conductance.

This section has shown how a thermo-mechanical model for coated contacts can be used to predict the enhancement in thermal contact conductance for a coated joint. For the particular case considered, an aluminum to aluminum joint, it was demonstrated that up to an order of magnitude improvement in the contact conductance is possible, depending upon the choice of coating material and the thickness employed. It should also be noted that aluminum substrates are relatively soft and have a relatively high thermal conductivity, and if the joint in question had been, for example, steel against steel, the improvement in the contact conductance would have been even more impressive.

5.8 Other Enhancement Techniques

Another method of enhancing the contact conductance, or perhaps in this case better said as minimizing the contact resistance, is to solder or braze the joint in question. Although these joining methods are strong, permanent interfaces, it is important to recognize that the resulting contact resistances may not be negligible and often may be significant.

Yovanovich (1970) examined the effect of soldering and brazing on contact resistance. It was observed that silver-soldered interfaces had contact resistances which varied considerably, and, which in all cases, exceeded the theoretical value by at least an order of magnitude. Correlation of the lowest contact resistance data from Yovanovich (1970), which corresponds to the best interfaces measured, resulted in

$$R = \frac{\psi}{2} \left[\frac{1}{k_1} + \frac{2}{k_2} + \frac{1}{k_3} \right] \tag{2.57}$$

Table 2.6: Ranking the effectiveness of coatings $[k'/(H')^{0.93}]$

Coating Thickness (μm)	Lead	Tin	Silver
0	3.05	3.05	3.05
1	3.72	3.96	3.53
2	7.05	6.81	3.98
4	19.6	10.5	4.68
8	18.0	10.8	6.24
16	21.0	12.9	8.16
∞	19.9	12.2	8.38

$P = 2000 \, kN/m^2$, $\sigma = 4.0 \, \mu m$, $m = 0.20 \, radians$

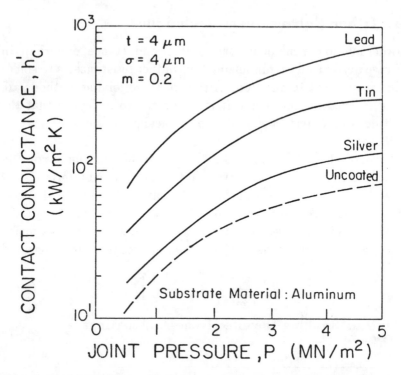

Figure 2.18: Effect of joint pressure upon joint conductance of a layered interface.

where R is the contact resistance, k_1 and k_3 are the thermal conductivities of the contact solids, k_2 is the conductivity of the interface material (solder), and ψ is the constriction parameter for the soldered interface. The constriction parameter is a complex function of the surface geometry and its thermal properties and cannot be predicted from first principles. The parameter, ψ, was empirically determined to be approximately $1.0\ m^{-1}$ for a 25 mm diameter cross section.

When a mechanical joint cannot be made permanent by soldering or brazing, the contact resistance can often be minimized by the use of thermal grease, or as shown by Yovanovich (1973), by inserting a soft foil or as discussed in the previous section by coating one or both surfaces with a relatively soft metallic coating. Any interstitial substance whose thermal conductivity is greater than air, and which will wet and/or fill the gap between the contacting solids, will have a significant effect on the contact resistance. Attempts to classify the effectiveness of greases and foils have been made by Fletcher (1973) and Snaith, O'Callaghan and Probert (1984).

With grease, the reduction in contact resistance occurs because the grease fills the otherwise void interstitial space which tends to alleviate the *pinching and spreading* of the heat flow lines at the interface. Thermal greases are available in various formulations having differing heat transfer effectiveness. Tests on thermal joint compounds by Feldman, Hong and Marjon (1980) and Oktay, Dessauér and Howath (1983) have shown that the best thermal compounds possess thermal conductivities as high as 1.25 $W/m\ K$, which is approximately 50 times greater than air.

Soft foils are coined when inserted between two harder surfaces. Again the interstitial spaces are partially filled, but this time with a soft metal whose thermal conductivity is orders of magnitude better than the best thermal grease. The result is an increased contact area which again tends to alleviate the pinching effect. Yovanovich (1973) has shown that the thermal performance of a foil depends primarily on the ratio of the foil thermal conductivity to hardness. The higher this index the better the foil performance. The thickness of the foil is also critical to the performance and for a given material, as shown in Fig. 2.19, there exists an optimum foil thickness.

6 SUMMARY

Spreading and constriction resistance concepts and solutions for half-spaces and semi-infinite flux tubes as they pertain to microelectronic thermal management have been reviewed. Recently developed dimensionless spreading resistance expressions have been presented which can be used to solve numerous thermal problems which frequently arise in microelectronics.

Thermal contact, gap and joint conductances for conforming, rough surfaces are reviewed. The effect of thin metallic layers bonded to one surface upon joint conductance is reviewed, and an example of the application of the theory to a practical microelectronics problem is presented to show how the various parameters effect the joint resistance.

An expression for the prediction of the minimum thermal resistance at soldered joints is also presented. The minimum resistance

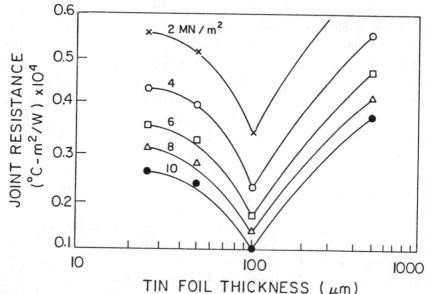

Figure 2.19: Effect of metallic foil thickness on joint resistance

applicable to joints which have been soldered with extreme care. Finally, the effect of very thin metallic foils to minimize joint resistance is considered, and the presence of an optimum thickness is reported.

7   NOMENCLATURE

a Contact spot radius, m

A Area, m^2

b Flux tube radius, m

C Constriction parameter correction factor, correlation coefficient, dimensionless conductance,
$$C = (h/ks)(\sigma/m)$$

c_p Specific heat at constant pressure, $kjoule/kg\ K$

c_v Specific heat at constant volume, $kjoule/kg\ K$

d Equivalent Vickers diagonal or indentation depth, μm

H Microhardness, Pa

H' Effective microhardness of soft layer on harder substrate, Pa

h Thermal contact conductance, $W/m^2 K$

I_g Gap integral

J_n Bessel function, first kind, order n

K Thermal conductivity ratio (substrate to layer)

k Thermal conductivity, $W/m \cdot K$

k_s Harmonic mean thermal conductivity,
$k_s = 2k_1k_2/(k_1 + k_2), W/m \cdot K$

k' Effective thermal conductivity, $W/m \cdot K$

M Gas parameter, $M = \alpha\beta\Lambda, m$

m Combined average absolute asperity slope,
$m = \sqrt{m_1^2 + m_2^2}$, radians

N Number of contact spots in apparent area A_a

n Number of contact spots per unit area, m^{-2}

P Pressure, apparent contact pressure

Pr Prandtl number

Q Heat flow rate, W

q Heat flux, W/m^2

R Resistance, K/W

T Temperature, K

t Local gap thickness, layer thickness, m

u Relative local gap thickness, $u \equiv t/\sigma$

x Parameter $x = Y/\sqrt{2}\sigma, m$

Y Separation distance of mean planes of contacting surfaces, m

Greek Symbols

α Accommodation coefficient

β Gas parameter, $\beta = (2\gamma)/(\gamma + 1)/Pr$
relative coating thickness, $\beta = \delta/a$

γ Specific heat ratio $(= c_p/c_v)$

γ_n Constriction parameter modification factor attributable
to heat flux distribution

δ Conductive layer thickness, m

δ_n Eigenvalues are the roots of $J_1(\delta_n) = 0$

ϵ Relative contact spot radius, $\epsilon = a/b$
or contact area
size, $\epsilon = \sqrt{A_c/A_t}$

κ Thermal conductivity ratio

Λ Molecular mean free path, m

ρ_n Constriction parameter modification factor attributable to contact temperature basis

σ Combined RMS roughness $\sigma = \sqrt{\sigma_1^2 + \sigma_2^2}, m$

τ Relative layer thickness, $\tau = t/a$

ϕ_n Constriction parameter modification factor attributable to the layer

ψ_∞ Dimensionless spreading resistance

ψ Constriction parameter

Superscripts

In general, a prime indicates the layer

Subscripts

a Apparent
b Bulk
c Contact or constriction
g Gas, gap
i Summation index
j Joint
L Layer
P Constant pressure
r Real
t Total
v Vickers; constant volume
1 One side of contact or layer
2 Other side of contact or substrate
∞ Continuum

8   REFERENCES

Al-Astrabadi, F R, O'Callaghan, P W, Snaith, B and Probert, S D (1981). Prediction of Optimal Interfacial Filler Thickness for Minimum Thermal Contact Resistance, AIAA-81-1166, presented at AIAA 16th Thermophysics Conference, Palo Alto, CA.

Antonetti, V W (1983). On The Use of Metallic Coatings to Enhance Thermal Contact Conductance, Ph.D. thesis, University of Waterloo.

Antonetti, V W and Yovanovich, M M (1984). Thermal Contact Resistance in Microelectronic Equipment, *International Journal of Hybrid Microelectronics*, pp. 44-50, Vol. 7, No. 3.

Antonetti, V W and Yovanovich, M M (1985). Enhancement of Thermal Contact Conductance by Metallic Coatings: Theory and Experiment, *Journal of Heat Transfer*, Vol. 107, pp. 513-519.

Bergles, A E, Chu, R C, and Seely, J H (1977). Survey of Heat Transfer Techniques Applied to Electronic Packages, *Proceedings of the Technical Program, National Electronic Packaging and Production Conference*, Anaheim, CA.

Ellison, G N (1984). Thermal Computation for Electronic Equipment, Van Nostrand Reinhold, New York, N.Y.

Feldman, K J, Hong, Y M and Marjon, P L (1980). Tests on Thermal Joint Compounds to 200°F., AIAA-80-1466, presented at the AIAA 15th Thermophysics Conference, Snowmass, Colorado.

Fletcher, L S (1973). Thermal Control of Materials for Spacecraft Systems, *Proceedings of the 10th Int. Symposium on Space Technology and Science*, Tokyo, Japan, pp. 579-586.

Fried, E and Kelley, M J (1965). Thermal Conductance of Metallic Contacts in a Vacuum, AIAA-65-661 AIAA Thermophysics Specialists Conference, Monterey, CA.

Hegazy, A A (1985). Thermal Joint Conductance of Conforming Rough Surfaces: Effect of Surface Micro-Hardness Variation, Ph.D. thesis, University of Waterloo.

Kadambi, V and Abuaf, N (1983). Axisymmetric and Three-dimensional Chip Spreader Calculations, AIChE Symposium Series on Heat Transfer, Seattle, Washington.

Kennedy, D P (1960). Spreading Resistance in Cylindrical Semiconductor Devices, *Journal of Applied Physics*, Vol. 31, pp. 1490-1497.

Kraus, A D and Bar-Cohen, A (1983). Thermal Analysis and Control of Electronic Equipment, *McGraw-Hill Co.*, New York, N.Y.

Kraus, A D, Chu, R C and Bar-Cohen, A (1982). Thermal Management of Microelectronics: Past, Present, and Future, Computers in Mechanical Engineering, pp. 69-79.

Madhusudana, C F and Fletcher, L S (1981). Contact Heat Transfer - The Last Decade, *AIAA Journal*, Vol. 24, No. 3, pp. 510–523.

Mal'kov, V A and Dobashin, P A (1969). The Effect of Soft-Metal Coatings and Linings on Contact Thermal Resistance, *Inzhenerno-Fizicheshii Zhurnal*, Vol. 17, No. 5, pp. 871-879.

Mikic, B B (1966). Thermal Contact Resistance, Sc.D. thesis, Department of Mechanical Engineering, MIT, Cambridge, MA.

Negus, K J and Yovanovich, M M (1983). Application of the Method of Optimized Images to Steady Three-Dimensional Conduction Problems, *ASME Paper* No. 84-WA/HT-110, ASME WAM, Boston, MA.

Negus, K J and Yovanovich, M M (1984). Constriction Resistance of Circular Flux Tubes With Mixed Boundary Conditions By Linear Superposition of Neumann Solutions, *ASME Paper* No. 84-HT-84, NHTC Niagara Falls, NY.

Negus, K J, Yovanovich, M M and Thompson, J C (1985). Thermal Constriction Resistance of Circular Contacts on Coated Surfaces: Effect of Contact Boundary Conditions, *AIAA Paper* No. 85-1014, AIAA 20th Thermophysics Conference, Williamsburg, VA.

Negus, K J, Yovanovich, M M and Beck, J V. On The Non-Dimensionalization of Constriction Resistance for Semi-Infinite Heat Flux Tubes, Submitted to ASME Journal of Heat Transfer.

Oktay, S, Dessauer, B and Horvath, J L (1983). New Internal and External Cooling Enhancements for the Air-Cooled IBM 4381 Module, IEEE International Conference on Computer Design: VLSI in Computers, Port Chester, NY.

Roess, L (1948). Theory of Spreading Conductance, Appendix to N.D. Weills and E.A. Ryder, *Thermal Resistance Measurements on Joints Formed Between Stationary Metal Surfaces*, presented at ASME Heat Transfer Meeting, Milwaukee, Wisc.

Simons, R E (1983). Thermal Management of Electronic Packages, Solid State Technology, pp. 131-136.

Snaith, B, O'Callaghan, P W and Probert, S D (1984). Interstitial Materials for Controlling Thermal Conductances Across Pressed Metallic Contacts, *Applied Energy*, Vol. 16, pp. 175-91.

Yovanovich, M M (1970). A Correlation of the Minimum Thermal Resistance at Soldered Joints, *Journal of Spacecraft and Rockets*, Vol. 7, No. 8, pp. 1013-1014.

Yovanovich, M M (1973). Effect of Foils Upon Joint Resistance: Evidence of Optimum Thickness, *AIAA Progress in Astronautics and Aeronautics, Thermal Control and Radiation*, Vol. 31, edited by Chang-Lin Tien, Cambridge, MA., pp. 227-245.

Yovanovich, M M (1976). General Expressions for Circular Constriction Resistances for Arbitrary Flux Distrubtion, AIAA Progress in Aeronautics and Astronautics, Radiative Transfer and Thermal Control, Vol. 49, edited by A.M. Smith, New York, pp. 381-396.

Yovanovich, M M (1976). Thermal Constriction Resistance of Contacts on a Half-Space: Integral Formulation, AIAA Progress in Astronautics and Aeronautics, Radiative Transfer and Thermal Control., Vol. 49, edited by A.M. Smith, New York, pp. 397-418.

Yovanovich, M M (1978). Thermal Contact Resistance in Microelectronics, *Proceedings of the Technical Program*, National Electronic Packaging and Production Conference, Anaheim, CA.

Yovanovich, M M (1981). New Contact and Gap Correlations for Conforming Rough Surfaces, AIAA-86-1164, presented at 16th Thermophysics Conference, Palo Alto, CA.

Yovanovich, M M and Burde, S S (1977). Centroidal and Area Average Resistances of Nonsymmetric, Singly Connected Contacts, *AIAA Journal*, Vol. 15, No. 10, pp. 1523-1525.

Yovanovich, M M, Burde, S S and Thompson, J C (1977). Thermal Constriction Resistance of Arbitrary Planar Contacts with Constant Flux, AIAA Progress in Astronautics and Aeronautics, Thermophysics of Spacecraft and Outer Planet Entry Probes, Vol. 56, edited by A.M. Smith, New York, pp. 127-139.

Yovanovich, M M, DeVaal, J and Hegazy, A H (1982). A Statistical Model to Predict Thermal Gap Conductance Between Conforming Rough Surfaces, *AIAA Paper 82-0888*, presented at the AIAA/ASME Third Joint Thermophysics, Fluids, Plasma and Heat Transfer Conference.

Yovanovich, M M, Hegazy, A and Antonetti, V W (1983). Experimental Verification of Thermal Conductance Models Based Upon Distributed Surface Microhardness, *AIAA-83-0532*, presented at AIAA 21st Aerospace Sciences Meeting, Reno, Nevada.

Yovanovich, M M, Hegazy, A and DeVaal, J (1982). Surface Hardness Distribution Effects Upon Contact, Gap and Joint Conductances, *AIAA-82-0887*, AIAA/ASME 3rd Joint Thermophysics, Fluids and Heat Transfer Conference, St. Louis, MO, June.

Chapter 3

DIRECT AIR-COOLING OF ELECTRONIC COMPONENTS

Robert J. Moffat
Department of Mechanical Engineering
Stanford University
Stanford, CA 94305

Alfonso Ortega
Department of Aerospace and Mechanical Engineering
University of Arizona
Tucson, AZ 85721

1 INTRODUCTION

The air-cooling problem will not have been "solved" until a designer can accurately predict the temperature of any element in an arbitrary, three-dimensional array with specified heat release on each element. Reference to a drawing showing the planform layout, the board spacing, the power distribution, and the cooling flow (if cooled by forced convection) should be the only data required.

There is little hope that correlations of heat transfer coefficients can accomplish all that would be desirable. Almost certainly, the ultimate solution will come from computational fluid mechanics programs coupled with conduction and radiation programs which acknowledge the detailed geometry and materials involved. Programs of such complexity will not be available for years, however, and there is an immediate need for more accurate design tools. These must be constructed from what is available now.

The conduction and radiation portions of the problem are well in hand as far as principles and programs are concerned. The conservation equations and rate equations for both types of heat transfer are well understood. There are large-scale programs which can deal with combined conduction, radiation, and convection in bodies of arbitrary shape. The weak link is in the handling of the convection process. Most current programs require, as input, specification of the heat transfer coefficient at each point on the surface and that information is not available.

Most situations of practical interest are beyond present analytical capabilities and must be dealt with experimentally. Significant problems arise in trying to assemble a coherent body of knowledge out of the results of independent experiments conducted in different laboratories, with different objectives, assumptions, and definitions. Convective heat transfer is very "situation specific" and results do not generalize well when the system geometry is one of the variables.

The heat transfer coefficient is a defined quantity, not a physically deterministic one. It is formed from the heat flux, the surface temperature, and a reference fluid temperature. It can be defined in terms of averages over large areas of the surface or at a point and there are several options for the reference temperature. From the phenomenological standpoint, the heat flux at a point on a surface depends on the distribution of temperature in the fluid above the point which, in turn, depends on the local fluid mechanics and the upstream distribution of heat flux. Thus the local heat transfer coefficient may change value when the upstream temperature or flow distribution is changed. In fact, the heat transfer coefficient at a point may change in response to any change in its upstream neighborhood. The extent of the neighborhood which affects a given point depends on the local fluid mechanics.

In an attempt at bringing order to the field in the face of this sensitivity, the historic tendency has been to simplify the situations studied experimentally and analytically and to establish some "benchmark" situations which are very well understood. Most of these benchmark situations involve uniform or stepped wall temperature or heat flux on smooth external surfaces or inside smooth walled channels with fully developed laminar or turbulent flow. These provide us with our library of established solutions.

This library serves well in heat transfer situations similar to those which lead to the selection of the existing set of benchmark solutions, but the electronics cooling situation has introduced some new problems. Geometry is one of the primary variables in electronics cooling. The most common situation, in the air-cooling area at least, involves elements of different shape arranged in an irregular array, between two walls with significant small-scale roughness, where the spacing between the channel walls must be considered a design variable. Dealing with this geometric variability is one of the primary problems of organizing cooling data.

We have no way of describing geometric shape in an orderly way, and one can never have a truly satisfactory one since "shape" is not a measurable property: two objects of different shape cannot be placed in any logical sequence involving a "greater than, equal to or less than" test. We are forced to fall back on drawing pictures to show shape, and quoting dimensions to describe size: size to be measured once "shape" is set. The inability to "rank order" shapes, plus the fact that shape affects the heat transfer coefficient, affects the approach to heat transfer research. Every new shape must be tested and the results obtained are specific to that shape alone, unless, by good fortune, two shapes act alike in their heat transfer behavior. Thus the heat transfer literature abounds with tests of particular shapes. It also abounds with attempts to describe the heat transfer behavior of different shapes by the same correlation through choice of some "critical dimension". There is no denying the utility of such collections, when they work, but it is important to keep in mind that one has no right to assume, except based on experience, that objects of different shape can be described by the same correlation. In the air-cooling environment, for example, simply changing the board-to-board spacing constitutes defining a new shape, even if the layout and shapes of the components is unchanged. One cannot guarantee to describe the heat transfer behavior resulting from those two different board spacings by different values of the same correlating parameter–they constitute two quite different geometries and may have quite different distributions of heat transfer.

So far, it has not been possible to really bring order to this situation and the present work certainly doesn't finish that job. We hope it helps, however. It does bring together, in an orderly way, a representative set of analytical and experimental results for free, forced, and mixed convection covering many situations of specific interest to the electronics cooling practitioner. It is as complete as our patience, and the time available, could make it. Certainly and inevitably, however, there are some fine and important works which have not been cited. This is particularly true of the contributions from the Japanese language literature, a situation which we hope will have improved before the next edition of this work.

The chapter is divided into two parts: Forced Convection and Natural Convection. These are presented as a single work by two authors, rather than as two separate works, since each of us contributed to the ideas in each section. However, the parts were separately written and are so treated as far as structure and nomenclature are concerned. Part I on Forced Convection (Sections 2 to 5) was composed by R. J. Moffat, and Part II on Natural Convection (Sections 6 to 9) by A. Ortega.

Part I
Forced Convection

2   ANALYTICAL AND NUMERICAL METHODS

2.1   Introduction to Analysis

Part I deals with calculating the heat transfer rate from a component to a cooling air stream under pure forced convection conditions. This restriction to "pure forced convection" implies that local buoyancy effects are negligible small. The situation may involve flow over components on isolated surfaces or within channels formed by two parallel planes. Not all of the heat released within a component is delivered directly to the cooling air. Some is first conducted to the board and some is radiated to other areas of the enclosure. An energy distribution analysis must be performed to determine what fraction of the power released by a component goes directly to the coolant by forced convection. Such an analysis requires a fully integrated program, simultaneously considering all three modes of heat transfer and all possible heat flow paths. Part I does not address the entire problem, but deals only with direct convection: it does not deal with conduction or radiation heat transfer.

Most thermal design is done on the basis of correlations which describe the heat transfer performance of a particular geometry in terms of the flow conditions, usually expressed as a Reynolds number. One of the major problems in electronic cooling is that practical geometries are so diverse that only rarely does a designer have a correlation for the actual geometry. The problem is one of choosing the best from among a few poor choices. This is unfortunate because there is little transferability of information from one geometry to another. If one dimension of a system is changed, every dimension must be changed in the same proportion. Only when strict geometric similarity is observed does Reynolds number become a robust parameter. It is essential to Reynolds number scaling that the two systems in question be of *exactly* the same shape–differing only in size. For example, a designer who wishes to estimate the effect of reducing the spacing between two boards, with no other change, finds that the new situation presents an entirely new geometry. Correlations describing the heat transfer coefficient as a function of Reynolds number for the original channel height may not apply to the new one, even though the board layouts remain the same and the only change is in the channel height! Faced with this unpleasant reality, and yet required to make decisions anyway, designers must make do with whatever help is available, and pass the burden to the experimentalists to 'develop' the hardware.

In the following sections, 28 analytical solutions and several sets of experimental results will be presented. These seem most relevant to the design of forced convection, air-cooled electronic systems. Presentation of the analytical solutions does not imply that they apply directly; in most cases, they will not. They will, however, show the expected behavior well enough to identify the trends and provide some guidance as to what should be measured and how to present the results.

2.2   Analytical Solutions for External Flow Situations: Average Heat Transfer

Boundary layers develop on any surface exposed to a flowing fluid. The boundary layers may be laminar, transitional, turbulent or "disturbed". The heat transfer from the surface depends on the characteristics of the boundary layer and that, in turn depends on the nature of the flow. All analytical solutions presently in the literature assume a disturbance-free flow over the surface and most deal either with laminar or turbulent states. Little is known about heat transfer from transitional or disturbed states. The customary criterion for describing the state of the boundary layer, for such situations, is the Reynolds number based on x, the distance along the surface. If the Reynolds number is less than 1×10^5, the boundary layer is assumed laminar. If above 2×10^5, it is assumed turbulent. In between, it is assumed transitional.

Laminar boundary layers are rarely found in practical situations related to electronic cooling and the laminar analyses presented here are chiefly useful as lower bounds for the heat transfer rates which can be expected.

Most practical situations involve flows which are highly disturbed – for example, the flow downstream of a fan or blower, or downstream of an obstruction. There are no accepted methods for predicting the heat transfer in those situations. The recommendation made here is to treat any boundary layer under a disturbed free-stream as turbulent, regardless of the Reynolds number, and even that may underestimate the heat transfer rates.

The equations presented in this section are all for air and other fluids of moderate Prandtl number ($0.5 < Pr < 15$).

For external flows, the characteristic dimension is either streamwise position, x, or length, L, depending on whether local or average information is sought. The reference gas temperature, T_∞, is the temperature of the flow far from the surface and is assumed to be uniform. The heat transfer coefficient is defined in terms of the convective heat flux and $(T_o - T_\infty)$ where T_o is the surface temperature.

When the surface temperature is not uniform, the definition of the average h can be ambiguous (see the discussion following eq (3.2).

Laminar boundary layer flow over an isolated, smooth, flat plate.

1. Uniform surface temperature, heated along the entire length (Kays and Crawford (1980) eq (9.14), p. 139).

$$\overline{Nu}_L = 0.664 \mathrm{Re}_L^{1/2} Pr^{1/3} \tag{3.1}$$

2. Uniform heat flux, heated along the entire length (from a local solution given by Kays and Crawford (1980)eq (9.40), p. 151).

$$\overline{Nu}_L = 0.906 Re_L^{1/2} Pr^{1/3} \tag{3.2}$$

In eq (3.2), Nu_L is defined as $\overline{h}L/k$ where $\overline{h}$ is found by integrating the expression for the local h in Kays and Crawford (1980), eq (9.40), p. 151. If, on the other hand, $\overline{h}$ is defined as $\overline{h} = \dot{q}_0''/\overline{\Delta T}$, then the constant would be 0.803 instead of 0.906.

3. Non-uniform temperature or non-uniform heat release situations should be dealt with using superposition to find the local values, following the methods outlined in Section 2.3, Parts 4 and 5. The average is then found by integration.

Laminar boundary layer up to $Re_x = 3 \times 10^5$ followed by transition to a turbulent boundary layer. Uniform temperature along the plate.

Recommended when the length Reynolds number exceeds 5×10^5 provided that there are no significant disturbances in the free stream. If there are significant disturbances, use eq (3.4) everywhere (Kraus and Bar Cohen (1983) p. 128).

$$\overline{Nu}_L = 0.036(Re_L - 14,251)Pr^{1/3} \tag{3.3}$$

Turbulent boundary layer from $x = 0$. Uniform temperature along the plate (Kraus and Bar Cohen (1983) p. 128).

$$\overline{Nu}_L = 0.036 Re_L^{4/5} Pr^{1/3} \tag{3.4}$$

2.3   Analytical Solutions for External Flow Situations: Local Heat Transfer

The local heat transfer coefficient is defined in terms of the local heat flux, local wall temperature and the gas temperature far from the surface. In general, the local heat transfer coefficient will be a strong function of the wall temperature distribution upstream of the point in question. Only the canonical cases (uniform wall temperature and uniform heat flux) have closed form explicit solutions. Non-uniform temperature distributions require the use of superposition.

Laminar boundary layers.

1. Uniform surface temperature, heating along the entire length (Kays and Crawford (1980) p. 139).

$$Nu_x = 0.332 Re_x^{1/2} Pr^{1/3} \qquad (3.5)$$

2. Uniform heat flux, heating along the entire length (Kays and Crawford (1980) p. 151).

$$Nu_x = 0.453 Re_x^{1/2} Pr^{1/3} \qquad (3.6)$$

3. Unheated from $x = 0$ to ξ, with uniform temperature thereafter where ξ is the location of the start of the heated portion of the surface (Kays and Crawford (1980), p. 147).

$$Nu_x = \frac{0.332 Re_x^{1/2} Pr^{1/3}}{[1 - (\xi/x)^{3/4}]^{1/3}} \qquad (3.7)$$

4. Arbitrary temperature distribution. The heat flux distribution, $q_o''(x)$, can be calculated by combining eq (3.7) with a description of the surface temperature distribution, which may include step changes as well as continuous changes (Kays and Crawford (1980) p. 149).

$$h(\xi, x) = \frac{(0.332k/x)\, Re_x^{1/2} Pr^{1/3}}{[1 - (\xi/x)^{3/4}]^{1/3}} \qquad (3.8)$$

$$q_o'' = \int_0^x h(\xi, x)(dT_o/d\xi)d\xi + \sum_{i=i}^{k} h(\xi_i, x)\Delta T_{o,i} \qquad (3.9)$$

5. Arbitrary heat flux distribution. The temperature distribution caused by a specified heat flux distribution can be calculated from (Kays and Crawford (1980) p. 151).

$$T_o - T_\infty = \frac{0.623}{k} Re_x^{-1/2} Pr^{-1/3} \int_0^x [1 - \left(\frac{\xi}{x}\right)^{3/4}]^{-2/3} \dot{q}_o''(\xi) d\xi \quad (3.10)$$

Turbulent boundary layers.

1. Uniform temperature, heating from the beginning of the plate (Kays and Crawford (1980) p. 213).

$$Nu_x = 0.0287 Re_x^{0.8} Pr^{0.6} \quad (3.11)$$

2. Unheated from x = 0 to x = ξ, with uniform temperature thereafter (Kays and Crawford (1980) p. 216).

$$Nu_x = \frac{0.0287 Re_x^{0.8} Pr^{0.6}}{[1 - (\xi/x)^{9/10}]^{1/9}} \quad (3.12)$$

3. Arbitrarily specified temperature or heat flux. Arbitrary temperature or heat flux distributions are dealt with using superposition, based on the step-temperature solution in eq (3.3) and a corresponding method for specified heat flux. The procedure is described by Kays and Crawford (1980, p. 216) for temperature boundary conditions and (p. 217) for heat flux.

2.4 Analytical Solutions for Heat Transfer with Fully Developed Flow and Heat Transfer Between Parallel Planes

In the analysis of channel flows, the heat transfer coefficient is almost always based on the difference between the wall temperature and the local mixed mean temperature of the flow in the channel at the location of the heat transfer measurement. The mixed mean temperature can be calculated by an energy balance from the inlet of the channel to the point in question, providing that the inlet temperature, the total heat load up to that point, the flow rate, and the fluid properties are all known. The hydraulic diameter, $D_h = 4r_h$, is conventionally used as the length scale in both the Reynolds number and the Nusselt number.

In fully developed channel flows (both flow and thermal) with constant temperature or constant heat flux boundary conditions, the heat transfer coefficient does not vary with streamwise position, hence there is no distinction between local and average values.

There are no general correlations for situations involving thermal boundary conditions which vary in the streamwise direction. With non-uniform wall temperature or heat flux, h may vary strongly in the streamwise direction - so strongly that it ceases to be useful as a convenient way of describing the processes. For example, just downstream of an abrupt drop in wall temperature, there can be heat transfer from the fluid to the wall, locally, even though the wall remains hotter than the mixed mean fluid temperature! In that region, h is negative. The method of superposition should be used in cases of variable wall temperature or heat flux, providing that the necessary data are available concerning the heat transfer and temperature wake functions for the geometry in question.

In the following sections, the classical solutions are presented for uniform boundary conditions. For a general treatment of superposition, the reader is referred to Kays and Crawford (1980). For a specific application to electronics cooling and an example of the data set required, the reader is referred to Arvizu and Moffat (1981).

Fully developed laminar flow with various thermal boundary conditions. Temperature field fully developed.

1. Uniform temperature, the same on both walls, aspect ratio of infinity (Kays and Crawford (1980) p. 103).

$$Nu = 7.54 \qquad (3.13)$$

2. Uniform temperature on one wall, the other insulated with an aspect ratio of infinity (Kays and Crawford (1980) p. 103).

$$Nu = 4.86 \qquad (3.14)$$

3. Uniform heat flux, the same on both walls, with a channel aspect ratio of infinity (Kays and Crawford (1980) p. 103).

$$Nu = 8.235 \qquad (3.15)$$

4. Uniform heat flux on one wall, the other insulated, with a channel aspect ratio of infinity (Kays and Crawford (1980) p. 103).

$$Nu = 5.385 \tag{3.16}$$

5. Uniform heat flux on each wall, but of different magnitudes: $\dot{q}_j''$ on the j-th wall, $\dot{q}_i''$ on the i-th wall. The Nusselt number must be calculated for each wall (Kays and Crawford (1980)p. 117).

$$Nu_i = \frac{5.385}{1 - 0.346(\dot{q}_j''/\dot{q}_i'')} \tag{3.17}$$

Fully developed laminar flow, with various thermal boundary conditions: in the thermally developing region

$$\text{for all equations } x^+ = \frac{2(x/D_h)}{RePr} \tag{3.18}$$

1. Both walls unheated until $x^+ = 0$ and both at the same uniform temperature thereafter (Kays and Crawford (1980) p. 116 with extrapolation).

Table 3.1: Nu_x versus x^+ in the thermal entrance region—parallel planes with both walls having the same uniform temperature for $x^+ > 0$. Fully developed velocity distribution throughout.

x^+	Nu_x
0.005	10.44
0.010	8.52
0.020	7.75
0.050	7.55
0.100	7.55

2. Both walls unheated until $x^+ = 0$ and both with the same uniform heat flux thereafter (Calculated from data in Kays and Crawford (1980) p. 117).

Table 3.2: Nu_x versus x^+ in the thermal entrance region—
parallel planes with both walls having the same uniform
heat flux for $x^+ > 0$. Fully developed velocity distribution
throughout.

x^+	Nu_x
0.005	11.86
0.020	8.80
0.100	8.25
0.200	8.23
0.300	8.23

3. Both walls unheated until $x^+ = 0$. One wall with uniform
   heat flux thereafter, the other wall insulated (Kays and Crawford (1980) p. 117).

Table 3.3: Nu_x versus x^+ in the thermal entrance region—parallel planes with one wall having uniform heat flux
for $x^+ > 0$, the other wall insulated. Fully developed velocity
distribution throughout.

x^+	Nu_x
0.0005	23.50
0.005	11.20
0.020	7.49
0.100	5.55
0.250	5.39
0.300	5.38

4. Both walls unheated until $x^+ = 0$. Both walls with uniform
   heat flux thereafter, but of different values: $\dot{q}_i''$ on one side
   and $\dot{q}_j''$ on the other. The Nusselt numbers must be calculated
   for each wall (Kays and Crawford (1980) p. 117).

$$Nu_i = \frac{Nu_x}{1 - \theta(\dot{q}_j''/\dot{q}_i'')} \tag{3.19}$$

Table 3.4: Nu_x and θ versus x^+ in the thermal entrance region–parallel planes with one wall having uniform heat flux $\dot{q}_i''$ and the other $\dot{q}_j''$ for $x^+ > 0$. Fully developed velocity distribution throughout.

x^+	Nu_x	θ
0.0005	23.50	0.01175
0.005	11.20	0.0560
0.020	7.49	0.1491
0.100	5.55	0.327
0.250	5.39	0.346
0.300	5.38	0.346

Fully developed turbulent flow, with various thermal boundary conditions, and fully developed temperature fields.

1. Uniform temperature on both walls (developed from data in Kays and Crawford (1980) p. 267).

$$Nu = 0.0379 Re^{0.755} Pr^{0.45} \tag{3.20}$$

2. Uniform heat flux on both walls, of equal value (developed from data in Kays and Crawford (1980) p. 267).

$$Nu = 0.0395 Re^{0.755} Pr^{0.45} \tag{3.21}$$

3. One wall with uniform heat flux, the other wall insulated (developed from data in Kays and Crawford (1980) p. 267).

$$Nu = 0.0334 Re^{0.755} \tag{3.22}$$

4. Both walls with uniform heat flux, but of different values: q_j'' on one wall and q_i'' on the other, positive if into the fluid. The Nusselt number must be calculated for each of the two walls (developed from data in Kays and Crawford (1980) p. 267).

$$Nu_i = \frac{0.0334 Re^{0.755}}{1 - 0.49 Re^{-0.1} \left(q_j'' / q_i'' \right)} \tag{3.23}$$

2.5 Analytical Solutions for Streamwise Variations in Heat Flux

When the wall temperature or heat flux is not uniform, the heat transfer from a spot on the surface cannot be determined from local conditions alone nor can h be specified in terms only of local properties.

The usual situation considered analytically is streamwise variation of either temperature or heat flux assuming spanwise uniformity. This situation can be dealt with by superposition providing the energy equation describing the situation is linear. The equation is inherently linear in form but becomes non-linear if significant variations in fluid properties or buoyancy effects are present. Without these complications, the energy equation can be considered linear.

For electronics cooling, the most relevant situation is that of variable heat flux at the wall rather than variable temperature. Therefore, only those solutions will be presented.

Fully Developed Laminar Flow in a Round Tube.

This problem was solved by Sellars, Tribus and Kline (1956) and is given here mainly to show the form of the solution (Kays and Crawford (1980) p. 124). Here T_e is the entrance temperature, at $x^+ = 0$.

$$T_o(x^+) - T_e = \frac{r_o}{k} \int_0^{x^+} g(x^+ - \xi)\dot{q}_o''(\xi)d\xi \qquad (3.24)$$

where

$$g(x^+) = 4 + \sum_{m=1}^{\infty} \frac{\exp\left\{-\gamma_m^2 x^+\right\}}{\gamma_m^2 A_m} \qquad (3.25)$$

The eigenvalues and constants are given in Table 3.5 on next the page.

Fully Developed Laminar Flow in an Annular Passage.

Lundberg, Reynolds, and Kays (1963) extended this approach to the annular flow passage including the case of two parallel planes. Their solution, eqs (3.26) and (3.27), allows prediction of the wall

Table 3.5: **Eigenvalues and constants for streamwise variation of heat flux: Round tube, fully developed laminar flow.
For larger m, $\gamma_m = 4m + 4/3$ and $A_m = 0.358\gamma_m^{-2.32}$.**

m	γ_m^2	A_m
1	25.68	7.630×10^{-3}
2	83.86	2.058×10^{-3}
3	174.20	0.901×10^{-3}
4	296.50	0.487×10^{-3}
5	450.90	0.297×10^{-3}

temperature distributions from a description of the wall heat flux
distributions (Kays and Crawford (1980) p. 126).

$$T_i(x^+) - T_m = \frac{D_h}{k}\left[\int_0^{x^+} \frac{1}{Nu_{ii}(x^+ - \xi)}d\dot{q}_i''(\xi) - \int_0^{x^+} \frac{\theta_i(x^+ - \xi)}{Nu_{ii}(x^+ - \xi)}d\dot{q}_o''(\xi)\right]$$

(3.26)

$$T_o(x^+) - T_m = \frac{D_h}{k}\left[\int_0^{x^+} \frac{1}{Nu_{oo}(x^+ - \xi)}d\dot{q}_o''(\xi) - \int_0^{x^+} \frac{\theta_o(x^+ - \xi)}{Nu_{oo}(x^+ - \xi)}d\dot{q}_i''(\xi)\right]$$

(3.27)

In these equations, the values of Nu_{ii}, Nu_{oo}, θ_i, and θ_o are taken
from Table 3.6.

Table 3.6: **Nusselt numbers and influence coefficients for
streamwise variation of heat flux on the walls of a parallel planes channel with fully developed laminar flow (see
Equations (3.26) and (3.27)).**

x^+	Nu_{ii}	Nu_{oo}	θ_i	θ_o
0.0005	23.5	23.5	0.01175	0.01175
0.005	11.2	11.2	0.0560	0.0560
0.02	7.49	7.49	0.1491	0.1491
0.1	5.55	5.55	0.327	0.327
0.25	5.39	5.39	0.346	0.346
∞	5.38	5.38	0.346	0.346

Equations (3.26) and (3.27) are for spanwise uniform heating at each x^+. This is equivalent to uniform heating in each row of an array. The constants given are for smooth wall passages in laminar flow with fully developed hydrodynamics before the heating begins. The streamwise location of the point in question is x^+, while the location of the heating is ξ. T_i and T_o refer to the "inner" and "outer" walls of the passage in the general annular situation. In the present case the subscripts serve only to identify which coefficient applies to which wall.

Hydrodynamically Developing Laminar Flow

When the flow is developing hydrodynamically as well as thermally, the problem is more complicated. No general solution can be presented for superposition within the hydrodynamically developing region because a separate step-function solution would have to be available for every point in the hydrodynamically developing region. These solutions do not exist.

Fully Developed Turbulent Flow Between Parallel Planes

Turbulent flow alters the apparent conductivity of the air but does not change the basic problem, hence, streamwise variations in wall heat flux can still be dealt with using eqs (3.26) and (3.27) but the coefficients describing the solution are different. One new variable is present (Reynolds number) and the new solution is presented in terms of x/D_h rather than x^+ but, other than that, the procedure for solving for the wall temperature is the same as for laminar flows. The appropriate values of Nu_{ii} and θ_i are presented in Table 3.7 taken from Kays and Crawford (1980), p. 267, for three Reynolds numbers and for Prandtl Number $=1$.

2.6 Analytical Solutions for Individually Heated Spots on a Wall

No analytical solutions were found for this case, which would represent an idealization of a single heated element on a printed circuit board. Such solutions, if they existed, would form the basis for superposition solutions for randomly distributed power on the elements of a board. At present, the only approach seems to be experimental.

Table 3.7: Nusselt numbers and influence coefficients for streamwise variation of heat flux on the walls of a parallel planes channel with fully developed turbulent flow (see Equations 3.26 and 3.27)

x/D_h	Nu_{ii}	θ_i	Nu_{ii}	θ_i	Nu_{ii}	θ_i
	Re = 7,096		Re = 73,612		Re = 494,576	
1	47.3	0.013	234	0.005	940	0.000
3	37.9	0.033	203	0.018	851	0.009
10	31.5	0.089	177	0.049	761	0.030
30	28.0	0.173	160	0.114	697	0.0077
100	27.1	0.200	152	0.155	661	0.123

2.7 Numerical Methods for Calculating Heat Transfer

There are two classes of numerical methods applicable to electronics cooling problems–"systems programs" aimed at integrating convection, conduction and radiation to find the temperature distribution on the board and "fluid mechanics programs" aimed at calculating the details of the flow field near the board, as a precursor to calculating the convective heat transfer.

Most of the "systems" programs use simple correlations for the convective heat transfer coefficient. These are embedded in a finite difference or finite element program which deals with conduction and radiation as well. These programs are not sources of information concerning convective heat transfer but users of that information.

The numerical fluid mechanics programs, on the other hand, are addressing the issue of learning about the heat transfer rates. Again, there are two classes of programs. These are distinguished by the level of heuristic modeling involved. On the one hand there are several proprietary codes available for calculating the flow distribution in complex passages. These are elliptic or semi-elliptic three-dimensional turbulent flow codes using either mixing length or turbulence kinetic energy models. These are all derivatives of programs which have been discussed in the open literature over the past ten years and can be presumed to have most of the same characteristics. Experience with such codes has shown them to be quite geometry-specific. Good results can be obtained for different geometries, but only with "feedback" from the user to adjust the constants in the models until usefully accurate results are obtained. There is little hope for true generality in such codes because no truly general turbulence model has yet been devised. However, with feedback and with limited variability in geometry, such codes could be useful. Because these codes are proprietary, it does not seem appropriate to name them individually. Most of them are not "open" to the user for modifications to turbulence modeling and users are not able to develop them for specific applications. It seems safe to guess, however, that most are derivatives of one or the other of the codes which emerged from the late 1960's work of Patankar and Spalding–finite difference solvers with modular fluid mechanics models.

Such programs have been discussed in connection with electronics cooling by Sparrow and Chukaev (1980) and Schmidt and Patankar (1986). Both of these report numerical results for regular arrays of obstructions in parallel-plane channels with steady, fully developed, laminar flow. The work of Sparrow and Chukaev considered ribs parallel to the direction of flow arranged in a spanwise regular array, while Schmidt and Patankar dealt with transverse elements. Both evaluated friction factors and heat transfer coefficients and investigated the effects of geometric variations.

Davalath and Bayazitoglu (1985) treated the two-dimensional conjugate heat transfer problem, using a program similar to TEACH-T (Gosman (1976)). They calculated the heat transfer coefficient through a channel containing two-dimensional ribs. The heat transfer coefficient was defined in terms of the difference between the wall temperature and the inlet temperature.

More recently, Asako and Faghri (1987) attacked the three-dimensional problem using a $16 \times 22 \times 30$ node representation of one of the periodic modules of a fully developed flow. The numerical method used prior developments by Patankar (1980 and 1981) but is claimed to be the first successful calculation of the fully three-dimensional problem producing values for the friction factor and the Nusselt number and their distributions. The average Nusselt numbers appear a bit low, compared with data from 3-D experiments, but the method is very promising. Values are given for the Nusselt numbers on each of the faces of the element, which should aid a designer.

The second category of numerical methods involves the direct solution of the Navier-Stokes equations, with no heuristic models. A research group headed by Mikic and Patera at MIT have shown that such solutions can predict usefully accurate heat transfer results and can lead to new insights in design (Kozlu et al (1987) and Karniadakis et al (1987)). Using a spectral element method of solution (Ghaddar et al (1986)), solutions have been obtained for two-dimensional situations in 20 to 30 minutes of Cray-time. Results have been limited to laminar flows but with Reynolds numbers near the values used in practice.

This method has proven itself as a reliable predictor of behavior, and is now being used by the MIT group as an investigative tool which seeks to optimize heat transfer enhancement for prescribed pumping power requirements. The program is used to predict geometries which "tune" the oscillations from bluff bodies in the flow, or grooves on the wall, to excite instabilities in the main flow field. These in turn increase the heat transfer but with only small increases in the pressure drop. This appears to be the first useful predictor program that does not depend on constants determined by experience.

3   THE HEAT TRANSFER COEFFICIENT

The convective heat transfer from a component is usually described in terms of a heat transfer coefficient, either calculated, measured, or estimated from a correlation. It is based on the fact that the heat flow from a bounding surface to a fluid stream is directly proportional to the surface area and the temperature difference between the surface and the fluid. The convective heat transfer coefficient is a defined quantity, not a physical one, and its definition involves some subtleties which are not always appreciated. The purpose of this section is to review some of the issues which can cause errors in application.

Thus, the heat transfer coefficient, h, is defined as the heat flux from the surface divided by the temperature difference between the surface and the fluid. Significant difficulties in application arise because there is more than one way to define the reference fluid temperature and the numerical value of h, the heat transfer coefficient depends on which option is chosen.

3.1   The Options

In the electronics cooling situation, there are four options for the reference temperature, each of which has been used by some segment of the industry. Thus, there are four options for the definition of h:

$$h = \frac{\dot{q}_o}{A(T_o - T_r)} \qquad (3.28)$$

where T_r may be

1. The adiabatic temperature of the element.
2. The mixed mean temperature of the air in the channel.
3. The inlet temperature.
4. The air temperature far from the element.

For a channel flow problem, it is not appropriate to speak of "the temperature far from the surface", hence option 4 is not available. The other three options are viable, however, and three different h values can be calculated from any set of experimental data. The three are: h_{ad}, h_m, and h_{in}. For low channel heights, where the mixing is good, and the fluid temperature nearly uniform, h_{ad} and h_m will have nearly the same value, but at high channel spacing, where the mixing is not good, there can be a large difference: h_{ad} might be twice as large as h_m and three times as large as h_{in}. All three values would yield the correct heat transfer rate, if used with their appropriate reference temperature, but if data which, in fact, represent h_{ad} are used as though they represent h_m, then significant errors can occur.

Option 1: h_{ad}

The heat transfer coefficient based on adiabatic temperature is the recommended option for the electronics cooling situation. This recommendation is based on the following four arguments:

First, h_{ad} is all there is! Substantially all of the experimental data available at the present time use this option. The question is not whether or not to use h_{ad}–there is no alternative–the question is "how to avoid mis-using the existing data base?"

Any experiment in which only one element at a time is heated is implicitly using the adiabatic temperature of the element as the reference gas temperature in the definition of h. The same is true of any experimental technique in which one element at a time is coated with napthalene, or with a swollen polymer - the implicit basis for heat transfer coefficients determined from such experiments is the adiabatic temperature of the element itself.

Second, h_{ad} is the "best behaved" of all of the options; it is a function only of geometry and flow. It is the only one of the four options whose value at a given location does not depend on the upstream heating pattern.

Its value can be measured in simple tests with only one element heated (the test specimen) and this same value will apply regardless of how many upwind elements are also heated. As long as the geometry and Reynolds number of the flow are the same, h_{ad} will be the same.

Third, using h_{ad} opens the door to superposition solutions for situations involving non-uniform heat release.

Fourth, any situation which can be handled accurately with h_m or h_{in} can be handled just as accurately with h_{ad}.

Option 2: h_m

This is the traditional heat transfer coefficient (based on the mixed mean temperature) used in heat exchanger design and process heat transfer calculations. This definition is used in substantially all of the analytical solutions presented in texts and reference works on convective heat transfer for channel-flow heat transfer. A review of the electronics cooling literature has revealed no experimental data for h_m.

There is no way to measure h_m experimentally without heating all of the elements in the entire channel, and then the distribution of h_m which results is valid only for the distribution of heat release which was used in the test. This means that each different distribution of heat release will yield a different distribution of h_m along the channel.

The usual correlations show h_m to be a function of Reynolds number and fluid properties. It is therefore constant along the length of the channel (once past the entrance region). This constancy is the result of the wall temperature boundary conditions used in the analysis or experiment. Only for "constant temperature" or "constant heat flux" conditions is h_m well behaved. Streamwise variations in temperature or heat release cause h_m to change drastically and h_m can take on any value ranging between plus and minus infinity.

Option 3: h_{in}

The channel inlet temperature is used as the reference temperature by some practitioners in the electronics cooling area because it is part of the environmental description and is unambiguous. This option has more utility in describing systems tests than in analyzing the details of the heat transfer process. When h is based on T_{in}, all variations in the heat transfer process are accounted for by variations in h itself. Thus, of the three options, h_{in} can be expected to vary most widely as the thermal conditions change.

Option 4: h_∞

In free convection situations or in forced convection with very large spacing between board, one might use h_∞, where the reference temperature was taken as the fluid temperature far from the surface. This is typical of boundary layer analyses.

3.2   The Consequences of Misinterpreting h

The most common error is using h_{ad} to calculate the temperature rise of a component with respect to the bulk mean fluid temperature $(T_o - T_m)$. This can lead to large errors in predicted element temperature, especially in relatively open channels populated with flatpacks, where the mixing is not good and where the real adiabatic element temperature will almost always be hotter than the bulk mean fluid temperature.

The size of the error will depend on how much the mean temperature has risen, and how good the mixing is in the channel. Typical values for the error are between one and four times the coolant temperature rise. That is, if the coolant rise is $10°$ C, an error of from 10 to $40°$ C in the temperatures of the elements in the last row might result if data for h_{ad} were used as though the definition was h_m.

Such errors are common because, while almost all of the published data are, in fact, values of h_{ad}, the usual assumption is that they represent h_m. In a representative situation, where the correct use of h_{ad} would predict a final temperature of about $115°$ C for the elements in the last row, using h_{ad} with $(T_o - T_m)$ underestimated the maximum temperature rise by $20°$ C while treating h_{ad} as though it was h_{in} underestimated T_{max} by $30°$ C.

3.3   Finding the Adiabatic Element Temperature

The adiabatic temperature of an element depends on the distribution of upstream heat release, and must be known before the h_{ad} data can be applied. The adiabatic temperature of an element can be calculated, for an arbitrary distribution of heat release, using a set of superpositional kernel functions, one for each of the upstream elements. The kernel functions describe the adiabatic wall temperature distribution downstream of a heated element in terms of the temperature rise of the heated element or its heat release. In the present state of the art, these kernel functions must be found experimentally.

The adiabatic temperature can be found by superposition, as demonstrated by Arvizu and Moffat (1982), Moffat, Arvizu, and Ortega (1985) and Bibier and Sammakia (1986). The superposition method accounts for the contribution from each heated element upstream of the element under scrutiny. Such a calculation is sequential, beginning in the first row and working through the array. The thermal wake from the first row of heated elements strikes the second row, raising the adiabatic temperature there, and passes on to heat the rows further downstream. Two types of information are needed: 1) the heat transfer coefficient pertaining to the element under scrutiny and 2) the distribution of adiabatic element temperatures downstream of a single heated element.

The correlation for the adiabatic temperature distribution downstream of an element depends on the element geometry and the Reynolds number–there is no general correlation. Arvizu and Moffat (1982) recommended a two-part correlation. The first, a correlation based on Reynolds number, describes the adiabatic temperature rise of the first element downstream of the heated element, θ_1. The second part relates the adiabatic temperature rise of the elements further downstream to the temperature rise of that first one.

A correlation for θ_1 was given by Moffat, Arvizu and Ortega (1985)

$$\theta_1 = C_1 Re_B^{-0.28} \tag{3.29}$$

where

$$C_1 = 0.83 \text{ for } S/B = 2.0 \text{ and } C_1 = 0.59 \text{ for } S/B = 3.0$$

Elements downstream of the first element display a reduced adiabatic rise. For an element which is N rows behind the heated element the adiabatic rise is $1/N$ times θ_1. This is an approximation which fits the data for cubes reasonably well, and, as shown by Wirtz and Dykshoorn (1984), is not too far off the mark for flatpacks.

3.4   Using h_{ad}

There are several formulations of the superposition equation, but typical of the application of h_{ad} is

$$T_i - T_{in} = \frac{\dot{q}_i''}{h_i A_i} + \sum_{j=1}^{J} \theta_{ij} \frac{\dot{q}_j''}{h_j} \qquad (3.30)$$

The first term calculates the temperature rise of the component above its own adiabatic temperature due to its own heat release.

The second term calculates the difference between the adiabatic temperature of the component and the inlet temperature of the coolant.

The common error of using h_{ad} data with T_m is tantamount to assuming that the second term vanishes and that the fluid is perfectly mixed. Experiments show that this is not a good assertion except for very small board spacing and very intrusive elements–flatpacks do not encourage mixing.

This issue has also been described by Nakayama (1986) who discusses the source of the problem (insufficient mixing) and presents data for the ratio of the adiabatic temperature rise to the mean temperature rise. He shows this ratio to be between 1 (perfect mixing) and 4, based on the data of Wirtz and Dykshoorn (1984) and Arvizu and Moffat (1982).

4   EXPERIMENTAL DATA:
TWO-DIMENSIONAL SITUATIONS

Whereas the empty, parallel planes channel offers only two length scales, plate spacing and streamwise distance, the parallel plane channel with two-dimensional, transverse ribs, shown in Fig. 3.1 has five. If the height of the rib, B, is taken as the scaling dimension, four dimensionless groups are required: $H/B, L/B, S/B$ and x/B (or N, the rib number). The geometric possibilities are bounded by the smooth walled channel (when $S/B = 1$ the ribbed wall is smooth again) and by isolated ribs (when $S/B > 20$ the ribs act independently).

The heat transfer behavior of a ribbed channel can be divided into two regions: the entrance region and the fully developed region. Furthermore, one must make the decision to study either the heat transfer from the rib alone, or from the rib-and-wall together. In the following work, heat transfer is measured from the rib alone and does not include the heat transfer from the wall.

4.1 The Entrance Region

The number of ribs which should be described as "in the entrance region", as far as heat transfer is concerned, varies between 2 and 6, depending on the rib spacing (S/B) and channel height (H/B). Figure 3.2 taken from Arvizu and Moffat (1981) shows the number of rows required to enter the fully developed region as a function of H/B, S/B and Reynolds number. Taken together with other (unpublished) data, the evidence suggests that the extent of the entrance region is not a strong function of Reynolds number but depends mainly on the geometry.

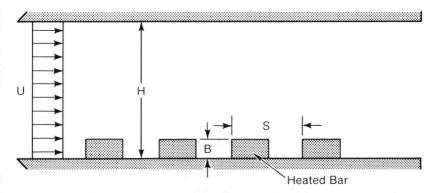

Figure 3.1: Schematic representation of a channel with two dimensional ribs. [Arvizu and Moffat (1981) Fig. 2.1]

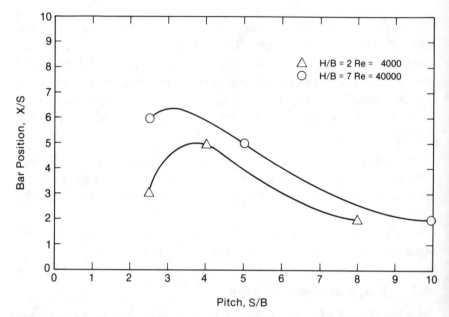

Figure 3.2: **The number of rows required to achieve fully developed heat transfer for two-dimensional ribs in a channel flow. [Arvizu and Moffat (1981) Fig. 2.5].**

The heat transfer from the first rib in an array of heated ribs can be larger than from an isolated rib by 20 percent or more if the presence of the next rib creates a Type I flow situation, as described in the next section.

4.2   The Fully Developed Region

The changing flow fields and heat transfer behavior as S/B increases are illustrated in Fig. 3.3 for H/B=7.0. Three flow regimes are identified, following the work of Liu, Kline, and Johnston (1966). These are identified as:

> Type I: The Driven Cavity Regime
> Type II: The Wake Interference Regime
> Type III: The Independent Roughness Regime

The line through the data is not extended to S/B = 1 to avoid dealing with the discontinuity in area when the gap between elements first appears. Channel height was the same for all runs shown in this figure and the Reynolds number, Re_H, was defined using the approach velocity and the height of the empty channel. When channel height is a variable, the data are better organized using Re_B as can be seen by examining the data in Fig. 3.4. Rib height was not a variable in this program, hence this presentation is no more general than simply plotting the data against approach velocity, a fact which should be kept in mind.

The data for relatively open channels, H/B = 4.6 and 7.0, form an ordered group as though there is no significant difference between these two passage heights. Within that group, increasing the rib spacing from S/B = 2.5 to 4.0 causes an increase in Nusselt number. Increases in S/B from 4.0 to 10 cause only a slight further increase in Nusselt number. The data for H/B = 2.5 form another group displaying a different order property. At this relatively close channel spacing, data for S/B = 4.0 and 8.0 show no difference. The heat transfer results in Figs. 3.2, 3.3, and 3.4 were taken on ribs 1.27 cm high and 2.54 cm wide (in the flow direction).

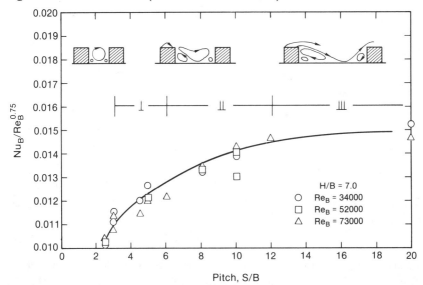

Figure 3.3: Variation of $Nu/Re^{0.75}$ with pitch, S/B, and Reynolds number, UH/ν, for fully developed flow. [from a student lab report, cited in Arvizu and Moffat (1981) as Fig. 2.4].

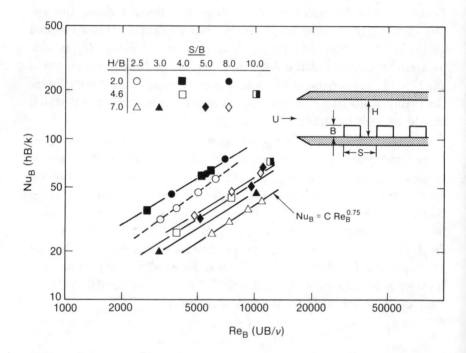

Figure 3.4: Element Nusselt number (hB/k) versus element Reynolds number (UB/ν) for large, rectangular ribs. [Arvizu and Moffat (1981) Fig. 2.6].

Results from a study by Lehmann and Wirtz (1984) on a different rib geometry, are shown in Fig. 3.5 in the coordinates reported by the authors: case-to-ambient convective resistance versus Reynolds number. The term ambient here refers to the fluid temperature at the location of the element. This temperature was uniform except for the disturbance caused by the heated element itself. The recommended correlation is:

$$R_{c-a} = C_2 Re_{H_1}^{-0.6} \qquad (3.31)$$

where $Re_{H_1} = U(H_1)/\nu$ and C_2 is given by Fig. 3.6 as a function of b/H_1 with b equal to wall-to-wall spacing and H_1 taken as the rib length in the flow direction.

The case-to-ambient resistance is equal to $1/Nu_{ad}$ where the Nusselt number is based on h_{ad} and the element length in the streamwise direction.

The Lehmann and Wirtz (1984) study used ribs 1.25 *cm* high and 5.0 *cm* in the flow direction, spaced 1.25 *cm* apart. This arrangement formed square grooves between the rows and simulated flatpacks in a moderately dense array.

The two experiments used different techniques, but both report the same kind of heat transfer coefficient, h_{ad}. Lehmann and Wirtz (1984) also reported local and average h values measured interferometrically on the top face of one rib in an otherwise passive array. They used a Plexiglas specimen, heated on its top face only, using a thin film heater. The study by Arvizu and Moffat (1981) used an energy balance technique on one heated aluminum rib in an otherwise passive array hence their reported values include the heat transfer from the edges. The edges are expected to have lower h values than the top face and they account for 50 percent of the total area of the specimens. This might explain some of the difference in h values.

Both studies measured h on a single heated specimen inside an array. As a consequence, both h values represent "adiabatic" values;

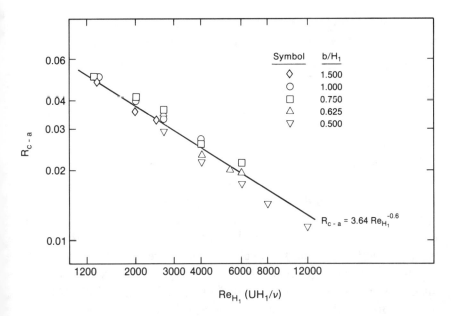

Figure 3.5: Case-to-ambient resistance versus Re_{H_1} for two-dimensional ribs in a channel flow. [Lehmann and Wirtz (1984) Fig. 10]. See Fig. 3.8 for nomenclature.

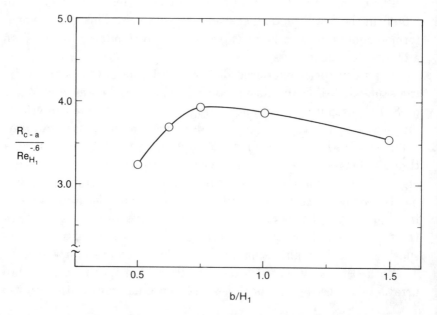

$$\frac{R_{c\text{-}a}}{Re_{H_1}^{-.6}}$$

b/H$_1$

Figure 3.6: Variation of the coefficient C_2 in the correlation of Equation 3.31 [Lehmann and Wirtz (1984) Fig. 11].

that is, to calculate the heat transfer rate from an element in a fully powered array, the appropriate equation would be:

$$q_{conv} = hA(T_o - T_{ad}) \qquad (3.32)$$

Use of these h values with either the mixed mean temperature or the inlet temperature will result in significant errors in the predicted heat transfer.

Because of the differences in geometry and definition, it is not appropriate to show both data sets on the same graph, but Figs. 3.7 and 3.8 were constructed to facilitate the comparison and it is hard to resist the temptation to "explain" the differences in behavior.

Both data sets represent the fully-developed heat transfer region–after the entrance effects have died off. The Arvizu-Moffat (1981) data, shown in Fig. 3.7, form two distinct groups, one corresponding to small channel spacing (H/B = 2) and one for large channel spacing (H/B = 4.6 and 7.0). In each group the Nusselt number increases as the streamwise spacing (S/B) increases up to limit at about S/B = 4.0. Further increases in streamwise spacing have no effect. This seems to be the lower limit of the Type III behavior, as far as heat transfer is concerned, referring to the classification proposed by Liu et al (1966). The increase may reflect the increasing activity on the upstream and downstream walls of the rib since, in these experiments, h represents the average over all three faces shown.

Figure 3.7 shows a drop in Nusselt number as the passage height opens up. This could mean that the S/B spacings defining the flow regimes in Fig. 3.3 change with passage height.

The Lehmann and Wirtz (1984) data, shown in Fig. 3.8, are also from the fully developed region–after nine or ten ribs. The data are (relatively) more coherent than the Arvizu-Moffat (1981) data, showing only a 25 percent drop in Nusselt number at constant Reynolds number, as the passage height is increased. This is a much smaller change than that shown in Fig. 3.7 for roughly the same

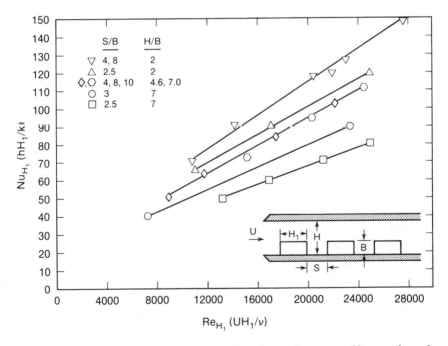

Figure 3.7: The Arvizu and Moffat data for two-dimensional ribs, plotted using Lehmann and Wirtz's recommended definition of Reynolds number. [Arvizu and Moffat, 1981, Fig. 2.6].

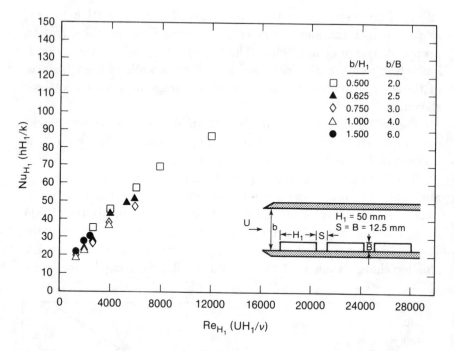

Figure 3.8: Lehmann and Wirtz's data for two-dimensional ribs, as Nusselt number versus their recommended definition of Reynolds number. [from the data in Lehmann and Wirtz (1984) Fig. 11].

range of ratios of channel height to element height. Reynolds number is defined in terms of velocity and element length, not passage height, hence these experiments at constant Reynolds number are also at constant velocity. This indicated that the drop in Nusselt number is not explainable by a velocity effect. The square cavity yields a Type I flow situation which, apparently, does not interact vigorously with the main stream. Smoke-wire studies by Lehmann and Wirtz (1984) support this notion and show the cavities to be relatively isolated from the main flow and not inducing much mixing.

The Lehmann and Wirtz (1984) data are about 15 percent higher than the Arvizu and Moffat (1981) data in the Reynolds number range where both sets have data. Both data sets were examined in several coordinates, i.e., different definitions of Reynolds number and Nusselt number. None collected the data better than those used in Figs. 3.7 and 3.8. The issue is still in doubt, however, as to the "best" scaling length since neither study varied the dimensions of the ribs–particularly their streamwise width. There is no evidence that the Reynolds number based on the streamwise length and the velocity over the top is a robust scaling parameter–at the moment, all that has been varied is velocity. Wirtz's argument in favor of using the length in the flow direction seems physically sound when the gap between elements is small enough so that a Type I flow is trapped in the cavity.

Figure 3.9 shows pressure drop data for three conditions in terms of the cumulative pressure coefficient C_p vs N, the number of ribs. The pressure coefficient C_p is defined as the difference between the upstream total pressure and the downstream static pressure divided by the velocity head in the channel approaching the array: the open channel velocity head.

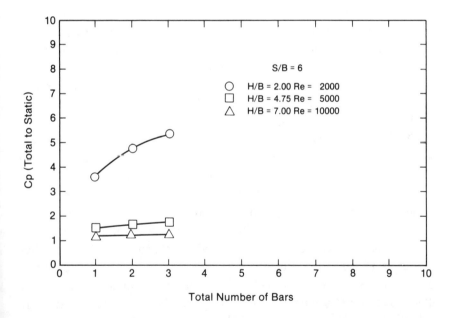

Figure 3.9: Overall pressure coefficient, total to static, for large, rectangular ribs. [Arvizu and Moffat (1981) Fig. 2.7]

5   EXPERIMENTAL DATA: THREE-DIMENSIONAL SITUATIONS

5.1   General Remarks

The most common geometries encountered in electronic cooling are rectangular blocks or short cylinders mounted on one of two parallel planes which define a flow passage. The block-like elements range from dual-in-line packages (DIPs) to flatpacks, and the cylinders may be mounted either parallel to the wall or perpendicular to it. The individual elements may be arranged in regular arrays, either staggered or row-wise, or dispersed in irregular patterns. On a given board, the sizes of the elements may be either uniform or non-uniform. Board spacing is usually kept to a minimum, to conserve volume and to minimize electrical transit time delays.

The flow is fan driven, but may be described as forced draft or induced draft depending on whether the fan is upwind or downwind of the boards. Frequently, fans are used in both locations at the same time, to balance case pressure against ambient (for control of dust ingestion) or, simply, to get a higher pressure ratio than one fan can produce. Close-coupled forced draft systems may suffer from non-uniform flow due to the discharge pattern of the fan. The flow exits in an annular jet, not a cylindrical one, and there is little or no flow along the centerline. If the fan is very close to the entrance to the flow passage, the lanes on the centerline may not receive sufficient flow. The discharge is turbulent, however, which will raise the heat transfer coefficient relative to undisturbed flow at the same mean velocity but only on elements in the first few rows of an array.

The most frequent experiment used to measure heat transfer behavior is the isolated element test in which a single component is placed in a small wind tunnel and tested over a range of air speeds. The components are usually used in arrays, however, where different values of h will exist in practice than in the isolated element tests.

This section presents heat transfer coefficient data both for isolated elements and arrays of elements and covers three geometries: cubes, flatpacks, and stub cylinders. Heat transfer coefficients are uniformly presented in terms of h_{ad}, which is the form used in the original publications.

In addition to the heat transfer coefficient, the thermal wake functions and the pressure drop per row are given where available because these data are necessary if the heat transfer coefficients are to be used with best accuracy. This point was emphasized by Arvizu and Moffat (1982) in a paper describing the use of superposition for calculating the operating temperatures of individual elements in an array with arbitrary power distribution.

5.2   Single Elements

Buller and Kilburn (1981) presented data for forced-convection heat transfer from single, rectangular modules. Five specimens were tested: three "plain" and two with finned heat sinks attached. Their results for all five geometries could be represented by a single correlation within ± 15 percent if the Reynolds number and the Nusselt number were defined using a characteristic length based on frontal area, perimeter, total surface area, and flow length.

$$L = (A_f/C_f)(A_t/l)^{1/2} \qquad (3.33)$$

Where:

A_f = Frontal surface area of the component.

A_t = Total wetted surface area of the component.

C_f = Circumference of the frontal area of the component.

l = Component length in the flow direction.

L = Characteristic length for Reynolds number and Nusselt number.

While this approach was very promising in its ability to describe data from different geometries with a single correlation, it does not appear to have been used by other workers until Chang, Shyu, and Fang (1987) applied it in a study of single and double elements.

Their elements were 6 × 6 cm squares, 2 cm high installed at the downstream end of rectangular channel operating at channel Reynolds numbers between 5,200 and 62,000. Results were reported in terms of the element Reynolds number, based on channel mean velocity (volume flow rate divided by open channel flow area) and element characteristic length. The results can be expected to be higher

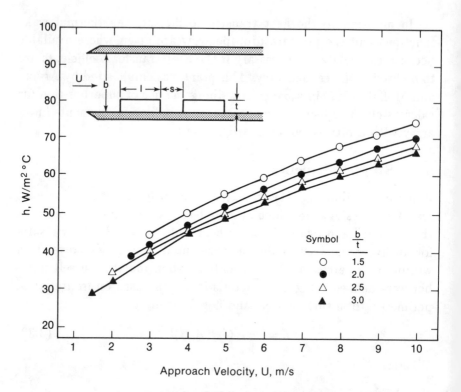

Figure 3.10: Heat transfer coefficient versus approach velocity for a single element in a fully developed channel flow. [Chang et al (1987) Fig. 4]

than would be obtained in tests with a low-turbulence free stream of the same nominal velocity. Two series of tests were reported, one with a single element installed and one with two elements separated in the streamwise direction.

Figure 3.10 shows h as a function of velocity for a single element tested with different channel heights. At a given mean velocity of approach to the element, h is higher for the smaller passage heights. This could represent either the effects of local accelerations around the block or changes in the scale and intensity of the free-stream turbulence.

The effect of flow blockage by an unheated upstream element is shown in Fig. 3.11 as the ratio of the heat transfer coefficient with two elements present, h_2, to that with only one present, h_1. The data are presented in terms of the ratio of the gap width, s, (measured between the two elements) to the height of the element, t, or to the streamwise length of the element, l.

The upstream element was never heated, hence the h value measured in these tests is h_{ad}. The effects of the thermal wake were not measured in this study, although the issue was discussed by the authors.

5.3   Arrays of Elements

The general situation to be discussed in the following sections is that of an array of elements on one wall of a parallel planes channel. The heat transfer behavior can vary from that of a smooth walled channel, where uniform elements are placed so close together that the new surface formed by their tops resembles a smooth wall with small grooves, to that of isolated elements. These bounding cases are now well enough understood so that at least some guidance is available over a useful range of conditions. The main problems arise in the middle ground with sparsely populated arrays, where the elements

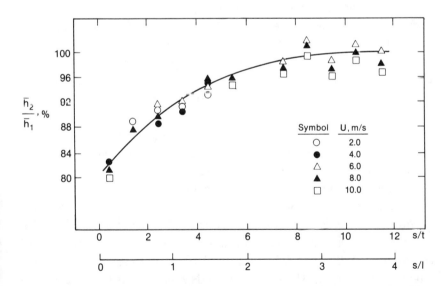

Figure 3.11: The ratio $\bar{h}_2/\bar{h}_1$ versus gap width, s, where $\bar{h}$ is the heat transfer coefficient on the downwind element: $\bar{h}_2$ with blockage, and $\bar{h}_1$ without. [Chang et al (1987) Fig. 7]

are too densely arranged to be treated as independent elements, yet not close enough together to form a smooth wall or a grooved wall.

The sparsely populated array cannot be treated by simple interpolation between the bounding cases. Element geometry, array pattern, and passage height each provide length scales to this problem and the dimensionality is high. Even if the elements were uniform in size and shape, which is unlikely, there are too many variables for an orderly study and all seem to be important. Complex geometries have been dealt with in compact heat exchanger analysis by defining the hydraulic radius of the flow passage as a volume- averaged property as, for example, by Kays and London (1984). This approach has been tried, and found to be unsuccessful (Moffat et al (1985)). The apparent reason for the failure is that a passage with an array of elements on one wall provides not one, but two possible paths for flow. Some of the flow stays inside the array and some bypasses it, running through the clearance gap without affecting the heat transfer from the elements. The flow appears to divide between the two paths according to the pressure drop characteristics of the array and the passage height. Because the flow partition changes as the passage height changes, no generality will be achieved by describing the hydraulic radius: each change in passage height defines an entirely new flow split. The compact heat exchanger situation is different in that the heat transfer core can be treated as homogeneous–the passages are small, there are many in parallel and every possible path through the core sees the same type of passage.

One must be prepared to find different responses to the Reynolds number and the passage height for different element geometries because these factors affect the flow partition as well as affecting the heat transfer from each individual element.

Strictly speaking, the heat transfer data obtained from systems of different geometry cannot be "compared" with one another. No universal length scale description has been found. This is in spite of the work of Buller and Kilburn (1981) which represented several different geometries on the same Reynolds number basis. Yet the urge to "compare and collect" is almost irresistible. The problem is to decide on what parameter to "pivot" in making the comparison, that is, what parameter should be held constant to make the comparison most meaningful.

Table 3.8 represents heat transfer results for seven situations, six experimental and one analytical. The experiments are discussed in more detail in the following sections. These situations represent increasingly intrusive protuberances, beginning with a smooth walled channel and proceeding to isolated rectangular elements. The heat transfer coefficient itself is presented, rather than any dimensionless correlation because the elements are of different shape and there is no universally suitable length scale.

Table 3.8: Heat-Transfer Coefficients for Seven Situations of Turbulent Flow With and Without Modules Between Parallel Planes [Moffat et al (1985).

Source	Case	Correlating Equation*	h	$C_P Row$†
Kays	Smooth planes	$h = 0.023\, Re^{0.78}$	14	–
Sparrow	Dense flatpacks	$h = 0.078\, Re^{0.72}$	29	0.034
Sparrow	With barrier	$h = 0.112\, Re^{0.70}$	43	–
Wirtz	Sparse flatpacks	$h = 0.324\, Re^{0.60}$	48	–
Arvizu	Sparse cubes (3/1)	$h = 0.600\, Re^{0.56}$	60	0.120
Arvizu	Dense cubes (2/1)	$h = 0.650\, Re^{0.56}$	65	0.160
Buller	Single flatpacks	$h = 0.722\, Re^{0.54}$	61	

*The form of each correlating equation was extracted from its parent data set as a whole while h was found specifically for a value of $Re = 3700$ either using that equation or from the data. Reynolds number is based on channel height and mean velocity in the empty channel upstream of the elements.

†C_P/Row is the pressure drop per row, made dimensionless by the velocity over the elements.

The entry labeled "Kays" represents the analytical results of Kays and Crawford (1980), p. 255 and are for turbulent flow between parallel, smooth planes when one wall is heated at constant heat flux and the other is adiabatic. The channel Reynolds number was used in their equation to calculate the heat transfer coefficient corresponding to the approach Reynolds number in a populated channel: the Reynolds number upstream of the first row of elements.

The entries labeled "Sparrow" represent the heat transfer and pressure drop results of Sparrow, Niethammer and Chaboki (1982) for a Reynolds number of 3700. That Reynolds number could be found in both of the other studies and was within the range covered.

The entry labeled "Wirtz" considers the data of Wirtz and Dykshoorn (1984) and are from their tests on flatpacks at a Reynolds number of 3700 with H/B = 3.6. The parameter H/B had very little effect on their results (contrary to the findings with tests on cubes) but this value resulted in a clearance gap over the elements which was the same as the gap used by Sparrow et al (1982).

The data from Moffat, Arvizu, and Ortega (1986) led to the entries marked "Arvizu" and were taken from their runs at H/B = 2.48 with an approach Reynolds number of 3700. For those build-ups, the free height above the elements was the same as in the Sparrow et al (1982) work (1.667 cm) and the ratio of channel height to element height was almost the same (2.48 compared to 2.67). The pressure drop per row was made dimensionless using the velocity above the elements (assuming all of the flow passed over the array) to be consistent with the definition used in the Sparrow et al (1982) work.

The entry attributed to "Buller" represents data taken from the work of Buller and Kilburn (1981) and was calculated by extracting h from their tabulated Stanton numbers and, in order to compare it with the other values, estimating the value it would have had at a Reynolds number of 3700.

There is a regular increase in the value of h, corresponding to a regular increase in C_P/Row. The fact that the heat transfer increases when turbulence intensity increases has long been known. As noted by Wirtz and Dykshoorn (1984), and others, there is a decrease in the Reynolds number exponent as the turbulence intensity increases. One feature of Table 3.8 which is of particular interest is the fact that the heat transfer coefficient increases roughly in proportion to the square root of C_P/Row. Qualitatively, this seems intuitively reasonable; an increased pressure drop per row means higher turbulence production which should lead to higher heat transfer coefficients. The increase would not be expected to be linear.

It appears, from Table 3.8 that one must be prepared to find any exponent on the Reynolds number from 0.54 to 0.78 and one may find an exponent which varies along the flow direction, even within a uniform array, due to changing values of turbulence.

5.4 The Implications of Non-Uniform Heat Release

When the heat release is not uniformly distributed over the elements of an array, there is no general method available for dealing with the heat transfer problem through correlations. The fluid mechanics are set by the geometry of the array and that determines the local velocity distribution and turbulence properties around the individual elements but the temperature distribution downstream of each heated element is also important. This introduces a new concern which involves the thermal wake function, $\theta_{i,j}$. This is just as important a property of the element as its heat transfer coefficient for without the thermal wake function, one cannot apply the heat transfer coefficient data. The thermal wake function is not, properly speaking, a property of the element itself, but of its downstream neighbors and the flow field. Nevertheless, it must be known, and it is associated with the position and geometry of the heated element as well as the downstream neighbors.

Whereas the uniform temperature and uniform heat flux situations can be dealt with by knowing only a properly defined heat transfer coefficient, the non-uniform heating situations can only be dealt with by superposition and superposition requires two types of information: h_{ad} and $\theta_{i,j}$, the thermal wake function.

At present, given the state of the art of numerical fluid mechanics, it is not possible to solve for the temperature distribution in the array with any assurance, except with a well-tuned code. The thermal wake functions, as well as the heat transfer coefficients, must be obtained experimentally.

The general method of superposition is as old as the art of heat transfer itself and is one of the classical analytical tools. Its application in experiments in electronics cooling is more recent. The technique was described by Arvizu and Moffat (1981) in a paper which reviewed the classical superposition approach and presented a preliminary data base by which its use could be demonstrated in an electronics cooling environment. The method has since been tested by Bibier and Sammakia (1986) on an array of flatpacks using an in-house generated data base and good results were obtained.

The thermal effects of non-uniform heating will become increasingly important as circuit designers come closer to the limit of ma-

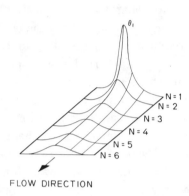

FLOW DIRECTION

Figure 3.12: The thermal wake function downstream of a single heated cube. [Moffat, Arvizu and Ortega (1985) Fig. 2]

terials. In preparation for this, it is important to assemble all of the information needed for heat transfer design. Thus, in the sections which follow, the thermal wake functions will be described wherever possible.

5.5   Arrays of Cubical Elements

Moffat, Arvizu, and Ortega (1985) presented heat transfer coefficients and thermal wake functions measured on a square, in-line, array of 112 cubical elements. The elements were arranged in fourteen columns with eight elements in each row with the spacing between rows the same as between columns. The flow was parallel to the columns. Only one element was heated at a time and both h and the thermal wake function were measured as a function of channel spacing and approach velocity for two spacings, $S/B = 2.0$ and 3.0. A series of tests was conducted with the heated element in different rows to determine the effect of position within the array.

The heat transfer coefficients reported from this study are based on the adiabatic element temperature, and the significance of this choice was emphasized by the authors. The objective of the work was to lay the groundwork for, and to demonstrate the practicality of, superposition solutions for the temperature distribution within an array. To that end, the thermal wake functions were as important as the heat transfer coefficients, and they played a prominent role in that report.

5.5.1   The Temperature Wake Function

Figure 3.12 shows the general shape of the thermal wake function behind a single heated element, is represented as an operating surface: temperature rise as a function of position downwind of a heated element. The value θ_1, the peak on this distribution, is not the temperature of the heated element, but the adiabatic temperature rise of the first element behind the heated element. The adiabatic rise of the first downstream element depends on the streamwise spacing, as

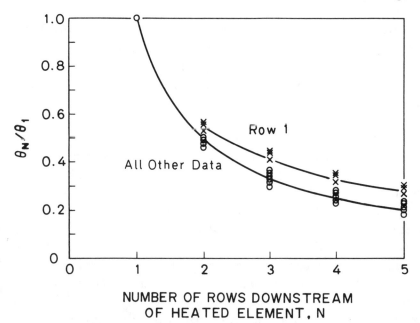

Figure 3.13: Variation of θ/θ_1 along the elements in the column containing the heated element. [Moffat, Arvizu, and Ortega (1985) Fig. 3]

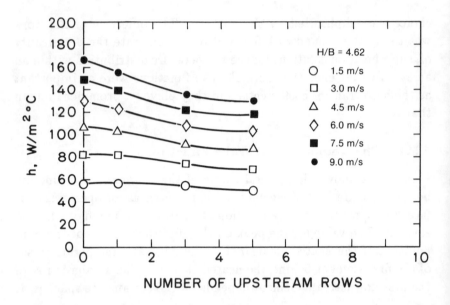

Figure 3.14: Heat transfer coefficient versus row number in the entrance region of a channel with $H/B = 4.62$. [Moffat, Arvizu, and Ortega (1985) Fig. 4]

seen by the variation of C_1 in eq (3.34) and, to a lesser extent, on the row in which the heated element was located.

Along the column containing the heated element, the adiabatic temperature rise drops off sharply in the downstream direction. Figure 3.13 shows a representative sample of the data. To a first approximation, this roll-off goes like 1/N, where N is the number of rows behind the heated element. It spreads only slightly to the laterally neighboring columns. If the lateral temperature rise is expressed as a fraction of the rise along the centerline, the fraction is found to increase linearly with row number. This has the net effect of yielding an approximately constant temperature rise in the flanking columns. There is likely a limit to the downstream extent of this simple model but it seems acceptable within the first five or six rows. The shape of the wake function surface is described by three parameters: the initial value, θ_1, the roll-off along the centerline and the spanwise spreading rate as a function of the centerline value. Again, to a first approximation, this shape can be modeled by the simple set of functions:

$$\theta_1 = \frac{T_1 - T_{in}}{T_{heated} - T_{in}} = C_1 Re_B^{-0.28} \qquad (3.34)$$

where:

$$C_1 = 0.83 \text{ for } S/B = 2.0$$

and

$$C_1 = 0.59 \text{ for } S/B = 3.0$$

and

$$\theta_N = \frac{T_N - T_{in}}{T_{heated} - T_{in}} \cong \frac{1}{N}\theta_1 \qquad (3.35)$$

5.5.2 The Heat Transfer Coefficient

Figures 3.14 and 3.15 show the distribution of heat transfer coefficients in the entrance region for two channel heights, H/B = 4.62 and 1.00. For large H/B (Fig. 3.14), the heat transfer coefficient is highest in row 1, and decreases monotonically with penetration into the array, reaching a "fully developed" value after three of four

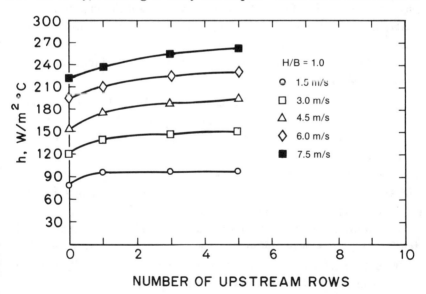

Figure 3.15: Heat transfer coefficient versus row number in the entrance region of a channel with **H/B = 1.0**. [Moffat, Arvizu, and Ortega (1985) Fig. 5]

rows. In the worst case, at 9.0 m/s, the drop was about 30 percent relative to the initial value. The general appearance is that of a classical entrance region heat transfer coefficient distribution. A similar response (although the variation was less pronounced) was seen by Sparrow, Niethammer and Chaboki (1982) and by Wirtz and Dykshoorn (1984) in tests on flatpacks.

The value of h in the first row can be affected by the turbulence level in the approaching stream. The first row values shown here for H/B = 4.62 are probably high due to a separation bubble at the leading edge of the test surface upstream of the first row. Subsequent repeat runs for the situations shown here have yielded values about 20 percent lower in the first row when the flow is known to have low turbulence (Anderson (1987)). Because the flow condition in service may not be well known, one must be prepared for variation to this extent in the first row due to separations which might exist.

However, the behavior in Fig. 3.14 is not representative of all situations. Figure 3.15 shows data from another entrance region situation, but now h rises by about 20 percent with penetration into the array, instead of falling by 30 percent. The difference in response is attributed to the effect of the channel height on the flow distribution inside the channel. In either case, the entrance region appears to include only the first three or four rows, after which h_{ad} reaches a fully developed state.

From Figs. 3.14 and 3.15, it is clear that the entrance region of a channel containing an array of protrusions cannot be dealt with in the same way as the entrance region of an empty channel. There appear to be two opposing mechanisms at work in the three-dimensional array, one resulting in a tendency for h to go up with distance into the array (H/B = 1.0), and one resulting in a tendency for it to go down (H/B = 4.62). The balance between the two shifts as H/B changes and, probably, depends on the height of the elements and the density and pattern of the array. Tests with an array of cubes at an intermediate H/B (2.25) showed h to remain substantially constant with distance into the array over the first six rows.

Arvizu and Moffat (1981) explained these results by postulating hat h responded to both the mean velocity and the turbulence intensity around the elements. For H/B = 1.0, there can be no change in mean velocity with distance into the array but the turbulence intensity will rise during the first few rows. This would explain the increase in heat transfer coefficient with row number for H/B = 1.0. For H/B greater than 1.0, there exists a bypass path over the tops of the elements and, within the first few rows, fluid which approached the array below the crests of the elements would be extruded into the clearance gap. This would reduce the average velocity around the elements causing h to drop, as seen in the data for H/B = 4.62.

Heat transfer in the fully developed region of the array is also strongly affected by H/B, even in tests conducted at constant approach velocity, as shown by the data in Fig. 3.16. Here h (measured in row 6) drops significantly as the channel height increases for every value of approach velocity. At 7.5 m/s, for example, h drops from 265 $W/m^2\ K$ to 120 $W/m^2\ K$ as H/B increases from 1.0 to 4.62.

Figures 3.17 and 3.18 show the effect of array density on this sensitivity to H/B. It is apparent that the more dense spacing (S/B = 2) is far more sensitive to channel height. At an approach velocity of 6 m/s the heat transfer coefficient drops more than 50 percent when H/B changes from 1.0 to 4.62 while it drops only 30 percent

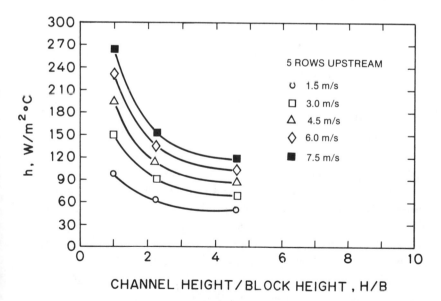

Figure 3.16: Fully developed heat transfer coefficient (Row 6) versus channel height for $S/B = 2.0$, parametric in velocity. [Moffat, Arvizu, and Ortega (1985) Fig. 6]

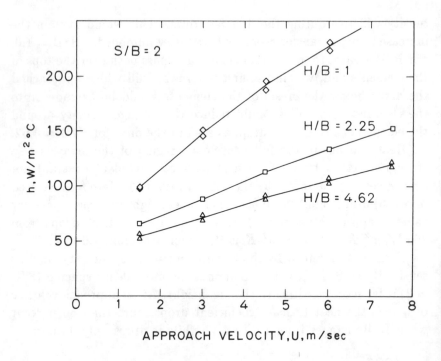

Figure 3.17: Fully developed heat transfer coefficient (Rows 5 and 6) versus approach velocity for $S/B = 2.0$. [Moffat, Arvizu, and Ortega (1985) Fig. 7]

for the less dense array (S/B = 3). This behavior makes it unlikely that any simple correlation will be found which describes the effect of H/B for arrays of different densities.

The data in these figures were examined in other sets of coordinates, seeking a set which produced better ordering but this examination met with no success. In particular, the two candidates most often suggested, Reynolds number based on channel height and Reynolds number based on the hydraulic radius of the flow path, were notably unsuccessful. Figure 3.19 shows all of the data of this study, in terms of Nusselt number versus Reynolds number using channel height, h, as the length scale. The data collect within $\pm$ 30 percent around a mean correlation of

$$Nu_H = 0.77 Re_H^{0.57} \qquad (3.36)$$

The conclusion which must be drawn from the data presented so far is that even a simple system like a square, in-line, array of cubes requires at least five parameters to specify its heat transfer behavior. These are approach velocity, channel height, element size, element spacing and position within the array.

Arvizu and Moffat (1981) proposed to reduce the dimensionality of this problem by eliminating H/B as a parameter. They assumed that the heat transfer to an element depended mainly on the velocity around it and that the flow bypassed the array whenever a clearance gap existed over the tops of the elements, that is, whenever H/B was larger than 1.0. Velocity measurements made with a hot-wire anemometer confirmed this, as shown in Fig. 3.20. The velocity distribution changes dramatically inside the array, with the aver-

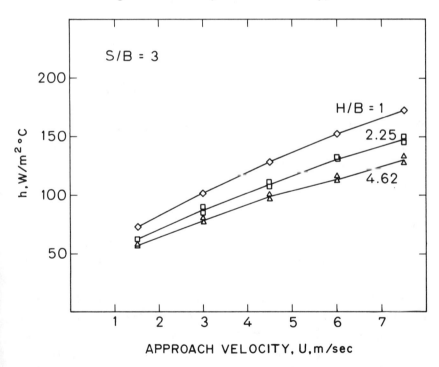

Figure 3.18: Fully developed heat transfer coefficients (Rows 5 and 6) versus approach velocity for *S/B* = 3.0, parametric in H/B. [Moffat, Arvizu, and Ortega (1985) Fig. 8]

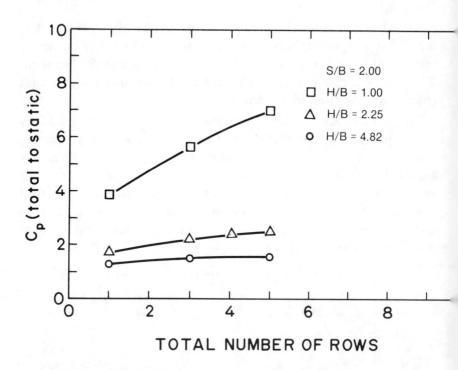

Figure 3.19: Nusselt number based on channel height versus Reynolds number based on channel height and approach velocity, all cases tested. [Moffat, Arvizu, and Ortega (1985) Fig. 9]

age velocity below the crests of the elements (at $Y/H = 0.22$) only about 60 percent of the average velocity in the undisturbed channel. Accompanying the decrease in velocity is an increase in turbulence intensity, as shown in Fig. 3.21, which would increase h for a given mean velocity, but this effect is not considered in the model.

In the Arvizu and Moffat (1981) model, the flow is presumed to partition itself between the two paths because the pressure drop per unit length must be the same for the flow in the array-space and in the bypass gap. This has the effect of sharply increasing the flow in the bypass path and reducing the flow around the elements as H/B is increased. This, in turn, lowers h. The basic assumption was that the flow within the array depended only on the pressure drop per unit length applied to the array. Thus the pressure drop versus flow characteristics of the array at $H/B = 1.0$ could be used to estimate the flow which stayed in the array region at any other value of H/B. From this assumption, the average array velocity can be calculated by eq (3.37). The velocity ratio, $U_{array}/U_{approach}$, is called the partition function:

$$\frac{U_{array}}{U_{approach}} = \frac{\sqrt{C_P(N,H/B)}}{\sqrt{C_P(N,H/B=1.0)}} \quad (3.37)$$

There is clear evidence of the bypass mechanism in smoke-wire photographs presented by Moffat et al (1985) shown in Figs. 3.22 and 3.23.

The partition function depends on the pressure drop characteristics of the array. Figures 3.24 and 3.25 show the overall pressure coefficient, total to static, as a function of row number for two array densities, $S/B = 2.0$ and 3.0. These data enable the calculation of the partition function as a function of position (row number) and

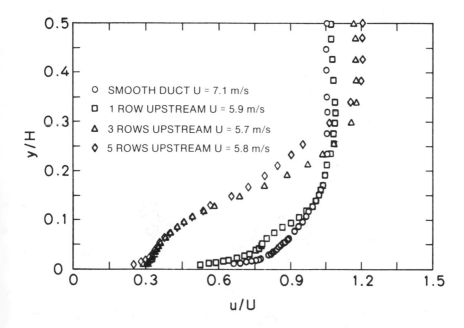

Figure 3.21: The U-component of turbulence intensity, measured within and upstream of the array for $H/B = 4.62$ and $S/B = 2.0$. [Moffat, Arvizu, and Ortega (1985) Fig. 11]

H/B. From the partition function and the approach velocity, the array velocity can be calculated, again as a function of position and H/B.

The array velocity concept provides a qualitative, physical basis for the trends shown in the data, but does not entirely collapse the data. Figure 3.26 shows all of the data from that study expressed in terms of the Reynolds number based on array velocity and element size. Most of the data, entrance region and fully developed, for all H/B and all S/B, are contained within the cross-hatched band. The one data set which lies distinctly outside the band is the fully developed data for $H/B = 1.0$ and $S/B = 2$. There is no residual trend with H/B within the cross-hatched region. This suggests that the array velocity concept has potential value for reducing the dimensionality of this problem. It remains to identify the reason for the exceptional behavior of the fully developed data from the case of

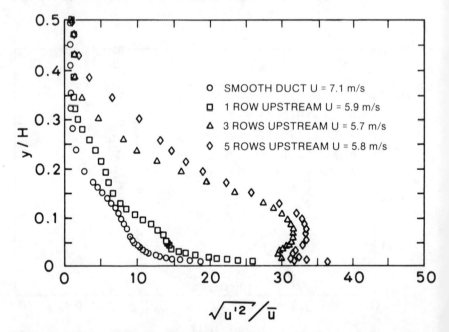

Figure 3.20: Velocity profiles within and upstream of the array for $H/B = 4.62$ and $S/B = 2.0$. [Moffat, Arvizu, and Ortega (1985) Fig. 10]

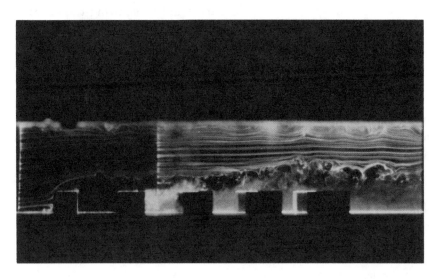

Figure 3.22: Smoke-wire visualization of the flow within the array, side view, for $U_{approach} = 5.0$ m/s and $H/B = 4.0$. [Moffat, Arvizu, and Ortega (1985) Fig. 12]

Figure 3.23: Smoke-wire visualization of the flow within the array, plan view, for $U_{approach} = 5.0$ m/s with $S/B = 3.0$ and $H/B = 4.0$. [Moffat, Arvizu, and Ortega (1985) Fig. 13]

$H/B = 1.0$ and $S/B = 2$. Those data could be showing the effect of turbulence intensity on h.

The study by Wirtz and Dykshoorn (1984) examined the array velocity concept in connection with heat transfer tests on flatpacks and concluded that it did not help much. Their study used elements $2.54 \times 2.54 \times 0.64$ *cm* spaced 2.54 *cm* apart. This spacing may have caused the elements to act as independent elements rather than an array, in terms of the classification scheme of Liu et al (1966). It is also possible that the average heat transfer was dominated by the top face, which accounts for nearly half of the total area. The top face is exposed to the bypass velocity rather than the array velocity. Their evidence is clear, however, that the array velocity concept is of little value in dealing with flatpacks.

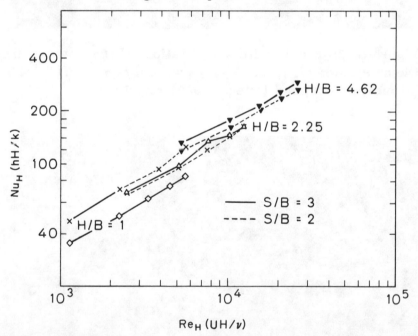

Figure 3.24: **Overall pressure coefficient, total to static, versus position within the array for** $S/B = 2.0$. [Moffat, Arvizu, and Ortega (1985) Fig. 14]

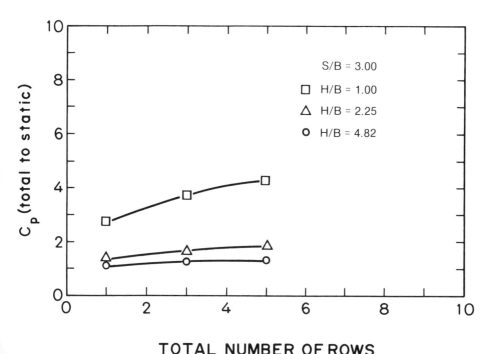

Figure 3.25: Overall pressure coefficient, total to static, versus position within the array for $S/B = 3.0$. [Moffat, Arvizu, and Ortega (1985) Fig. 16]

5.6   Arrays of Cylindrical Elements

Piatt (1986) measured the heat transfer coefficients and temperature wake functions for a square, in-line array of stub cylinders on an adiabatic wall in channel flows with various heights. The cylinders were 1.0 cm high and 1.9 cm in diameter, mounted with their centers 3.8 cm apart in each direction. Reynolds number and Nusselt number were defined using the channel height as the length dimension. The approach velocity was used in calculating Reynolds number.

Figure 3.27 shows Nusselt number versus Reynolds number for the cylindrical elements. These data are fit within a few percent by

$$Nu_H = 0.16Re^{0.75} \tag{3.38}$$

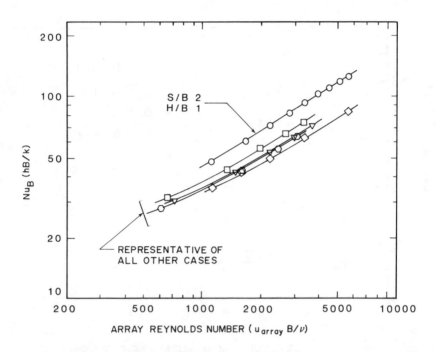

Figure 3.26: Nusselt number based on element height versus array Reynolds number, all data from this series. [Moffat, Arvizu, and Ortega (1985) Fig. 16]

The legends are interpreted as, for example, $H/B = 1$, heated element in Row 1, Approach velocity = 1.5 and 3.0 m/s.

The thermal wake function tests showed that the adiabatic temperature rise of the element directly behind the heated element was considerably lower than for cubes. The comparison is difficult to make explicitly because the stub cylinders have two possible length scales (height and diameter) while the cubes have only one. Piatt did not investigate this issue in enough depth to define a correlating function in terms of both H and D but some useful insight can be drawn from the data available. The adiabatic temperature rise of the first unheated stub cylinder was 0.085 compared with 0.155 for a cube whose side equaled the height or 0.128 for a cube whose side equaled the diameter, for the same approach velocity. The smaller temperature rise on the cylinders could be explained by the reduced degree of recirculation which would be expected behind the cylinders as compared with the cubes.

Elements further downstream show diminished adiabatic rise in a pattern which agreed very well with the 1/N dependence reported for cubes.

5.7   Arrays of Flatpacks

Sparrow, Niethammer, and Chaboki (1982) measured the heat transfer and pressure drop characteristics of an in-line array of rectangular blocks with the added complication of "missing elements" and added obstructions. The basic elements were 2.667 *cm* square, and 1.0 *cm* high and the elements were spaced 0.67 *cm* apart in each direction. The grooves thus formed had an aspect ratio (depth to width) of 3:2 and would thus be considered "narrow". The passage height was fixed at 2.667 *cm* (measured wall to wall) yielding a ratio of channel height to element height of 2.667.

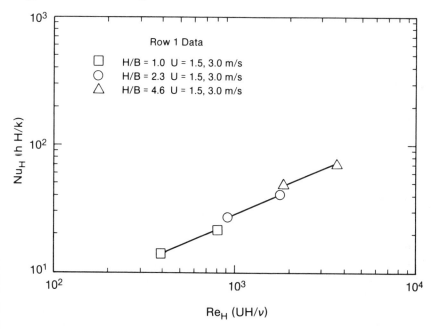

Figure 3.27: **Nusselt number versus Reynolds number for short cylinders, 1.0 *cm* high and 1.9 *cm* diameter, in a square, in-line array. [Piatt (1986) Fig. 4.6]**

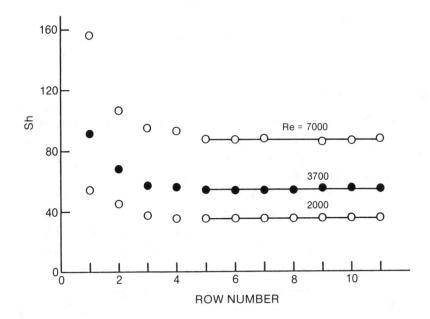

Figure 3.28: Sherwood number versus Row number for flat-packs in a square, in-line array, for three values of Reynolds number. Napthalene sublimation technique. [Sparrow, Niethammer, and Chaboki (1982) Fig. 3].

The napthalene sublimation technique was used with one element at a time being made of napthalene. The heat transfer Nusselt numbers were deduced from the mass transfer Sherwood numbers by the relationship:

$$Nu = (Pr/Sc)^m Sh \tag{3.39}$$

where Sh is the Sherwood number. With $Pr = 0.7$ for air, $Sc = 2.5$ for napthalene and $m = 0.36$ following a recommendation by Zukauskaus (1972) based on flow normal to tube-banks

$$Nu = 0.632 Sh \tag{3.40}$$

The Sherwood number was defined using the element length, 2.667 cm, as the characteristic length and the reference velocity was calculated from the volume flow rate and the flow area between the

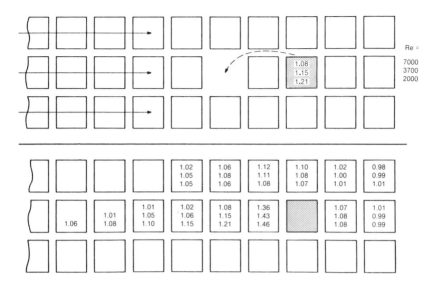

Figure 3.29: The effect of a missing module on the heat transfer coefficient at the observed location, for Reynolds numbers of 7000, 3700, and 2000. [Sparrow, Niethammer, and Chaboki (1982) Fig. 4]

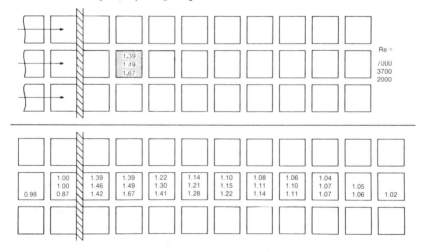

Figure 3.30: The effect of an added barrier of height (b-t)/H=1/5 on the heat transfer coefficients on the downstream modules, for Reynolds numbers of 7000, 3700, and 2000. [Sparrow, Niethammer, and Chaboki (1982) Fig. 6].

tops of the elements and the opposite wall with no allowance for flow in the axial grooves. The resulting heat transfer coefficients are to be interpreted as h_{ad} because only one element at a time was active.

Figure 3.28 shows the reported values of Sherwood number for three different Reynolds numbers. The pronounced drop in Sherwood number over the first four rows more likely represents the conventional entrance region behavior of a smooth walled channel than any effect of flow partitioning. This deduction is based on the fact that the elements are short and closely packed. The authors recommend the following correlations:

$$Sh = 0.148 Re^{0.72} \tag{3.41}$$

or

$$Nu = 0.0935 Re^{0.72} \tag{3.42}$$

where

$$Re = UL/\nu$$
$$Sh = KL/D$$

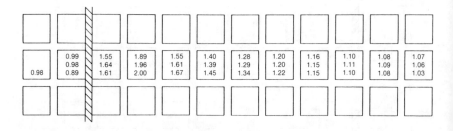

Figure 3.31: The effects of a taller barrier, (b-t)/h =2/5, on the heat transfer coefficients on the downstream modules, for Reynolds numbers of 7000, 3700, and 2000. [Sparrow, Niethammer, and Chaboki (1982) Fig 6.]

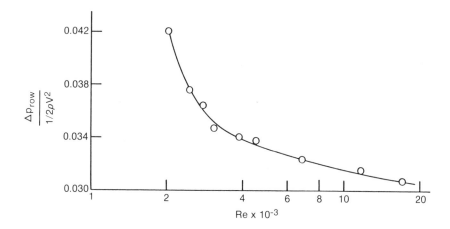

Figure 3.32: The per-row pressure drop, static to static, for a fully populated array with no added barriers. [Sparrow, Niethammer, and Chaboki (1982) Fig. 9].

$U = $ Velocity over the crests of the elements.

$K = $ Mass transfer coefficient based on the napthalene density differences between the surface layer and the free stream

$L = $ Element length in flow direction.

$D = $ Napthalene diffusion coefficient, calculated from the Schmidt number of napthalene in air (2.5) and the kinematic viscosity of air.

No measurements were made of the napthalene vapor concentration on the downstream passive elements. Hence, there is no way to deduce the equivalent concentration wake function. The heat transfer data reported here could be incorporated into design studies using temperature wake functions deduced from the study by Wirtz and Dykshoorn (1984).

The effects of missing modules were reported in terms of the ratio h/h^*, where h is the heat transfer coefficient on the observed module when one module is missing (somewhere in the array) and h^* is the heat transfer coefficient on the observed module when none are missing. The value of h/h^* is a function of the relative position of the missing module from the observed module. Figure 3.29

shows the effect of a single missing module when both the observed and the missing module are in the fully developed region. The figure requires careful interpretation. It represents a commentary on the values of h at the observed location when one of the other modules is missing. The observed location is indicated here by the square which does not contain any numbers. The numbered squares are possible locations of the missing module. The three numbers in each numbered square indicate the heat transfer coefficients on the observed module when the numbered square is empty. The three values represent Reynolds numbers of 7000, 3700, and 2000 from top to bottom. Thus, in Fig. 3.29, if the module immediately upstream of the observed module is missing, the heat transfer coefficient on the observed module would

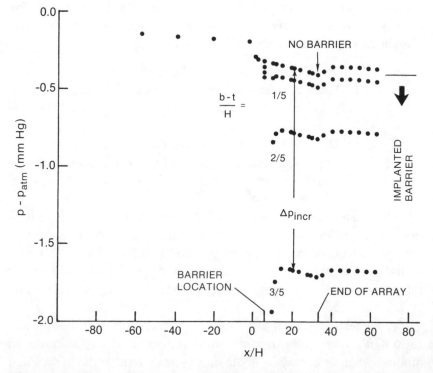

Figure 3.33: Axial pressure distributions for arrays with and without barriers, at Reynolds number of 6900. [Sparrow, Niethammer, and Chaboki (1982) Fig. 10].

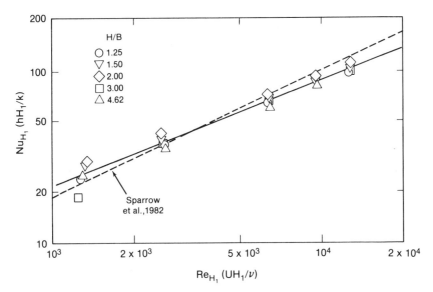

Figure 3.34: Nusselt number, hL/k, versus Reynolds number, $V\,L/\nu$, for flatpacks of length L. [Wirtz and Dykshoorn (1984) Fig. 6]

module would be increased by 46 percent at a Reynolds number of 2000. An empty location two rows downstream of the observed module has essentially no effect on h at the observed location.

The pattern of these effects will probably be different when the missing module is in the developing region. Thus, side-by-side missing modules have essentially the same effect as individual missing modules within their own columns.

If a barrier is introduced into the array, it tends to increase h on the downstream modules as shown in Fig. 3.30. The effect is largest on the second row downstream of the barrier, possibly because the sheet-jet produced by the barrier skips over the first row, re-attaching to the module wall at or before the second row. The effect persists for 10 rows downstream. The increase in heat transfer coefficient is largest at low Reynolds numbers, 67 percent at $Re = 2000$, which is consistent with other observations concerning the effect of high turbulence on heat transfer. The barrier used in this study was 1.33 cm high, half the height of the empty channel, and protruded 0.33 cm above the tops of the modules. There is no way to accurately assess the effect of the barrier on the temperature wake function since no adiabatic concentrations were measured. It does seem reasonable to assume complete, or nearly complete, mixing in the first row downstream of the barrier with the temperature distribution re-starting from there and following the same distribution observed by Wirtz and Dykshoorn (1984) in the far field.

Figure 3.31 shows the effects of a taller barrier, 1.668 *cm* high, in the same channel. The effect is, once again, largest on the second row downstream and is largest at the low Reynolds number (100 percent at $Re = 2000$). A measurable increase was still visible ten rows downstream. Figures 3.32 and 3.33 show the per-row pressure drop for the fully populated array and the array with barriers. These values represent static-to-static pressure drops made dimensionless using the reference velocity based on free height above the modules.

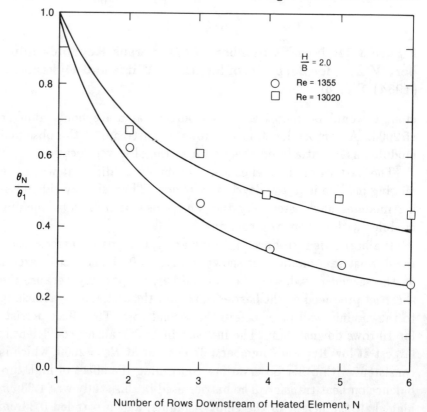

Number of Rows Downstream of Heated Element, N

Figure 3.35: A correlation for θ_i/θ_1 sensitive to H/B and *Re*, for flatpacks. [Wirtz and Dykshoorn (1984) Fig. 9].

In Fig. 3.33, the barrier heights are referred to by their dimensionless heights, $(b - t)/H$. The two cases for which the heat transfer disturbances were shown are $(b - t)/H = 1/5$ and $(b - t)/H = 2/5$. The barrier height b is related to the free height above the modules, $H(1.667\ cm)$, and the thickness of the modules, $t(1.00\ cm)$.

The study of flatpacks conducted by Wirtz and Dykshoorn (1984) used a conventional energy balance technique to measure the heat transfer coefficient on individually heated elements. It documented the thermal wake functions as well as the adiabatic heat transfer coefficients.

This study used a square, in-line array (8 rows and 5 columns) of modules 2.54 cm square and 0.635 cm thick, spaced 5.08 cm apart, center to center, in both directions. The channel height was variable but was found to have little effect on the results. Good correlation was obtained using Nusselt number against Reynolds number defined using the approach velocity and the module length in the flow direction. Figure 3.34 shows the Wirtz and Dykshoorn (1984) data, eq (3.43) and compares it with that of Sparrow, Niethammer and Chaboki (1982).

Wirtz and Dykshoorn (1984) tested the array velocity concept and found that it did not help. Using the array velocity caused the data to disperse more than using approach velocity.

The correlation

$$Nu = 0.348 Re^{0.6} \tag{3.43}$$

is recommended. Here

$$
\begin{aligned}
Nu &= hH_1/k \\
H_1 &= \text{module length in the flow direction} \\
Re &= UH_1/\nu \\
U &= \text{approach velocity}
\end{aligned}
$$

The adiabatic temperature rise of the first module downstream of the heated module for $H/B > 1.5$, is represented by

$$\theta_1 = 1.83(H/B)^{-0.5} Re^{-0.3} \tag{3.44}$$

For $H/B = 1.25$, Wirtz and Dykshoorn (1984) reported θ_1 to be essentially independent of Reynolds number but did not quote a value.

For $H/B = 1.25$, Wirtz and Dykshoorn (1984) reported θ_1 to be essentially independent of Reynolds number but did not quote a value.

The behavior of rows further downstream was dependent on Reynolds number and, to a lesser extent, on the ration H/B. The behavior was described in terms of the ratio θ_N/θ_1, following the earlier work of Arvizu and Moffat (1982). They recommended the correlation illustrated in Fig. 3.35:

$$\theta_N/\theta_1 = (1/N)^m \qquad (3.45)$$

where

$$m = -0.28\log_{10}(Re/10^6) \qquad (3.46)$$

They reported only a small effect on the temperatures in flanking columns and did not present a correlation for it.

Part II
NATURAL CONVECTION

6 INTRODUCTION

Passive cooling of electronic components by natural convection is the least expensive, quietest, and most reliable method of heat rejection. It therefore is often the preferred method of cooling telecommunications devices such as telephone switching units, avionics packages, computers, and many custom electronic units housed in equipment frames, racks or enclosures. The great variability in electronic packaging at the component, board and system levels has deterred systematic study of natural convection phenomena in electronic cooling. A second deterrent has been the highly coupled nature of passive cooling applications in which substrate conduction and longwave radiation to the environment are as important as convection.

Figure 3.36 shows examples of natural convection cooling in various electronic devices. In large frames or cabinets, as in Fig. 3.36a, cards or printed circuit boards (PCBs) containing heat dissipating components are typically mounted vertically in rows. Cooling air is induced to flow through the cabinet by the density imbalance between the heated, warm air inside the cabinet, and the cooler air outside, a situation known as induced or natural draft cooling. In other applications, particularly when devices are expected to operate in non-benign environments, the electronics may be located within a sealed enclosure, as shown in Fig. 3.36b. In this case, the electronics induce a recirculating flow which convects heat from the hot surfaces to the cooler enclosure surfaces. Longwave radiation may be as important as convection in such configurations. Enclosures are frequently equipped with extended surfaces to increase the surface area and enhance the external dissipation of heat, as shown in Fig.

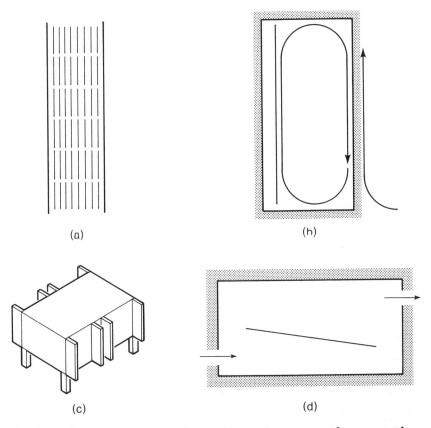

Figure 3.36: Typical configurations for natural convection cooling of electronics. (a) Vertical cabinet with array of PCB's, (Aung (1973)), (b) sealed enclosure, (c) externally finned enclosure, (Steinberg (1980)), (d) vented enclosure with horizontal PCB, [Johnson and Torok (1985)].

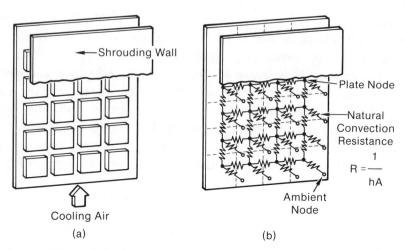

Figure 3.37: (a) Typical arrangement of electronic components on a PCB (b) Finite node representation [Steinberg
(1980)].

3.36c. Vented enclosures (Fig. 3.36d) are commonly used to house
electronic instruments and small computers that operate in clean
environments. In some exceptional cases, direct analytic or numerical
techniques can lead to immediately useful results as, for example,
in the case of densely packed components on PCBs with uniform
heat dissipation, or isolated components that do not interact with
other components. Generally, natural convection from surfaces with
heat dissipating components is influenced by the discrete nature of
the local heat sources, their interaction with other heat sources by
substrate conduction, radiation, thermal wake/plume interactions,
and the highly irregular, three-dimensional surface roughness caused
by the component array. Most of these phenomena have not been
extensively studied.

Modeling of natural air cooling of complicated electronic systems
has progressed along two fronts. The first is in the application of
finite-difference or finite-element techniques to analyze complex systems as discretized, interacting nodes. The basis of these techniques
is discussed in most undergraduate heat transfer textbooks, and
in electronics cooling references such as those by Steinberg (1980),
Kraus and Bar-Cohen (1983), and Ellison (1984). Figure 3.37 shows
a vertical PCB populated with electronic packages, and a typical fi-

nite node representation. The specification of thermal resistances is basic to this approach and requires the use of careful judgment by the analyst in determining convective heat transfer coefficients, conduction properties, and radiation view factors. A common approach in the industry is to lump all external thermal resistances into an "external" thermal resistance R_{ext}, defined as

$$R_{ext} = \frac{T_c - T_a}{Q_c}$$

which is directly measured in the configuration of interest. This approach requires the compilation of data for every configuration and is therefore not of general utility. The second area in which progress has been made is in the modelling of fluid flow and heat transfer in basic problems which are of specific importance to electronic cooling. These include vertical heated channels, open and partitioned enclosures with uniformly and discretely heated walls, surfaces with discrete sources of heat, and surfaces with heated roughness elements. The two recent reviews of Jaluria (1985) and Aihara (1987) provide broad coverage of most topics pertinent to natural convection air cooling of electronics. The aim of the present treatment is to focus on recent developments in topics of direct importance to this area.

7 NATURAL CONVECTION IN VERTICAL CHANNELS

7.1 Laminar Flow in Smooth Channels

As shown in Fig. 3.36a, in large cabinets or frames, flow is induced into the enclosure and partitioned into the various vertical channels which are formed by adjacent boards. Nominal spacing between boards (on the order of one to two *cm*) and relatively low flow rates insure that the flow is laminar. When components are closely spaced, the hydrodynamic effects of the cavity between elements are negligible and for practical purposes the board surface may be considered to be hydrodynamically smooth. Radiation and conduction into the board tend to render the surface heat flux uniform if there are not large variations in the power dissipated by each electronic component. Finally, if the boards are wide or if the edges of the channel

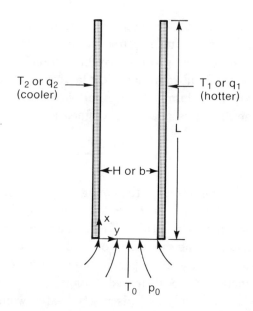

Figure 3.38: Nomenclature for vertical smooth walled channel.

are sealed by the enclosure, the flow in the channel will be nominally two-dimensional. There has been considerable work in the problem of laminar, buoyancy-induced flow between heated, vertical plates. The geometry and nomenclature are given in Fig. 3.38. Aung (1972) found analytic solutions for the case of fully-developed flow for symmetrically and asymmetrically heated channels with both constant temperature and constant heat flux boundary conditions. Aung, Fletcher, and Sernas (1972) numerically solved the problem in the developing region using a finite-difference technique and the results were compared to temperature measurements made using a Mach-Zender interferometer technique. Subsequently, Aung, Kessler, and Beitin (1973) compared these results with measurements made in real printed circuit board channels, with and without surface components. The earliest work in this problem was an experimental study conducted by Elenbaas (1942) on a channel formed by parallel plates at constant temperature. Early numerical work was performed by Bodoia and Osterle (1962) on a channel with uniform wall temperature, and an analytical integral solution was derived by Engel and Mueller (1967). Many subsequent investigators examined the effects of inlet boundary conditions, radiation, asymmetric heating, and mixed convection effects. A summary of investigations in this area is given by Ortega (1986), and expanded reference lists are given in the papers by Jaluria (1985), and Aihara (1987), as well as at the end of this chapter.

For laminar, constant property, boundary layer flow, Aung, Fletcher and Sernas (1973) give the following form of the conservation equations for two-dimensional buoyancy-induced flow in the vertical channel:

$$\frac{\partial U}{\partial X} + \frac{\partial V}{\partial Y} = 0 \qquad (3.47)$$

$$U\frac{\partial U}{\partial X} + V\frac{\partial U}{\partial Y} = \frac{\partial^2 U}{\partial Y^2} - \frac{dP}{dX} + \theta \qquad (3.48)$$

$$U\frac{\partial \theta}{\partial X} + V\frac{\partial \theta}{\partial Y} = \frac{1}{Pr}\frac{\partial^2 \theta}{\partial Y^2} \qquad (3.49)$$

where

$$U = \frac{b^2 u}{L\nu Gr}, \quad V = \frac{bv}{\nu}, X = \frac{x}{LGr}, Y = \frac{y}{b},$$

$$P = \frac{(p - p_0)b^4}{\rho L^2 \nu^2 Gr^2}, \quad Pr = \frac{\mu c}{k}$$

and where the dimensionless flow rate is given by

$$M \text{ or } Q = \frac{u_0 b^2}{L\nu Gr} = \int_0^1 U\, dY \qquad (3.50)$$

Boundary conditions are given for either uniform heat flux (UHF) or uniform wall temperature (UWT) conditions as:
Uniform Heat Flux:

$$r_H = \frac{q_2}{q_1}, \theta = \frac{T - T_0}{q_1 b/k}, Gr = \frac{g\beta q_1 b^5}{L\nu^2 x}$$

$$U = Q, \ V = 0, \ \ \theta = 0 \qquad\qquad \text{for} \ \ X = 0, \ \ 0 < Y < 1 \qquad (3.51)$$
$$U = 0, \ \ V = 0, \ \ \partial\theta/\partial Y = -r_H \quad \text{for} \ \ X \geq 0, \ \ Y = 0 \qquad (3.52)$$
$$U = 0, \ \ V = 0, \ \ \partial\theta/\partial Y = 1 \quad\ \text{for} \ \ X \geq 0, \ \ Y = 1 \qquad (3.53)$$

and for $P = 0$

$$X = 0 \ \text{ and } \ X = 1/Gr \qquad (3.54)$$

Uniform Wall Temperature:

$$r_T = \frac{T_2 - T_0}{T_1 - T_0}, \quad \theta = \frac{T - T_0}{T_1 - T_0}, \quad Gr = \frac{g\beta(T_1 - T_0)b^4}{L\nu^2}$$

$U = Q, \quad V = 0, \quad \theta = 0 \quad$ for $\quad X = 0, \quad 0 < Y < 1$

$U = 0, \quad V = 0, \quad \theta = r_T \quad$ for $\quad X \geq 0, \quad Y = 0$ $\qquad$ (3.55)

$U = 0, \quad V = 0, \quad \theta = 1 \quad$ for $\quad X \geq 0, \quad Y = 1$ $\qquad$ (3.56)

and for $P = 0$

$$X = 0 \quad \text{and} \quad X = 1/Gr$$

It is to be noted that Aung and co-workers, Miyatake et al (1972, 1973, 1974), Bodoia and Osterle (1962), and others have assumed that the static pressure at the channel inlet is equal to the ambient pressure at the same elevation, that is

$$u = u_0, \quad p - p_0 = 0 \quad \text{at} \quad X = 0 \qquad (3.57)$$

whereas Quintiere and Mueller (1973) and Aihara (1973) independently found that a more appropriate inlet condition accounts for the pressure drop that is incurred in accelerating quiescent fluid to an inlet velocity of u_0. Aihara (1973) found the most practical condition to be given by

$$u = u_0, p - p_0 = -\rho u_0^2/2 \quad \text{at} \quad X = 0 \qquad (3.58)$$

which assumes a uniform inlet velocity, and a pressure defect given by inviscid flow considerations. Equations (3.47), (3.48) and (3.49) and the associated boundary conditions form a system which is inherently elliptic in nature. It is not possible to specify the inlet flow rate and the channel length independently because the induced flow rate is set by the global channel buoyancy and the resistance to flow, both of which are integral properties. The methods used for numerically solving the equations reflect the character of the equations; typically, one may fix the inlet flow rate, integrate the equations until the pressure defect goes to zero and establish this as the channel exit. Alternatively, one may fix the channel length and iterate on the inlet flow rate until the pressure defect goes to zero at the exit. Kettleborough (1972), Nakamura et al (1982), and Ormiston et al (1985) have numerically solved the problem without invoking the boundary layer approximations in eqs (3.47), (3.48) and (3.49) and have extended the solution domain to include the region far from the channel entrance.

A local Nusselt number based upon the inlet air temperature as a reference may be defined on either wall of the channel as:

$$Nu = \frac{hb}{k} = \frac{qb}{(T_w - T_o)k} \qquad (3.59)$$

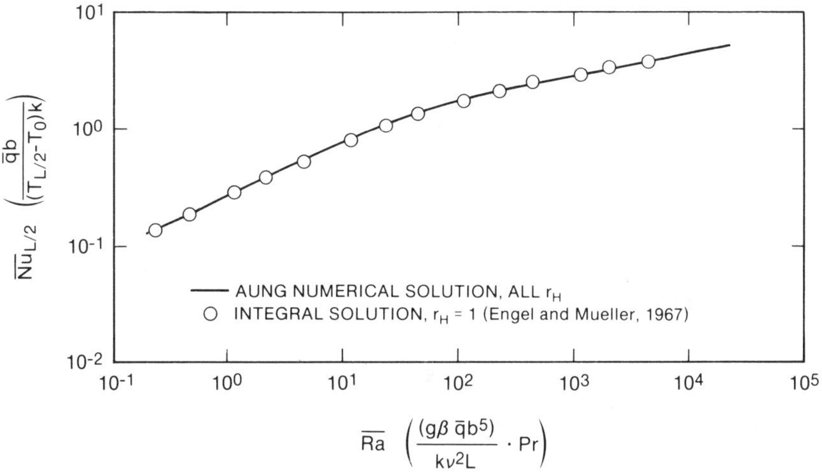

Figure 3.39: **Average Nusselt number as a function of Rayleigh number for UHF [Aung et al. (1972.)]**

where q is the local heat flux on that wall. Because of their importance for thermal design, the quantities that have received the most consideration are the maximum wall temperature, in the case of UHF conditions, or the total or average heat dissipation rate, for UWT. Both are represented non-dimensionally by Nu. Plate or channel average Nusselt numbers were computed by Aung et al (1972) using a finite difference technique and are shown in Figs. 3.39 and 3.40 for UHF and UWT boundary conditions respectively, for both symmetrically and asymmetrically heated channels. Figure 3.39 also shows results from the integral solution of Engel and Mueller (1967). Figures 3.41 and 3.42 show the computed induced flow rate as a function of dimensionless channel length. The average Nusselt number was taken to be the value of the Nusselt number evaluated at plate mid-height. Remarkably, the average Nusselt numbers are well represented by a single curve for either symmetrically or asymmetrically heated channels as long as the average of the two plate temperatures and heat fluxes evaluated at the plate mid-height are used in both Nu and the Rayleigh number, Ra. The average Nusselt number demonstrates asymptotic behavior at both large and small Rayleigh number. For large plate spacing where $Ra \to \infty$, the heat transfer approaches the boundary layer behavior of an isolated plate. For very small plate spacing or for very long channels where $Ra \to 0$, the hydrodynamic and thermal boundary layers develop and merge, resulting in fully developed flow which is invariant in the flow direction. This asymptotic behavior was first observed experimentally by Elenbaas (1942) using isothermally heated plates with open edges. Figure 3.43 shows Elenbaas' average plate Nusselt numbers as replotted by Aihara (1973). Wirtz and Stutzman (1982) presented interferograms which span the entire range of plate spacing and clearly show the transition from isolated plate boundary layer behavior to buoyancy-induced channel flow behavior.

7.2 Limiting Relations for Isolated Plate and Fully-Developed Channels

Both the isolated plate and fully-developed channel flow limits have been well developed. Table 3.9 is a summary of analytically derived limits for a constant property, incompressible, fluid with $Pr = 0.7$. Shown are the dimensionless flow rate, Q, defined by eq (3.50), the average Nusselt number (evaluated at plate mid-height), and the Nusselt number at the channel exit. The latter Nusselt number is the inverse of the dimensionless maximum channel temperature as defined by Aung et al (1972). The fully-developed channel flow results with UHF boundary conditions, eqs (3.65) and (3.66) and (3.70) and (3.71), follow the derivations of Miyatake et al (1973, 1974) which assume fully-developed, parabolic velocity profiles just as in a conventional forced convection situation. By similar arguments, Bar-Cohen and Rohsenow (1984) derived the equivalent results but neglected the constants 17/70 and 13/35, as did Aung (1972). Ortega (1986) showed that the Nusselt number based on the local channel

Table 3.9a: **A summary of asymptotic limiting relations for buoyancy induced channel flow.**

	Fully Developed Channel			Isolated Plate	
	Q	$Nu_{L/2}$	Nu_L	$Nu_{L/2}$	Nu_L
Isothermal-Symmetric	(3.59)	(3.60)	N/A	(3.61)	N/A
Isothermal-Assymmetric	(3.62)	(3.63)	N/A	(3.61)	N/A
Isoflux Symmetric	(3.64)	(3.65)	(3.66)	(3.67)	(3.68)
Isoflux-Asymmetric	(3.69)	(3.70)	(3.71)	(3.67)	(3.68)

Notes: $Q = Re/Gr$ for UWT
$Q = Re/Gr^*$ for UHF
N/A = Not applicable

Table 3.9b: **Equations for asymptotic limiting relations for buoyancy induced channel flow (Isothermal).**

Isothermal-Symmetric

$$Q = \frac{1}{12} \tag{3.59}$$

$$Nu_{L/2} = \frac{Ra}{24} \tag{3.60}$$

$$Nu_{L/2} = 0.519 Ra^{1/4} \tag{3.61}$$

Isothermal-Asymmetric

$$Q = \frac{1}{12} \tag{3.62}$$

$$Nu_{L/2} = \frac{Ra}{12} \tag{3.63}$$

Table 3.9c: Equations for asymptotic limiting relations for buoyancy induced channel flow (Isoflux).

Isoflux-Symmetric

$$Q = (12Ra^*)^{-1/2} \tag{3.64}$$

$$Nu_{L/2} = \left[\left(\frac{12}{Ra^*} \right)^{1/2} + \frac{17}{70} \right]^{-1} \tag{3.65}$$

$$Nu_L = \left[\left(\frac{48}{Ra^*} \right)^{1/2} + \frac{17}{70} \right]^{-1} \tag{3.66}$$

$$Nu_{L/2} = 0.596 Ra^{*1/5} \tag{3.67}$$

$$Nu_L = 0.519 Ra^{*1/5} \tag{3.68}$$

Isoflux-Asymmetric

$$Q = (24Ra^*)^{-1/2} \tag{3.69}$$

$$Nu_{L/2} = \left[\left(\frac{6}{Ra^*} \right)^{1/2} + \frac{13}{35} \right]^{-1} \tag{3.70}$$

$$Nu_L = \left[\left(\frac{24}{Ra^*} \right)^{1/2} + \frac{13}{35} \right]^{-1} \tag{3.71}$$

mixed mean temperature, Nu_m, is related to the conventional Nusselt number referenced to the inlet temperature through the expression:

$$Nu_m = \frac{qb}{(T_w - T_m)k} = Nu \left\{ 1 - \frac{T_m - T_o}{T_w - T_o} \right\}^{-1} = Nu \left\{ 1 - \frac{x/b}{RePr} Nu \right\}^{-1} \tag{3.72}$$

Using the expressions for induced channel flow rate given by eqs (3.64) and (3.69), one finds that $Nu_{m,L} = 70/17$ and $Nu_{m,L} = 35/13$ for the symmetric and asymmetric isoflux channels respectively. These fully developed local Nusselt numbers are identical to those for forced convection, Shah and London (1978), and reflect the invariant, self-similar nature of the fully-developed temperature profiles. The agreement of the analytical results of Aung (1972),

and the numerical results of Miyatake et al (1972, 1973, 1974) and
Aihara (1973), and others with the fully-developed asymptotic results, eqs (3.64) to (3.66) and (3.69) to (3.71) demonstrate that,
at least for constant property assumptions, there is very little difference between buoyancy-induced convection and forced convection
for fully-developed laminar flow. For isothermal boundary conditions this is readily apparent, since the mean fluid temperature approaches   the wall temperatures thereby diminishing any potential
for local buoyancy effects on the buoyancy-induced forced convection
flow. From Figs. 3.41 and 3.42, it is apparent that the induced flow
rate has very different behavior for UHF or UWT conditions. For
fully developed long channels with uniformly heated walls, the dimensionless flow rate, Q, asymptotes to the single expressions given
by eq (3.64) (symmetric, $r_H = 1$) or (3.69) (asymmetric, $r_H = 0$)
which shows that the channel Reynolds number is proportional to
$[Gr^*L/b]^{1/2}$ or $q^{1/2}L^{1/2}$. For a fixed heat flux, the Reynolds number varies as the square-root of the channel length. On the other
hand, for walls at constant temperature, the asymptotic flow rate is

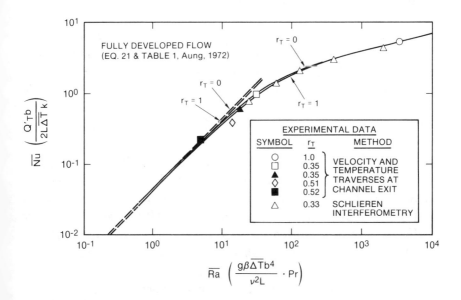

**Figure 3.40: Average Nusselt number as a function of Ray-
leigh number for UWT [Aung et al. (1972)].**

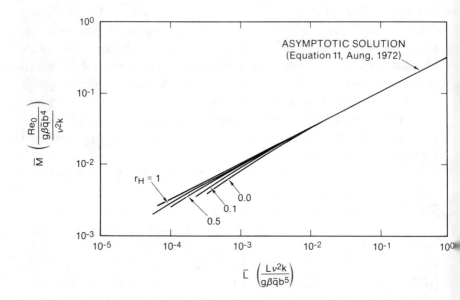

Figure 3.41: Dimensionless flow rate for asymmetric UHF channel [Aung et al. (1972)].

given by eq (3.59) which indicates that Re_o is proportional to Gr, or for fixed properties, Re_o is directly proportional to ΔT, independent of the channel length. The difference in the two limiting behaviors can be readily appreciated if one notes that for UWT, the bulk fluid temperature and hence the density approach a constant, whereas for UHF the bulk fluid temperature and density continuously change. In either case, the flow rate departs from the limiting behavior for increasing Rayleigh number first, in response to the increasing proportion of the channel which is not fully-developed, then increasingly in response to the proportion of the channel in which the velocity and temperature fields depart from the normally developing state of a global buoyancy-induced forced convection flow due to accumulative buoyancy effects. These phenomena are discussed further in the next section.

There are discrepancies in the reported experimental and numerical evidence with regard to the isolated plate asymptotic limits. For the isoflux plate, the data of Wirtz and Stutzman (1982) are asymptotic to the line given by

$$Nu_L = 0.577 Ra^{*1/5} \tag{3.73}$$

based on measurements for $Ra^* < 2414$. This also corresponds approximately to the limit found numerically by Aung et al (1972), and is parallel to, but roughly 11 percent higher, than the theoretical limit given by eq (3.68). Aung et al (1972) attributed this to the uniform velocity entrance condition not being realistic at large plate spacing (large Ra^*). The same difficulty was found by Aung et al (1972) and Bodoia and Osterle (1962) for the isothermal plate. Aihara (1973) performed a thorough study of the effects of inlet conditions for the isothermal channel and concluded that the error is, in part, due to the assumption of a uniform inlet velocity profile, and, in part, to neglecting the pressure defect at the inlet due to the acceleration of quiescent fluid to velocity u_o as given by eq (3.58). His results, shown in Figs. 3.43 and 3.44, for the inlet conditions of

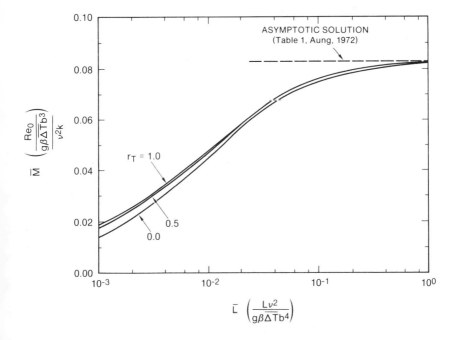

Figure 3.42: Dimensionless flow rate for asymmetric UWT channel [Aung et al. (1972)].

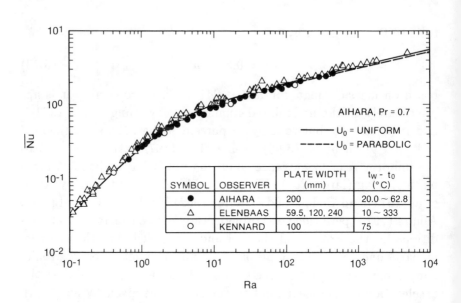

Figure 3.43: Average plate Nusselt number for isothermal plates [Aihara (1973)].

eq (3.58) compare favorably with the data of Elenbaas (1942). Bar-Cohen and Rohsenow (1982), citing the work of Aung and Miyatake et al (1973) recommended the correlations

$$Nu_L = 0.630 Ra^{*^{1/5}} \qquad (3.74a)$$

and

$$Nu_{L/2} = 0.730 Ra^{*^{1/5}} \qquad (3.74b)$$

for the isoflux plate limit which are parallel to but 21% higher, than the theoretical boundary layer limit for $Pr = 0.7$, but these correlations are apparently supported only by the aforementioned erroneous numerical calculations. Similarly, they recommend the correlation

$$Nu_L = 0.590 Ra^{1/4} \qquad (3.75)$$

for the isothermal plate limit which is parallel to and 14 percent higher than the theoretical limit given by eq (3.61). Aihara's (1973) results, and recent measurements by Sparrow and Azevedo (1985) on asymmetric, isothermal plates in water, show that the channel interval up to infinite spacing (isolated plate) and showed near perfect agreement with the theoretical isolated plate limit, $Nu_{L/2} = 0.613Ra^{1/4}$ for $Pr = 10$. In light of the available numerical and experimental data, there is no reason to believe that the theoretical isolated plate limits given by eq (3.61), for the isothermal plate, and (3.67) and (3.68), for the isoflux plate are not suitable for design. The data of Wirtz and Stutzman (1982), which apparently show a higher large spacing asymptote, demonstrate that it is difficult to achieve true isolated plate behavior, even at moderately large plate spacings. Their recommendation for modified isolated plate limits reflect channel behavior up to their largest plate spacing, $Ra < 2414$, and thus may be useful for design within these limits.

7.3 Local Heat Transfer Behavior

The manner in which the velocity and temperature fields develop are illustrated by the calculations of Aung et al (1972) in their Fig. 3 which is reproduced in Fig. 3.45. On the left are dimensionless temperature and axial velocity profiles for a dimensionless flow rate of $\bar{M} = 0.15$ ($Q = 0.075, Ra^* \approx 10$, and on the right for $\bar{M} = 0.005$ ($Q = 0.0025, Ra^* \approx 5000$), both cases for an asymmetrically heated, isoflux channel. For small Ra^* the velocity and temperature distributions develop quickly and agree very well with the fully-developed solutions previously discussed. Despite the asymmetric heating, the velocity profile is a normal fully-developed laminar profile, and the local Nusselt number, as noted earlier, is indistinguishable from the forced convection value. For a relatively feeble flow rate, as in the figures on the right, which would arise in a short channel, or one with relatively large plate spacing, the flow is gradually entrained towards the hotter wall and the profiles at the exit begin to resemble an isolated plate boundary layer flow, with most of the temperature increase occurring in only half of the channel. Because of the highly skewed temperature profile, only part of the fluid is buoyant, thereby reducing the overall global buoyancy and hence the flow rate. The elliptic nature of buoyancy-induced flow is apparent. Aung et al (1972) note that for such low flow rates, the specification of uniform entrance velocity profile is incorrect, and this assumption leads to the observed overprediction of the isolated plate Nusselt number. This seems highly plausible, and conclusions about local results must be qualified with the knowledge that they are constrained by the imposed inlet conditions. Miyatake et al (1973) have computed the local Nusselt number distributions using the identical formulation and boundary conditions as Aung et al (1972) and results from their Fig. 5 are shown in Fig. 3.46. Here the local Nu, defined by eq (3.59) is plotted against the dimensionless axial dimension, $(b/x)Gr^*Pr$, parametric on the dimensionless flow rate. For $Q = 0.3(Ra^* = 0.53)$, which is a similar condition found in Fig 3.45a, the Nusselt number decreases in the developing region and asymptotes to the fully-developed local Nusselt number given in Table 3.9, eq (3.71), with $x = L$. For $Q = 0.01$ the local Nusselt number no longer approaches the fully-developed asymptote but instead ends with its value equal to that given by the isolated plate expression given by eq (3.68), again with $x = L$. The departure of the local Nusselt number from the isolated plate limit for the low flow rate reflects the combined forced-free nature of the developing flow wherein a predominantly forced channel flow at the inlet undergoes transition to a predominantly buoyant boundary layer flow at the exit. Again it is noted that these results are valid only as long as the assumed flat entrance velocity profile is reasonably accurate. It was previously found that the fully-developed, constant property, laminar channel flow is identical for either buoyancy-induced or forced flow. A pertinent question is whether this is also the case in the hydrodynamic and thermal development region. Ortega (1986) replotted the Miyatake et al (1973) local Nusselt number results in conventional 2-D channel forced convection coordinates and compared them to the forced convection solution of Heaton et al (1964) for simultaneously developing velocity and temperature fields. The result is shown in Fig. 3.47. Here Nu_m is the Nusselt number based on the difference between the wall and local mixed-mean temperature, computed from eq (3.72), and the axial coordinate has been normalized as a conventional Graetz number by the conversion

$$\frac{x/b}{RePr} = \frac{1}{Q[(b/x)Gr^*Pr]}$$

At the lowest Rayleigh number, the local Nusselt number is indistinguishable from the forced convection solution both in the developing and fully-developed region. Note that the fully-developed Nusselt number has the value 35/13. For increasing Ra^* (decreasing Q), Nu_m develops as the forced convection case but then departs before reaching the forced convection asymptote. The departure is proportional to the extent to which the early, predominantly forced, flow is perturbed by accumulative buoyancy effects. At increasingly large channel spacing it is anticipated that the inlet flow is entrained towards the hot wall almost immediately, so that the flow is no longer a predominant forced channel flow at the entrance, but rather a developing buoyant boundary layer flow which is perturbed by the weak channel flow. Yao (1983) has examined this transition at the entrance of the channel with an initially uniform entrance velocity profile analytically and found that local buoyancy effects are negligible if

$$Gr^2 < Re, \quad \text{for uniform heat flux, and} \quad (3.76a)$$

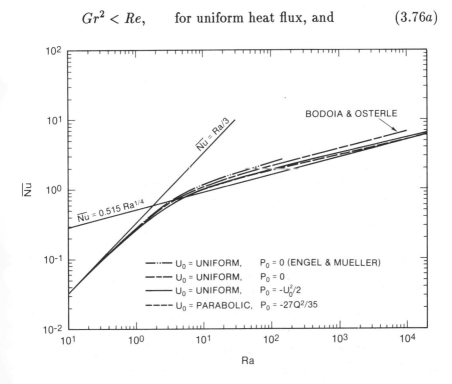

Figure 3.44: **Comparison of predicted average Nusselt number for isothermal plates with various inlet conditions** [Aihara (1973)].

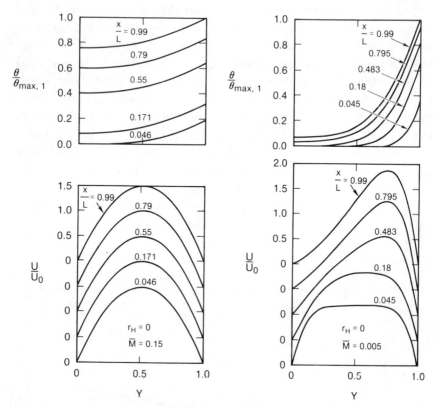

Figure 3.45: Calculated velocity and temperature fields for asymmetric UHF channel [Aung et al. (1972)].

$$Gr < Re, \qquad \text{for uniform wall temperature} \qquad (3.76b)$$

where $Gr = g\beta\Delta T a^3/\nu^2, a$ is the channel half-width, and ΔT is given by $\Delta T = T_w - T_o$, for uniform wall temperature, and $\Delta T = aq/(kRe^{1/2})$, for uniform heat flux. He concluded that when buoyancy effects are important, they are negligibly weak up to a predictable distance after which they are significant and continue to grow in importance. These tendencies are confirmed by Fig. 3.47. Miyatake and Fujii (1974) noted the remarkable similarity of their results for asymmetrically heated parallel plane channels to forced flow and attempted to collapse all of their local Nusselt number results to one universal line by defining a type of buoyancy-driven channel flow Graetz number which they give as

$$\phi = \frac{(b/x)Gr^*Pr}{[(b/L)Gr^*Pr]^{1/2}}$$

The denominator is proportional to the inverse of the dimensionless flow rate, Q, for fully-developed flow as given by eq (3.69). Their results, reproduced in Fig. 3.48a for a symmetric, isoflux channel, show remarkable universality for a large range of Rayleigh numbers. Aihara (1986) has found that using the improved inlet boundary condition that takes acceleration into account, it is not possible to correlate all the data using the parameter ϕ. He proposed an alternative, more empirical, definition which correlates his results very well. He showed that the results, correlated in this way, are indistinguishable from forced convection results correlated in the same manner. Furthermore he demonstrated that techniques used to correct for temperature-dependent fluid properties in forced convection are equally useful for buoyancy-driven channel convection.

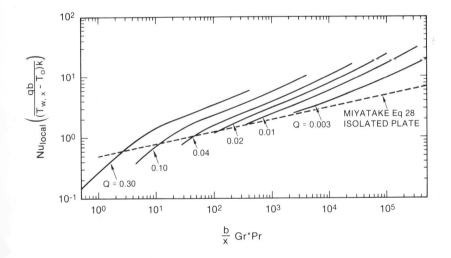

Figure 3.46: Variation of local Nusselt number for asymmetric UHF channel [Miyatake et al. (1973)].

7.4   Composite Relations for Thermal Design

For design purposes it is helpful to express the average Nusselt number, or maximum dimensionless plate temperatures, as composite, algebraic expressions that span the asymptotic limits and the intermediate transition regime as well. The method of Churchill and Usagi (1972) has been widely used for this purpose. Their technique recommends a correlation of the form

$$Nu^{-n} = Nu_1^{-n} + Nu_2^{-n}$$

Table 3.10a: Summary of relations for vertical plate arrays (after Bar-Cohen and Rohsenow (1984)).

Nu (Average)	Nu_L (Channel exit)	Optimum[a] Spacing	Optimum Nu[a] and Ra
Isothermal-Symmetric			
(3.77)	N/A	(3.78)	(3.79)
Isothermal-Asymmetric			
(3.80)	N/A	(3.81)	(3.82)
Isoflux-Symmetric[b]			
(3.83)	(3.84)	(3.85)	(3.86)
Isoflux-Asymmetric			
(3.87)	(3.88)[c]	(3.89)	(3.90)

[a] For negligibly thick plates.
[b] Based on T_w at $L/2$
[c] Derived by the present author from eqs (3.71) and (3.74a), $n = 2$
   N/A Not applicable.

Table 3.10b: Relations for vertical plate arrays [Bar-Cohen and Rohsenow (1984)].

Isothermal-Symmetric

$$Nu = \left[\frac{576}{Ra^2} + \frac{2.873}{Ra^{1/2}}\right]^{-1/2} \tag{3.77}$$

$$b_{opt} = 2.714P^{-0.25} \tag{3.78}$$

$$Nu_{opt} = 1.31$$

$$Ra_{opt} = 54.3 \tag{3.79}$$

Isothermal-Asymmetric

$$Nu = \left[\frac{144}{Ra^2} + \frac{2.873}{Ra^{1/2}}\right]^{-1/2} \tag{3.80}$$

$$b_{opt} = 2.154P^{-0.25} \tag{3.81}$$

$$Nu_{opt} = 1.04$$

$$Ra_{opt} = 21.5 \tag{3.82}$$

Isoflux-Symmetric

$$Nu_{L/2} = \left[\frac{12}{Ra^*} + \frac{1.85}{(Ra^*)^{0.4}}\right]^{-1/2} \tag{3.83}$$

$$Nu_L = \left[\frac{48}{Ra^*} + \frac{2.51}{Ra^{*0.4}}\right]^{-1/2} \tag{3.84}$$

$$b_{opt} = 1.472R^{-0.2} \tag{3.85}$$

$$(Nu_{L/2})_{opt} = 0.62$$

$$Ra^*_{opt} = 6.9 \tag{3.86}$$

Isoflux-Asymmetric

$$Nu_{L/2} = \left[\frac{6}{Ra^*} + \frac{1.88}{(Ra^*)^{0.4}}\right]^{-1/2} \tag{3.87}$$

$$Nu_L = \left[\frac{24}{Ra^*} + \frac{2.51}{(Ra^*)^{0.4}}\right]^{-1/2} \tag{3.88}$$

$$b_{opt} = 1.169R^{-0.2} \tag{3.89}$$

$$(Nu_{L/2})_{opt} = 0.49$$

$$Ra^*_{opt} = 2.2 \tag{3.90}$$

where Nu_1 and Nu_2 are the asymptotic relations and n is a constant chosen to match available data in the transition regime. Bar-Cohen and Rohsenow (1984), Wirtz and Stutzman (1982), Sparrow and Azevedo (1985), and Churchill (1977) have derived such expressions for parallel plate channels. Table 3.10 reproduces the results derived by Bar-Cohen and Rohsenow (1984). It is to be noted that the Bar-Cohen-Rohsenow results are all for $n = 2$. Wirtz and Stutzman (1982), using their eq 28 and $n = 3$, derived

$$Nu_L = \frac{0.144(Ra^*)^{1/2}}{[1 + 0.0156(Ra^*)^{0.9}]^{1/3}} \tag{3.91}$$

for symmetric, isoflux plates as compared to the Bar-Cohen-Rohsenow eq (3.84). Sparrow and Azevedo's (1985) expression for asymmetric,

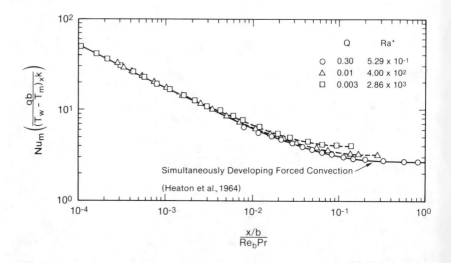

Figure 3.47: Miyatake et al. (1973) local Nusselt number for asymmetric UHF channel in conventional forced convection coordinates compared to forced convection solution for simultaneously developing thermal and hydrodynamic flow [Ortega (1986)].

isothermal plates is given by

$$Nu = \left\{ \frac{144}{Ra^2} + \frac{2.61}{(Ra)^{1/2}} \right\}^{-1/2} \tag{3.92}$$

and is the same as eq (3.80) except for the constant 2.61 which reflects their use of the theoretical limit for $Pr = 10$. It is worth repeating that the Nusselt numbers predicted by the composite expressions of Table 3.10 overpredict the theoretical isolated plate limits by 14 percent for the isothermal plate, and by 21 percent for the isoflux plate, but show excellent agreement with data and numerical calculations in the transition and fully-developed flow regimes.

Johnson (1986) has compared eqs (3.84) and (3.91) to experimental data for uniformly heated printed wiring boards (PWB) mounted vertically in a frame, similar to the configuration of Fig 3.36a. These data were taken on boards which ranged from quasi-smooth model plates to real circuit boards with dummy components mounted on the surfaces. Representative data are shown in Fig. 3.49 from the original paper, plotted in conventional channel coordinates for isoflux conditions. Most of the data lie in the range $Ra^* < 300$ which, according to the author, represent most applications for AT&T Bell Labs assemblies which are densely populated, narrow-channeled, and intensely powered. There are six points in the range $Ra^* > 550$ and few points in the transition regime. The correlations of Bar-Cohen and Rohsenow (1984) and Wirtz and Stutzman (1982) and the asymptotic limits given by eq (3.66) (neglecting the constant 17/70) and eq (3.68) (using a constant equal to 0.524 rather than 0.519) are shown in Fig. 3.50. Johnson concluded that both the composite correlations of Wirtz and Stutzman (1982) and Bar-Cohen and Rohsenow (1984) agree very well with the AT&T data for $Ra^* < 300$, but because of the sparsity of the data at higher Ra^*, cannot confirm the accuracy of the expressions for $100 < Ra^* < 1000$. In view of the limited data available for large Ra^*, Johnson recommended that the most conservative correlation, that attributed to Aung et al (1972) for the isolated plate limit be employed.

7.5   Optimum Spacing for Arrays of Vertical Heated Plates

Following Aung (1973) and Bar-cohen and Rohsenow (1984), the maximum power dissipated from an array of plates in a given cabinet volume is given as

$$Q_T = 2LSNq \tag{3.93}$$
$$= 2LSN(T_w - T_o)Nu(k/b)$$

where $Nu = Nu_L$ for $\Delta T = \Delta T_L, Nu = Nu_{L/2}$ for $\Delta T = \Delta T_{L/2}$ and N is the number of thermally active plates with a heat dissipation rate equal to (qLS) per side. For plates of negligible thickness, $N = W/b$ for symmetric heating (every plate heated), and $N = W/2b$ for asymmetric heating (alternating heated and insulting plates) where W is the total cabinet width. Since Nu decreases for decreasing b, and N increases, there is clearly an optimum spacing for given volume and allowable maximum or average plate temperature.

7.5.1   Isothermal Plates

For isothermal conditions the optimum spacing may be found by substituting eqs (3.77) or (3.80) for Nu in eq (3.93) and finding b that satisfies

$$\frac{d}{db}[Q_T/(T_w - T_o)] = 0 \tag{3.94}$$

to maximize the total heat dissipation per unit temperature increase. Using the composite eqs (3.77) and (3.80), Bar-Cohen and Rohsenow (1984) found the results which are summarized in Table 3.10, eqs (3.79) and (3.80) and (3.82) and (3.83), where $P = \rho^2 c_p g \beta (T_w - T_o)/(\mu k L) = Ra/(b^4)$. They found that for a given volume, an optimum array of isothermal plates alternating with insulating plates (asymmetric array) cannot dissipate more than 63 percent of the heat dissipated by an optimum array of isothermal plates. Hence, whenever possible, arrays should be arranged with all boards heated.

In order to dissipate the maximum heat from a confined space with isothermal plates, Bar-Cohen and Rosenow (1984) recommend finding the combination of parameters that will yield $Nu_{L/2}$ as close as possible to the isolated plate value. By arbitrarily setting $Nu_{L/2}$

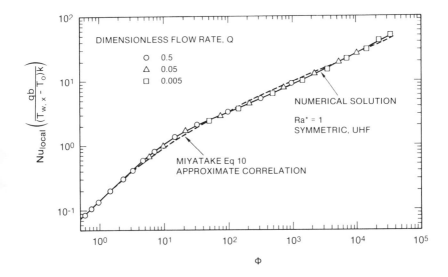

Figure 3.48: Correlation of computed local Nusselt number for asymmetric UHF channel [Miyatake and Fujii (1974)].

to 99 percent of the isolated plate value, they found the following parameters for maximum dissipation:

$$b_{max} = 4.64/P^{1/4}, \quad Ra_{max} = 463$$

for symmetric, isothermal plates, and

$$b_{max} = 3.68/P^{1/4}, \quad Ra_{max} = 184$$

for asymmetric, isothermal plates. The maximum spacing corresponds roughly to the boundary layer thickness at $x = L$ for an equivalent isolated plate times the number of heated walls. It is to be noted here that Aihara (1973) and Sparrow and Azevedo (1985) found that, in fact, the isolated plate does not form the upper bound for Nusselt number, but that it peaks at an intermediate plate spacing. This intermediate spacing is the true b_{max}, but it does not differ appreciably from the Bar-Cohen-Rohsenow (1984) values.

7.5.2   Isoflux Plates

Using eq (3.94), the equivalent optimum spacing for isoflux plates can be found, but the result depends on whether one maximizes the total

heat dissipation per unit "average" temperature increase, or per unit
"maximum" temperature increase. Bar-Cohen and Rohsenow (1984)
derived the results for the former by maximizing $Q_T/(T_w - T_o)_{L/2}$
and these results are summarized in Table 3.10, eqs (3.85) and (3.86)
and (3.89) and (3.90), where $R = \rho^2 c_p g \beta q / \mu L k^2 = Ra^*/b^5$. For a
given average plate temperature increase, eqs (3.86) and (3.90) may
be used to determine the optimum spacing, b_{opt} and the allowable
uniform heat flux, q to yield the maximum possible heat dissipation
for the array. Alternatively, Aung (1973) derived results for optimum
spacing based on the maximum allowable temperature increase following the analysis of Bodoia and Osterle (1962). Referring to Fig.
3.51 he shows that the optimum Nusselt and Rayleigh number occur
where the slope of a log-log plot of $\overline{Nu}$ versus Ra is equal to one-half;
here $\overline{Nu}$ and Ra are defined as

$$\overline{Nu} = \frac{\bar{q}b}{(T_1 - T_o)_L \, k}$$

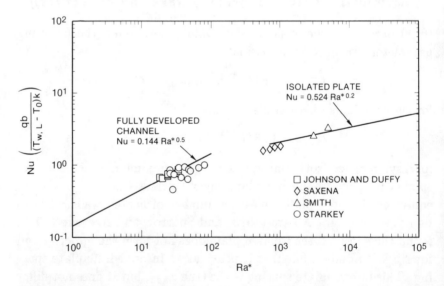

Figure 3.49: Measured Nusselt number at channel exit for
real PCB's compared to asymptotic relations for smooth
surfaces [Johnson (1986)].

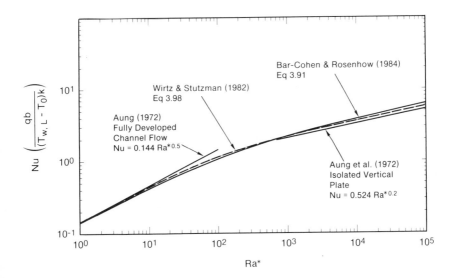

Figure 3.50: Comparison of composite relations of Wirtz and Stutzman (1985) and Bar-Cohen and Rohsenow (1984) for symmetric UHF conditions [Johnson (1986)].

and

$$Ra = \frac{g\,(T_1 - T_o)_L\,b^4}{\nu^2 L}Pr$$

where $\bar{q}$ is the average of the heat flux on both walls and T_1 is the temperature of the hotter wall. At this location, $\overline{Nu}_{opt} = \bar{p}Ra_{opt}^{1/2}$, and it is straightforward to show that for fixed volume and given maximum temperature increase

$$Q_T/(Q_T)_{symmetric} = \bar{p}/(\bar{p})_{symmetric} = E$$

where E, the heat transfer effectiveness of an optimally spaced array, is the ratio of total heat dissipated compared to the heat dissipated by a symmetrically heated channel at the same maximum temperature increase, both optimally spaced. Table 3.11 summarizes Aung's (1973) results and in addition, shows the optimum Nusselt and Rayleigh numbers using the conventional definitions used in Table 3.10. It is observed that the maximum heat dissipation from an optimally-spaced asymmetrically heated isoflux array is 65 percent of that for its symmetrically heated counterpart. This is much like the result for isothermal arrays in that asymmetry degrades the maximum heat dissipation for a given maximum temperature increase.

8 MODELING FLOW AND HEAT TRANSFER FROM COMPLEX SURFACES AND CHANNELS

8.1 Isolated Thermal Sources on Vertical Surfaces

The presence of a bounding surface affects natural convection from a surface mounted component only insofar as it restricts the entrainment of air. The characteristics otherwise are the same as for unconfined surfaces. Figure 3.52 presents the dependence of the Nusselt number on Grashof number for a single 1.27 *cm* cube on a vertical

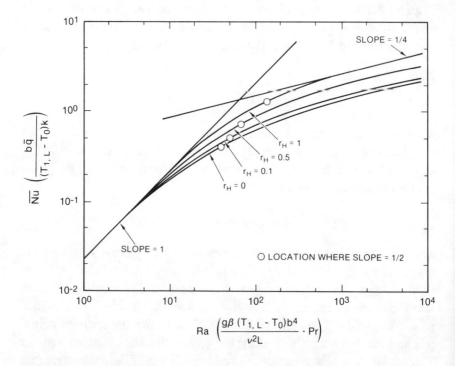

Figure 3.51: Optimum Nusselt and Rayleigh numbers for asymmetric UHF channel [Aung (1973)].

Table 3.11: Optimum Nusselt and Rayleigh number based on maximum wall temperature for asymmetrically heated UHF channels (after Aung (1973)).

r_H	Ra_{opt}	$\overline{Nu}_{opt}$	E	$(Nu_L)_{opt}$	Ra^*_{opt}
0	42	0.43	65%	0.86	48.84
0.1	51	0.51	70%	–	–
0.5	70	0.73	86%	–	–
1.0	135	1.18	100%	1.18	114.41

insulated surface, Ortega (1986), and for a cylinder of aspect ratio one mounted on a heated vertical plate, Sparrow and Chrysler (1981). The data are practically indistinguishable when the length scales used are the cube height and the cylinder diameter. They are well correlated by the expression

$$Nu = 0.426 Ra^{1/4} \qquad (3.95)$$

derived by Sparrow and Chrysler (1981). The one-fourth power dependence of Nusselt number on Grashof number is universal for many distinct geometries in laminar, external natural convection. Universal methods for correlating data for arbitrary geometries have been proposed by Raithby and Hollands (1975). Sparrow and Ansari (1983) have recently commented on a traditional correlating method attributed to King. Carey and Mollendorf (1977) have studied the thermal field above flush-mounted circular and square sources of heat on an insulated vertical surface and find that in the far-field, the wall-bounded thermal plume decay is bounded by that for theoretical point and line sources on vertical surfaces. They also found that the local Nusselt number on the heated, two-dimensional spot has the familiar one-fourth power dependence on the Rayleigh number. A similar investigation was carried out by Ravine and Richards (1984) who experimentally investigated the heat transfer from a flush-mounted, horizontal strip source of heat on a vertical surface with and without opposing, shrouding surfaces. Surprisingly, they found that the effect of a channel wall was to reduce the local heat transfer from

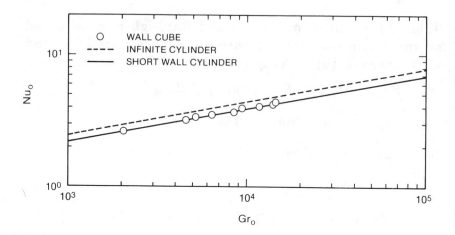

Figure 3.52: Nusselt number for a single cube on an insulated vertical surface and a short heated cylinder on an isothermal vertical surface [Ortega (1986)].

the heater as compared to a heater on an isolated plate. Apparently the enhancement due to the induced duct flow does not overcome the degradation of heat transfer due to the loss of uninhibited entrainment of air. This is partially attributable to the heater being located two heater widths downstream from the entrance. When their results are compared to those for isothermally heated plates of the same length as the heater, they find that the average heat transfer coefficient for the two-dimensional source is consistently 20 to 30 percent lower than that for the isolated plates. They observed that the additional viscous drag of the unheated parts of the channel decrease the flow rate compared to a heated channel without the unheated "chimney", but they did not comment on possible heat loss from the unheated chimney which would reduce the overall buoyancy and therefore the induced flow rate.

In contrast to Ravine and Richards' findings, Sparrow, Cook and Chrysler (1982) have shown that heat transfer from a short, heated cylinder mounted on an isothermal plate may be enhanced by as much as 60 percent by the effects of induced duct flow. They found the enhancements to be greatest when the duct is as narrow as possible and when the cylinder is nearest the duct entrance. In the external natural convection mode, the cylinder heat transfer is due

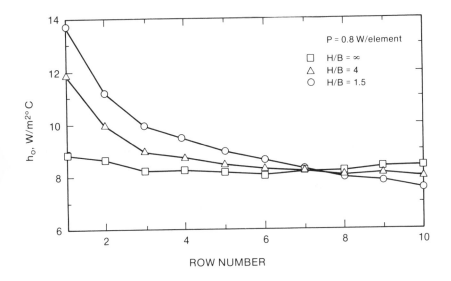

Figure 3.53: Variation of heat transfer coefficient based on inlet temperature in an array of heated cubes with and without an opposite wall (Ortega (1986)).

primarily to its locally generated buoyancy, but in the channel mode, the heat transfer is enhanced by the combined effects of the local cylinder buoyancy and the forced convection from the induced channel flow. Sparrow and Pfeil (1984) demonstrated this induced draft effect by comparing the heat transfer from a long, horizontal cylinder with and without unheated, vertical shrouding surfaces. Similar dramatic enhancements were reported by Ortega and Moffat (1985) on an array of cubical elements mounted on a vertical, insulated plate. Figure 3.53 shows the element heat transfer coefficient based on $(T_e - T_\infty)$ as a function of position in the cube array, parametric on plate-spacing to element height ratio. The heat transfer enhancement is greatest on row one, increasing monotonically with decreasing plate spacing. The enhancement is 54 percent for the smallest plate spacing. For elements in the array interior, the enhancement due to the channel forced convection effect is offset by the increasing mean channel temperature, and the net effect on the last two element rows is to decrease the heat transfer compared to the isolated array. It will be seen in the next section that it is beneficial to define a heat

transfer coefficient based on wall to adiabatic temperature difference
in order to examine the local convective heat transfer separate from
the effects of mean fluid temperature.

8.2   Interacting Thermal Sources on Vertical Surfaces

When thermal sources are in close proximity to each other on a
surface they may mutually interact, and the degree of interaction
naturally declines proportionally to the distance separating them.
Jaluria (1982) has numerically investigated the interaction of flush-
mounted, strip sources of heat on a vertical insulated surface using
the boundary-layer form of the governing equations. Figure 3.54a
shows the computed dimensionless temperature rise on a pair of
heated strips where the distance between the end of the first source

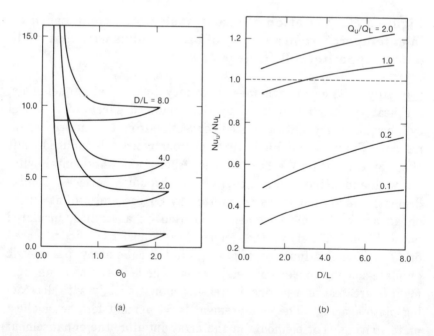

(a)   (b)

Figure 3.54: (a) Surface temperature variation on two,
one-dimensional heat sources on an insulated vertical surface, (b) Average Nusselt number of upper heater to lower
heater at various spacings, (Jaluria (1982)).

and the beginning of the second source, D, is normalized on the strip length, L, as is the axial position X. The dimensionless temperature and Grashof number are defined as

$$\theta = \frac{T - T_\infty}{qL/kGr^{1/5}}$$

and

$$Gr = \frac{g\beta qL^4}{k\nu^2}.$$

The effect of the thermal plume from the first source on the local heat transfer from the downwind source depends on the spacing. For $D/L = 2$, the dimensionless temperature is slightly greater than for the first, indicating a degradation of heat transfer, whereas a modest improvement is obtained for D/L of 4 and 8. Figure 3.54b shows the Nusselt number on the downstream source normalized on that for the upstream source. Of particular interest is the line for equal heating on both sources, $Q_u/Q_L = 1.0$, which shows that the cross-over spacing is roughly $D/L = 3$. Jaluria (1982) showed that the maximum velocity in the boundary layer flow induced by the first source <u>increases</u> downstream, due to buoyancy, and the adiabatic wall temperature <u>decreases</u>, due to entrainment of unheated fluid, thereby leading to an enchancement of heat transfer above that due solely to source buoyancy. Clearly the convective heat transfer from the second source has two components, one due to its own locally generated buoyancy, and the second due to convection by the predeveloped boundary layer flow to which it is exposed. Anderson and Moffat (1987) demonstrated that the local heat transfer from a cubical element in an array of heated elements is entirely due to the convective flow in the predeveloped boundary layer in which it is embedded. Figure 3.55 shows their local adiabatic Nusselt number, Nu_{ad}, plotted against an adiabatic Grashof number, Gr_{ad}, where the adiabatic temperature difference, $(T_e - T_{ad})$ and the element height, B, is used in both. For zero array heating, the heat transfer is due solely to the buoyancy generated by the element and therefore Nu_{ad} is strongly dependent on Gr_{ad}. When the element is exposed to a pre-developed boundary flow by heating the array of elements upstream of it, the element achieves a higher adiabatic temperature but the Nusselt number dependence on the local Gr_{ad} decreases in proportion to the level of upwind heating.

8.3   Effects of Real Surface Geometries

Whereas laminar flow results have been used successfully to design plates or channels with closely spaced surface components, the effects of surface irregularities and channel obstructions on overall heat dissipation and maximum card temperatures may be pronounced. Because these effects are not readily modeled there has not been much effort to characterize them.

8.3.1   Boundary Layer Flows

In a viscous dominated laminar boundary layer or channel flow, it is generally true that surface irregularities, such as protrusions or grooves, will not significantly affect the local convective heat transfer if the significant dimension is much smaller than the boundary layer

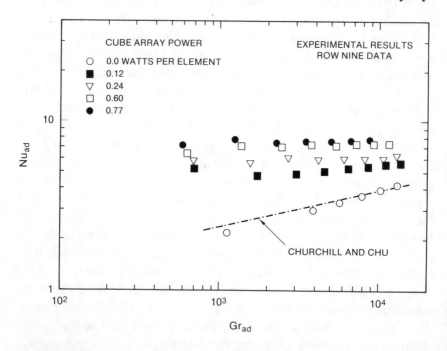

Figure 3.55: Adiabatic Nusselt number for a cubical element in an array of cubical elements on a vertical surface–dependence on local buoyancy (Anderson and Moffat (1987)).

thickness. This has been demonstrated by Fujii et al (1973) on vertical cylinders with horizontal grooves and sparse, three-dimensional protrusions in oil and water. Anderson and Bohn (1986) have noted that large, isolated, protrusions or grooves on heated vertical surfaces may alter the heat transfer in the local area about the element and may also affect the heat transfer downstream by inducing premature transition to turbulence on the plate which would normally be exposed to a laminar boundary layer. Figure 3.56 reproduces the numerically derived results of Shakerin et al (1986) which show the effects of a two-dimensional protrusion on one wall of a two-dimensional enclosure (with one wall at T_H and the other at a lower temperature T_C) where the roughness height is of the order of the boundary layer thickness on the heated wall. The effects of the roughness on the plate heat transfer are localized to within two roughness heights before and after the element, and the reduction of heat transfer on the plate is roughly balanced by the increased surface area of the roughness element. When there are repeated irregularities on a heated plate, the effects on both the local and the overall plate heat transfer are strongly dependent on the size and spacing of the protrusions (or depressions) relative to the boundary layer. Figure 3.57 shows the data and calculations of Horton (1981) showing the effect of horizontal and vertical grooves on the average heat transfer coefficient from a vertical, isothermal, plate with and without an opposite heated plate. For large plate spacings, the thermal boundary layer grows unconstrained and is thick compared to the grooves. The grooves do not appreciably affect the heat transfer. For small plate spacings, the clear channel spacing is of the order of the grooves and an increasing enhancement with decreasing spacing, especially for the large two-dimensional grooves is observed. Ortega and Moffat (1985, 1986) performed extensive experiments on a heated array of 1.27 cm cubical elements, spaced three element heights apart on a vertical, insulated surface with and without an opposing insulated surface. Because most of the heat is dissipated from the roughness element surfaces, the thermal boundary layer thickness is always of the order of the element height and spacing. Since the plate is not directly heated, they could not strictly comment on the effect of the roughness on the plate heat transfer. However they did compare the relative effectiveness of heat transfer from the element surfaces compared to that from a smooth plate dissipating the same average heat flux. Figure 3.58 shows this comparison, based on the behavior of the local Nusselt number, Nu_x, and the local Grashof number, Gr_x. The equivalent smooth plate correlations are

$$Nu_x = 0.406Gr_x^{1/4}, \; Pr = 0.72, Gr_x < 10^9 \qquad (3.96)$$

and

$$Nu_x = 0.096Gr_x^{1/3}, \; Pr = 0.72, Gr_x > 10^9 \qquad (3.97)$$

for laminar and turbulent regimes respectively (Kays and Crawford (1980) and Churchill and Chu (1975)). From Fig. 3.58, the local Nusselt number is from 40 to 50 percent greater than for the smooth plate everywhere. It is parallel to the result for turbulent plate convection despite the fact that the Grashof number range considered would place the equivalent smooth plate in the laminar regime. There are not sufficient data to attribute the increase in heat transfer to turbulence. The flow may be a complex, wall-bound, but laminar flow, with the increased fluid mixing greatly enhancing the heat transfer compared to that in a laminar boundary layer.

There are no comprehensive theories for the effects of large roughness on boundary layer flows, but the previously discussed contributions indicate that the effects are important only for sparse arrays or isolated elements. In these cases, the flow develops from the complex interaction of wall-bounded thermal plumes, and the most important consideration is the extent to which heat transfer from a source is affected by the boundary layer flow in which it is embedded.

8.3.2 Vertical Channel Flow

Surface irregularities or roughness may fundamentally alter the structure of the flow field near the wall and influence both the heat transfer rate and flow drag in a channel, as shown in the results of Horton (1981) previously discussed. For a buoyancy induced channel flow the total heat dissipated for a specified temperature rise depends on both. Their potential effects are illustrated by the measurements of Birnbreier (1981) on specially designed arrays of printed circuit boards, each roughly 10 cm high by 16 cm long. The boards were heated by 144 resistors uniformly distributed on the board, and large flow barriers (8 × 8 × 48 mm) were mounted on several locations to simulate the effects of larger components. Figure 3.59 shows the Nusselt number at the channel exit plotted against the symmetric, isoflux channel Rayleigh number for this array and for the smooth channel. The two behave very differently. For $Ra^* > 300$, the data follow the line

$$Nu_L = 0.20 Ra^{*^{0.31}} \tag{3.98}$$

which is steeper than the smooth isolated plate asymptote given by eq (3.68). The PCB Nusselt numbers greatly exceed the plate values at very large plate spacings. Birnbreier (1981) attributes this to the beneficial effect of radiative heat transfer. For smaller board spacing and decreasing Ra^* the effects of radiation are small and the Nusselt numbers undergo a transition to fully developed channel flow, below Ra^* of roughly 300. The data fall below the smooth channel fully-developed asymptote and show a distinctive slope, between 1/3 and 1/2, as compared to 1/2 for eq (3.66). That this behavior may be at least partly due to the effects of the surface roughness is readily demonstrated for both the isolated plate and fully-developed channel flow limits. The data of Ortega and Moffat (1985), shown in Fig. 3.58, for an isolated array of cubes, show behavior parallel to that of a fully turbulent boundary layer, where the surface temperature is independent of position for UHF. Nusselt number scales like $Nu_x = C_1 (Gr_x Pr)^{1/3}$, or in terms of the modified Grashof number for constant heat flux, $Nu_x = C_2 (Gr_x^* Pr)^{1/4}$. The same result transformed to channel coordinates is

$$Nu_L = C (Ra^*)^{1/4} (L/b)^{1/4} \tag{3.99}$$

In the other limit, for fully-developed channel flow and constant heat flux, the Nusselt number based on wall to inlet air temperature difference is given by

$$[Nu_L]^{-1} = [Nu_{m,L}]^{-1} + [Re\ Pr\ b/L]^{-1} \tag{3.100}$$

where $Nu_{m,L}$ is the local mixed mean Nusselt number. For large b/L

$$Nu_L = Re\ Pr\ b/L \tag{3.101}$$

For fully-developed, buoyancy-induced channel flow the induced Reynolds number was derived by Ortega and Moffat (1985) as

$$Gr^*L/b = 2Re^3Pr\left[\frac{1}{L}\int_0^L C_d dx\right] \quad (3.102)$$

where C_d is the local wall drag per unit area, normalized on the dynamic pressure, $1/2\rho u_o^2$. For laminar flow, $C_d = 12/Re$ and the familiar result given by eq (3.64) is obtained. When the channel drag is due primarily to form drag from obstructions or roughness elements the local drag coefficient is practically independent of Reynolds number, and the induced Reynolds number is given instead by

$$Re = \left\{\frac{Gr^*\,Pr\,L/b}{2\,Pr^2\,\bar{C}_d}\right\}^{1/3} \quad (3.103a)$$

and therefore

$$Q = Re/Gr^* = \left(\frac{Pr}{2\bar{C}_d}\right)^{1/3}\left(\frac{L}{b}\right)^{2/3}(Ra^*)^{-2/3} \quad (3.103b)$$

Substitution of this into eq (3.103) yields

$$Nu_L = C(Ra^*)^{1/3}(b/L)^{1/3} \quad (3.104)$$

The Birnbreier (1981) data show a slope of 0.31, compared to 0.25 predicted by eq (3.101) for large Rayleigh numbers, and a slope between 1/3 and 1/2 for small Rayleigh number, compared to 1/3 from eq (3.106). The data of Ortega and Moffat (1985) for the cube array also show a departure from smooth plate behavior. The Ortega and Moffat (1985) data are compared to an asymmetrically heated UHF channel, whereas the data of Birnbreier, Fig. 3.59, are compared to a symmetric, UHF configuration. Their data plotted in channel coordinates is shown in Fig. 3.60. As shown from the lines drawn through the data at the largest and smallest plate spacing, both limits follow the behavior derived above for large roughness effects and additionally demonstrate separate dependence on the channel aspect ratio, b/L, as predicted by the derived models. As expected, the data of Birnbreier (1981) for realistic PCB's, demonstrate behavior which is intermediate between the smooth plate models and the large roughness models.

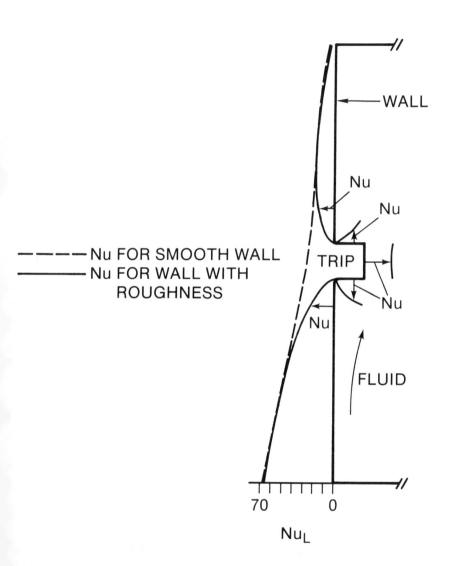

Figure 3.56: Effect of a two-dimensional protrusion on Nusselt number on a vertical enclosure wall at constant temperature (Shakerin et al. (1986)).

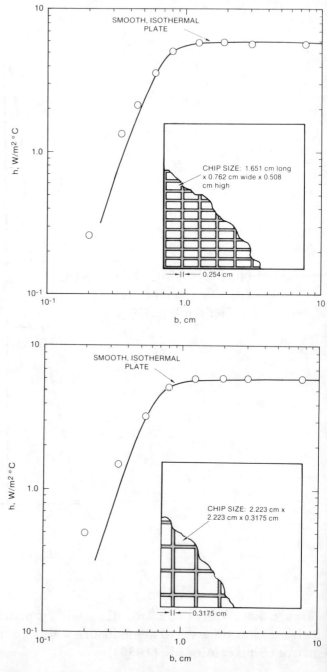

Figure 3.57: Effects of two types of surface grooves on average heat transfer coefficient from an isothermal channel wall [Horton (1981)].

9 MODELING NATURAL CONVECTION COOLED SYSTEMS USING FORCED CONVECTION PRINCIPLES

9.1 Effects of Local Buoyancy on Buoyancy-Induced Forced Flow in Channels

Analysis of complex electronic systems requires systematic evaluation of the heat transfer local to electronic components, the manner in which it is affected by other components, and the manner by which it in turn contributes to the overall, global properties of the system such as the maximum air temperature rise and induced flow rate. A typical example is shown in Fig. 3.61 in which the objective might be to predict the temperature rise of each component on a PCB, or the average temperature of a PCB in an array of PCBs in a cabinet, when the heat dissipation is not uniform. If one attempts to evaluate the temperature elevation of a discrete part of the system, the

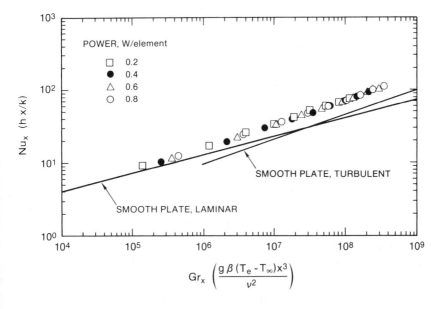

Figure 3.58: Local Nusselt number for an unshrouded array of cubical elements on a vertical insulated surface [Ortega and Moffat (1985)].

fundamental problems are to decide whether the surface in question is in "forced" or "natural" convection, and then to account for the thermal interaction of components.

The convective heat transfer from a discrete part of the system, such as a heated component, is driven by its own locally generated buoyancy and by the forced convection from the flow field in which it is immersed. The latter is due to the cumulative effects of other parts of the system. For external boundary layer flows, only the upwind components affect local heat transfer. For elliptic flow fields, such as vertical channels and sealed enclosures, the non-local forced convection arises from the combined effects of all parts of the system.

As demonstrated in Part I on forced convection cooling, it is fruitful to define a local adiabatic heat transfer coefficient when considering situations with non-uniform boundary conditions such that the heat dissipation from an element is given by

$$Q_e = h_{ad} A_e \left(T_e - T_{ad}\right) \tag{3.105}$$

where T_{ad} is the average temperature of a discrete element or part of a system with no internal heating. The meaning of this coefficient, and its use in forced convection situations has been discussed in Part I. Ortega (1986) measured the local adiabatic heat transfer coefficient in the same array of elements as used by Anderson and Moffat (1987), but with an opposing channel surface, and in buoyancy-driven and forced channel flows. In Fig. 3.62, the adiabatic heat transfer coefficient has been normalized on its forced convection value and plotted as a function of its own adiabatic Grashof number divided by the square of the measured inlet Reynolds number. This parameter may be thought of as a measure of the relative importance of local buoyancy to local forced flow to the heat transfer. All quantities are based on the local element length scale, B. Below Gr_{ad}/Re_B^2 of 0.3, the Nusselt number does not exceed its forced convection asymptote by more than 5 percent, and is not more than 10 percent greater for Gr_{ad}/Re_B^2 up to 0.5, for either buoyancy-driven or forced channel flow. The results for large plate spacing, $H/B = 4$, generally fall above those for the narrow channel, $H/B = 1.5$. This is because the average channel velocity was used in the normalizing Reynolds number. Since the ratio Gr_{ad}/Re_B^2 compares local mechanisms, it would be more appropriate to use the array velocity, as in

Table 3.12: Criteria for establishing the relative importance of local natural convection to induced forced convection.

$$Gr_{ad}/Re_B^2 \quad < 0.3, \quad \text{local natural convection negligible,}$$

$$0.3 < \quad Gr_{ad}/Re_B^2 \quad < 10.0 \quad \text{local forced and local natural convection equally important}$$

$$10.0 < \quad Gr_{ad}/Re_B^2, \quad \text{local natural convection dominant}$$

Section 5.7, as a measure of the local velocity about an element. As a general design principle, in the Reynolds and Grashof number regimes encountered in passive air cooling, the criteria displayed in Table 3.12 may be used to determine the relative importance of local natural convection, compared to induced forced convection.

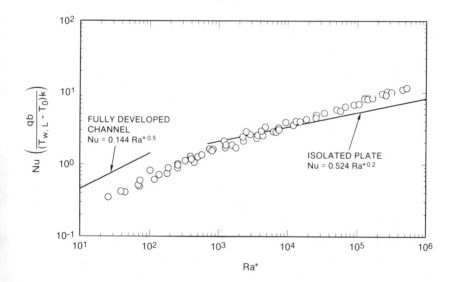

Figure 3.59: Nusselt number at channel exit for a vertical array of printed circuit boards [Birnbreier (1981)].

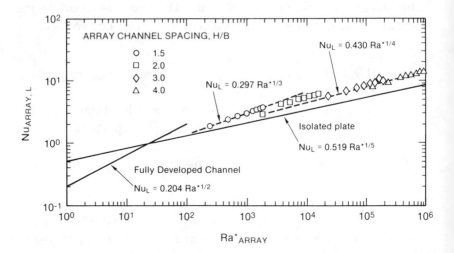

Figure 3.60: Nusselt number at channel exit for an array of cubical elements on a vertical, insulated channel wall with an insulated opposite wall [Ortega and Moffat (1985)].

9.2 Buoyancy-Induced Forced Convection in Vertical Channels

As a focus for discussion, consider the hypothetical channel containing heat dissipating electronic components depicted in Fig. 3.61. Because buoyancy is a cumulative property, it may be rationalized that if Gr_{ad}/Re_B^2 is small for every element in the channel, local buoyancy effects are negligible and the flow can be characterized as a "buoyancy-induced forced convection" flow. Conversely, if the parameter is non-negligible everywhere, it can be anticipated from previous discussion of the local behavior of smooth walled channels that the effects of buoyancy accelerate fluid towards the heated surfaces and transform the channel flow at the entrance to a buoyant boundary layer or wall jet downstream. In the sense that channel buoyancy evolves from the contributions of individual elements, the parameter Gr_{ad}/Re_B^2 embodies the same principles as Yao's criteria, eq (3.76), for smooth channels. Figure 3.63 shows the measured adiabatic Grashof number for the cube array of Ortega and Moffat plotted against the measured element Reynolds number squared, for naturally induced flow in their narrowest channel and for representative elements on the first row and in middle of the array. Such a representation may be though of as a domain map to show regimes of predominantly forced, mixed, or predominantly buoyant behavior.

For all array power dissipation levels, the ratio Gr_{ad}/Re_B^2 is less than 0.4, leading to the conclusion that localized buoyancy effects are negligible for the narrow-channel. Examination of local temperature profiles through the array, Fig. 3.64, indicate that after a short development length, the channel flow becomes fully-developed as is evidenced by the self-similarity of the profiles, with no indication

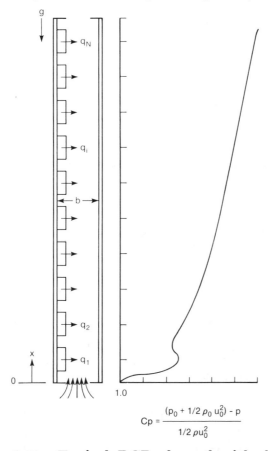

$$Cp = \frac{(p_0 + 1/2\,\rho_0\,u_0^2) - p}{1/2\,\rho u_0^2}$$

Figure 3.61: **Typical PCB channel with discrete components and variable heating showing local drag coefficient variation.**

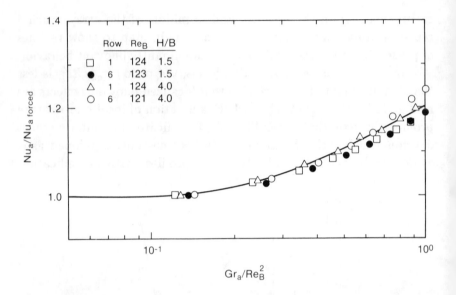

Figure 3.62: Adiabatic Nusselt number for an array of cubical elements normalized on forced convection value showing dependence on local buoyancy parameter Gr_{ad}/Re_B^2 [Moffat and Ortega (1986)].

of preferential boundary layer flow on the heated array. Following the approach demonstrated in Fig. 3.47, in which the local mixed mean Nusselt number for buoyancy-induced and forced convection in smooth channels is compared, a locally defined Nusselt number for the array for both situations can be compared and in this case, experimentally measured values of Nu_{ad} rather than Nu_m are chosen. In the buoyancy-induced flow, Nu_{ad} is measured in a fully heated array, by first measuring the element adiabatic temperature with no element power and then the element temperature with power dissipation. For forced flow, Nu_{ad} is measured by heating only the element in question with flow induced by an external blower. The two are compared in Figs. 3.65 and 3.66 for the same narrow channel spacing as the temperature profiles of Fig. 3.64.   On row one, the buoyancy-induced values are within 5 percent for all Re_B, well within experimental uncertainty for the experiment, and within 10 percent on row six, in the center of the array. Channel velocities are on the order of 20 cm/s at the entrance.

For large channel spacing, the results are summarized in Figs. 3.67, 3.68 and 3.69. As expected, the temperature profiles indicate a transition to array boundary layer flow where buoyancy effects accumulate near the heated array. Note that there is very little temperature rise past the channel center. From previous discussion, it is anticipated that the Nusselt number will not be appreciably perturbed from forced convection near the channel entrance, but will increasingly depart from the forced convection behavior further into the array. This is confirmed by Figs. 3.68 and 3.69 which show no departure from the forced convection data at row one, but a large increase at row six. From the discussion of array forced convection in Section 5.7, heat transfer from an element in a cube array is governed by the flow velocity local to the element (the "array velocity") because of the large lateral surface area. For large plate spacing, the array velocity decreases in the flow direction because

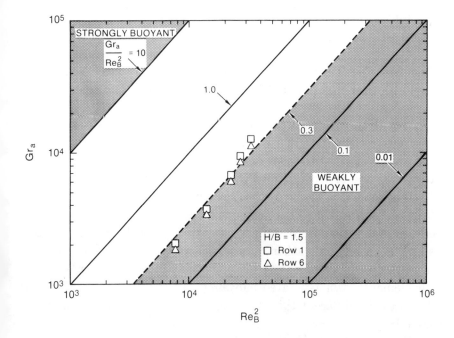

Figure 3.63: Heat transfer domain map for array of cubical elements with data for narrow channel spacing at various power levels [Moffat and Ortega (1986)].

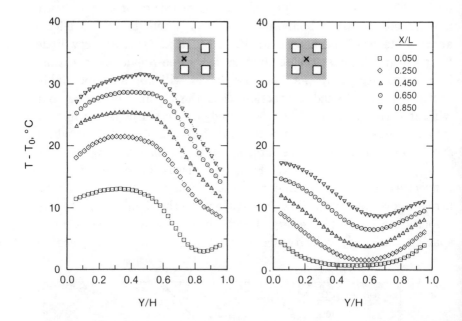

Figure 3.64: Temperature profiles for cube array with narrow channel spacing, H/B = 1.5 [Moffat and Ortega (1986)].

of array drag with a corresponding drop in the local adiabatic heat transfer coefficient. In contrast, the preceding results reveal that the array velocity increases in the direction of flow because of array buoyancy and is accompanied by a similar increase in Nu_{ad}.

It is important to clarify that the departure of the Nusselt number from the forced convection asymptote is not due to local buoyancy effects, but rather from the accumulation of local buoyancy effects which is an integral rather than a local property. The heat transfer from the row six element in Fig. 3.69 may indeed be unaffected by the element's own generated buoyancy, but the local "forced" flow is dramatically different than that for a forced channel flow with the same inlet Reynolds number.

For lack of better quantitative criteria at present, a good rule of thumb is that both local and accumulative buoyancy are unimportant in the absence of large temperature gradients normal to the direction of flow. These conditions are encountered for well-mixed, fully-developed flow in narrow channels.

10 A DESIGN EQUATION FOR INDUCED FLOW RATE IN CHANNELS WITH ARBITRARY HEATING AND FLOW RESTRICTION

For electronics channels that are in buoyancy induced forced flow, data and design principles discussed in part one of this chapter are equally applicable. In particular, the method of linear superposition for modelling a randomly heated channel is immediately applicable, as are local adiabatic heat transfer coefficients, thermal wake decay . functions, and array drag coefficients measured in "cold" forced flow bench tests. Referring again to the complex channel shown in Fig. 3.61, a useful, albeit approximate, design equation for predicting

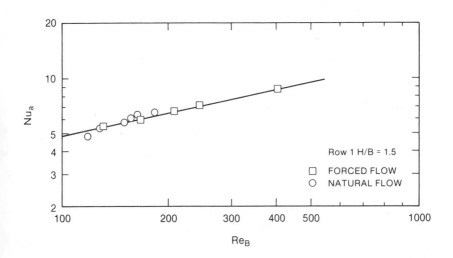

Figure 3.65: Comparison of adiabatic Nusselt number for forced and naturally-induced channel flow, $H/B = 1.5$, row one element [Moffat and Ortega (1986)].

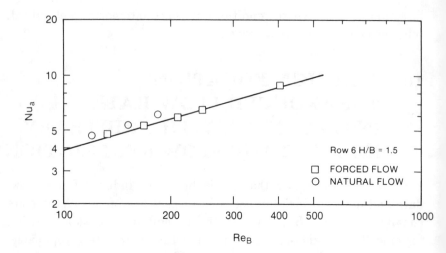

Figure 3.66: **Comparison of adiabatic Nusselt number for forced and naturally-induced channel flow, H/B = 1.5, row six element [Moffat and Ortega (1986)].**

induced flow rate in complex channels is derived here in brief. A complete derivation is given by Ortega (1986).

Consider a differentially small control volume including a portion of the air in a vertical passage whose overall height is L and whose wall-to-wall spacing is H. Assuming steady state conditions, the conservation of momentum in the $+x$ direction is expressed as:

$$\frac{d}{dx}\int_A \rho u^2 dA = g_c\left[-\frac{d}{dx}\int_{A_{in}} p dA - F_d' - \frac{g}{g_c}\int_A \rho dA\right] \qquad (3.106)$$

where F_d' is the total drag per unit length and A is the channel cross-sectional area, assumed to be constant in this treatment. Integrating from the bottom to the top of the channel yields

$$\int_{A_{in}} p_o dA - p_L A = F_d + \frac{g}{g_c}\int_o^L \bar{\rho} A dx + \frac{1}{g_c}\int_{A_{exit}} \rho u^2 dA - \frac{1}{g_c}\int_{A_{in}} \rho u^2 dA \qquad (3.107)$$

where

$$\bar{\rho} = \frac{1}{A}\int_A \rho dA$$

is the area averaged density at a vertical location in the channel. The pressure term at the exit is taken simply as p_L, assuming the air discharges as a low velocity jet into a uniform pressure region. At the inlet, only the total pressure is known and the inlet static pressure is expressed as

$$p_o(o,y) = p_L + \frac{g}{g_c}\rho_o L - \frac{1}{2g_c}\rho u^2$$

and hence the static pressure at the entrance can be evaluated in terms of the stagnation pressure and the entering momentum flux

$$\int_{A_{in}} p_o dA = \int_{A_{in}} \left(p_L + \frac{g}{g_c}\rho_o L \right) dA - \int_{A_{in}} \frac{1}{2} g_c \rho u^2 dA. \quad (3.108)$$

Substitution into eq (3.107) yields

$$\frac{g}{g_c} \int_0^L (\rho_o - \bar{\rho}) \, dx = \frac{F_d}{A} + \frac{1}{g_c A} \int_A \rho u^2 dA - \frac{1}{2g_c A} \int_A \rho u^2 dA \quad (3.109)$$

The term on the left is the buoyant pressure rise. The first term on the right is the total channel drag, and the second and third terms

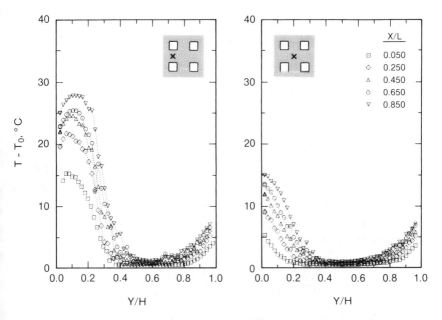

Figure 3.67: Temperature profiles for cube array with wide channel spacing, H/B = 4 [Moffat and Ortega (1986)].

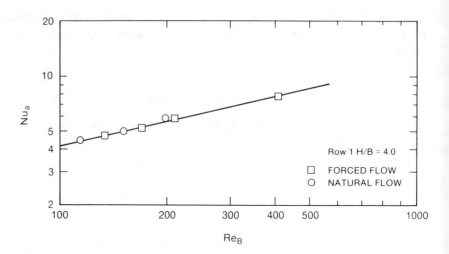

Figure 3.68: Comparison of adiabatic Nusselt number for forced and naturally-induced channel flow, H/B = 4, row one element [Moffat and Ortega (1986)].

represent two leaving and one entering velocity heads respectively. For applications in electronics cooling, the channel drag will be far greater than the difference in velocity heads. Thus, eq (3.109) can be simplified

$$\frac{g}{g_c} \int_0^L (\rho_o - \bar{\rho})\,dx = \frac{F_d}{A} + \frac{1}{2g_c}\rho_o u_o^2 \qquad (3.110)$$

which states that the total buoyant pressure rise, sometimes referred to as the "flotation" pressure, is balanced by the total channel drag plus the pressure loss due to the acceleration of quiescent fluid to velocity u_o. In the forced convection work of Part I an overall drag coefficient in terms of total-to-static pressure drop for the entire channel was defined as

$$C_p = \frac{(p_o + \rho_o u_o^2/2g_c) - p_L}{\rho_o u_o^2/2g_c} = \frac{(F_d/A) + (\rho_o u_o^2/2g_c)}{\rho_o u_o^2/2g_c} \qquad (3.111)$$

The key here is that the channel drag is the same whether the flow is buoyancy-induced or forced, and hence C_p can be measured in a cold-flow bench test. Figure 3.61 demonstrates the qualitative behavior

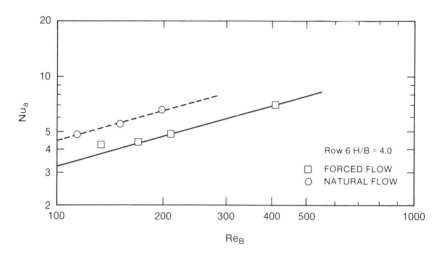

Figure 3.69: Comparison of adiabatic Nusselt number for forced and naturally-induced channel flow, H/B = 4, row six element [Moffat and Ortega (1986)].

of C_p in a channel populated with components, where an initial increase is incurred due to the entrance loss, followed by a consistent increase due to viscous and bluff-body drag of the components. For consistency, C_d is used as the symbol for the channel overall drag coefficient in this discussion, that is, $C_d = C_p$.

Assuming perfect gas behavior with constant specific heats allows the density difference in eq (3.110) to be expressed in terms of temperature difference. With this assumption and eq (3.111), the working form of the equation is

$$Gr_L = \frac{g \int_0^L \left[\left(\bar{T} - T_o \right) / \bar{T} \right] L^2 dx}{\nu^2} = \frac{C_d}{2} Re_H^2 \left(\frac{L}{H} \right)^2 \qquad (3.112)$$

and because $\bar{\rho}$ was defined as the area averaged density, $\bar{T}$ is the area averaged temperature. A shape factor can be defined for the temperature distribution, and this allows the integral to be replaced by information at the top of the channel:

$$S = \frac{Gr_L}{Gr_{L,max}} = \frac{\bar{T}_L \int_0^L \left[\left(\bar{T} - T_o \right) / \bar{T} \right] dx}{L \left(\bar{T}_L - T_o \right)} \qquad (3.113)$$

Two exact values for S are immediately derivable. For a uniform temperature channel, $\bar{T} = \bar{T}_L$ everywhere and therefore, $S = 1.0$. This situation would arise if all heating took place at the bottom of the channel, and the walls above were perfectly insulated. If it is now assumed that $\bar{T}$ is the fluid mixed-mean temperature, rather than the area averaged temperature, considerably more can be done. For uniform channel heating, the mixed mean temperature will rise linearly from T_o to $\bar{T}_L$, and for this case

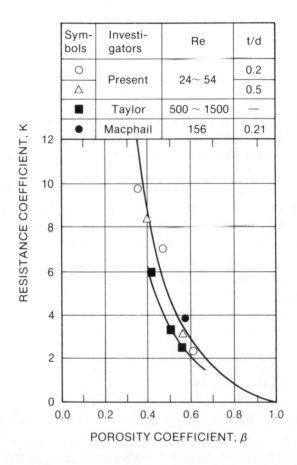

Figure 3.70: **Resistance coefficient for a perforated plate in a circular duct, data at high and low Reynolds number [Ishizuka et al. (1984)].**

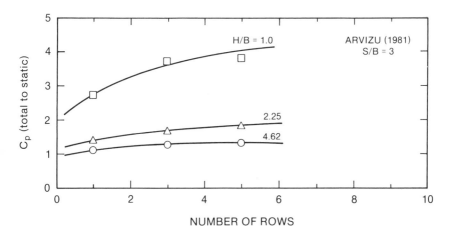

Figure 3.71: Drag coefficient for an array of cubical elements for three channel spacings measured in cold forced flow [Arvizu and Moffat (1981)].

$$S = \frac{1 - \left[\left(\bar{T}_L/T_o\right) - 1\right]^{-1} \ell n \left(\bar{T}_L/T_o\right)}{1 - \left(T_o/\bar{T}_L\right)} \qquad (3.114)$$

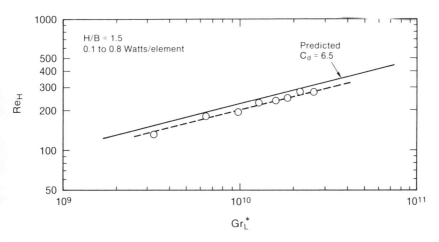

Figure 3.72: Predicted and measured channel Reynolds number for an array of cubical elements with narrow channel spacing, H/B = 1.5 [Ortega and Moffat (1986)].

Gr_L defined by eq (3.112) can be written in more conventional form
by combining with eq (3.113) to get

$$\frac{gS\left(\overline{T}_L - T_o\right)L^3}{\overline{T}_L\nu^2} = \frac{C_d}{2}Re_H^2\left(\frac{L}{H}\right)^2 \qquad (3.115)$$

It is also useful to express the result in terms of a Grashof number defined in terms of heat dissipation, rather than temperature
difference. From an energy balance, $\left(\overline{T}_L - T_o\right) = Q'/\dot{m}'c_p$, where
Q' is total power per unit channel width. Using this result, and
re-arranging terms, one obtains

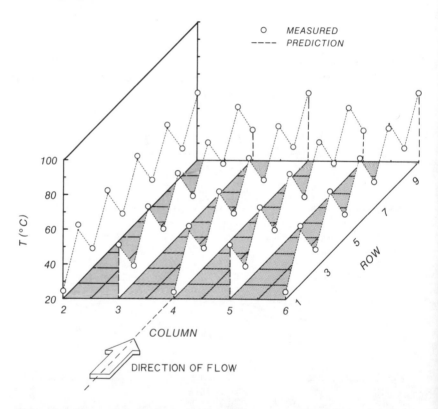

Figure 3.73: Predicted temperatures, using linear superposition and a forced convection data base, in a naturally
cooled array of cubical elements with narrow channel spacing, $H/B = 1.5$; alternate elements heated to 0.4 W/element,
others unheated [Ortega and Moffat (1986)].

$$Gr_L^* = \frac{gS\,(Q'/k)\,L^3}{\bar{T}_L \nu^2} = \frac{C_d}{2} Re_H^3 Pr \left(\frac{L}{H}\right)^2 \qquad (3.116)$$

For arbitrary heating distribution along the channel the shape factor can be defined alternatively by

$$S = \frac{T_o + (Q'/\dot{m}'c_p)}{(Q'/\dot{m}'c_p)} \int_o^L \frac{\theta(x)}{T_o + \theta(x)} dx \qquad (3.117)$$

where

$$\theta(x) = \bar{T}(x) - T_o = \int_o^x \frac{q''(\xi)}{\dot{m}'c_p} d\xi \qquad (3.118)$$

Here, q'' is the local heat release per unit width, per unit length. Equation (3.116) can thus be used directly, though iteratively, to predict the induced Reynolds number. Observe that the shape factor S will have a value close to 0.5 for uniform heating, but its exact value depends on the conditions. Steinberg (1980) Ishizuka, et al (1986) and others have derived simplified forms of eq (3.116) that are commonly used in complex, induced-draft systems.

Arvizu and Moffat (1981) have shown that the drag coefficient is independent of Reynolds number when it is due primarily from array drag. This is based on measurements in turbulent flows at much higher Reynolds number than are encountered in buoyancy- induced forced flows. However, both measurements and numerical computations for very low Reynolds number in channels with obstructions and constrictions indicate that this is true even in this regime. For example, Ishizuka et al (1984) found that the resistance coefficient for perforated plates shows a very weak dependence on Reynolds number over a very large range. Their data, shown in Fig. 3.70, for Reynolds numbers of the order of that encountered in electronics cooling applications, compare favorably to data at Reynolds numbers as much as 30 times greater than their base case. For the cube array used by Ortega and Moffat, measured drag coefficients are given in Fig. 3.71 for three plate spacings. Figure 3.72 shows the measured inlet Reynolds number for a narrow channel with plate spacing to element height of 1.5, compared to that predicted with eqs (3.116) through (3.118), and C_d estimated from the data of Fig. 3.71.

The data are roughly 15 percent lower than the predicted values, but they show the dependence Re_H proportional to $Gr_L^{*1/3}$ as predicted. Only modestly accurate values of the drag coefficient are required for an adequate prediction since Reynolds number varies as the cube root of C_d.

The ability to use forced convection data and design principles in natural convection cooled channels considerably increases the tools available to the designer. Of notable importance is the practical application of linear superposition principles in order to handle problems with non-uniform heat dissipation. The formal application of the method is outlined in Part I of this chapter, and it applies over exactly to buoyancy-induced channel flow. Two areas in which the method may find application are in the prediction of operating temperatures for individual components on a board, or the average operating temperature of a printed circuit board in a bank of boards, as in Fig. 3.36a. Ortega and Moffat (1986) demonstrated the technique by predicting the temperature of each element of an array of eighty elements, with nonuniform, but regular, heat dissipation. A typical comparison of measured versus predicted temperatures is shown in Fig. 3.73, where every other element was unpowered, and the rest dissipated 0.4 W/element. The adiabatic heat transfer coefficients and thermal wake functions used were measured in forced convection tests for the same geometry, and the induced Reynolds number was predicted using the method discussed in the preceding section. In a series of four cases with varying degrees of power nonuniformity, the temperatures were predicted to better than 15 percent of the absolute measured temperatures.

11   NOMENCLATURE

11.1   Part I

A Area, m^2

A_f Frontal area, m^2

A_i Area of the i-th element, m^2

A_m A constant in the eigenfunction expansion of eq (3.25)

A_t Total surface area, m^2

b	Barrier height in Sparrow, Niethammer and Chaboki, spacing between top of components and opposite wall, in Lehmann and Wirtz. m
B	Element height, m
C_f	Circumference of frontal area, m, see eq (3.33)
C_p	Pressure coefficient, may be total-to-static or static-to-static depending on author
D	Mass diffusion coefficient, m^2/s
D_h	Hydraulic diameter $D_h = 4r_h, m$
g	A weighting function defined by eq (3.25)
H	Channel height, in Arvizu and Moffat; free height above the modules in Sparrow, Niethammer and Chaboki, m
h_{ad}	Heat transfer coefficient defined using $(T - T_{ad}), W/m^2 K$
h_i	Heat transfer coefficient on the i-th element, $W/m^2 K$, see eq (3.30)
h_{in}	Heat transfer coefficient defined using $(T - T_{in}), W/m^2 K$
h_∞	Heat transfer coefficient defined using $(T - T_\infty)\ W/m^2 K$
h_m	Heat transfer coefficient defined using $(T - T_m), W/m^2 K$
$\bar{h}$	Average heat transfer coefficient, $W/m^2 K$
k	Thermal conductivity, W/mK
K	Mass transfer coefficient, m/s
L	Length, m; also characteristic length in eq (3.33)
N	A counter, usually for row number
$\dot{q}_0$	Heat transfer rate at the wall, W
$\dot{q}_i''$	Surface heat flux on the i-th surface, W/m^2
$\dot{q}_j''$	Surface heat flux on the j-th surface, W/m^2
$\dot{q}_0''$	Surface heat flux on the o-th surface, W/m^2
r_h	Hydraulic radius $r_h \triangleq$ flow area/wetted perimeter, m
r_0	Tube radius, m
R_{c-a}	Resistance, case to ambient, K/W, see eq (3.31)
S	Spacing between centers of elements, in the streamwise direction, m
t	Height (thickness) of modules in Sparrow, Niethammer and Chaboki, m
T_{air}	Air temperature, K

T_i	Temperature of the i-th wall, K
T_e	Entrance temperature in a channel flow, K
T_{heated}	Temperature of heated element, K
T_{in}	Temperature at channel entrance, K
T_o	Wall temperature, or temperature of the o-th surface, K
T_∞	Temperature of the flow, far from the wall, K
T_{ad}	Adiabatic temperature of a surface, K
T_m	Mean fluid temperature, K
T_{max}	Maximum temperature in eq (3.28), K
T_r	Reference gas temperature in eq (3.28), K
T_1	Adiabatic temperature of first element demonstration of heated element.
U	Channel approach velocity or average velocity between top of component and opposite wall, depending on author, m/s
U_{array}	Array velocity, eq (3.37), m/s
V	Velocity, m/s
x^+	Dimensionless distance, see eq (3.18) See eq (3.31).

Dimensionless Groups

Nu	Nusselt number, usually based on D_n for internal flows
Nu_{ad}	Nusselt number based on h_{ad}
Nu_i	Nusselt number on the i-th surface
Nu_B	Nusselt number based on element height, B
Nu_L	Nusselt number based on length, L
Nu_x	Nusselt number based on x
Nu_i	Nusselt number based on the i-th wall
Nu_{ii}	Nusselt number on the i-th wall when only the i-th wall is heated
Nu_∞	See Nu_{ii}
Pr	Prandtl number
Re	Reynolds number based on hydraulic diameter, for internal flows

Re_B	Reynolds number based on B
Re_{H_1}	Reynolds number based on element length in flow direction, H_1, and average channel velocity between the top of the elements and the opposite wall.
Re_x	Reynolds number based on x
Re_L	Reynolds number based on L
Re_B	Reynolds number based on element height B, and approach velocity
Re_H	Reynolds number based on channel height, H, and approach velocity
Re_{H1}	Reynolds number based on H_1, the length of an element, and V, the velocity over the crest of the elements.
Sc	Schmidt number
Sh	Sherwood number

Greek

$\Delta T_{o,i}$	The i-th step increase in wall temperature, K
ξ	Location of the beginning of heating, in eqs (3.7), (3.8), and (3.12)
	or dummy axial position relative to beginning of heating, m
γ_m	An eigenvalue as, for example, in eq (3.25)
$\theta, \theta_1, \theta_0$	Temperature weighting functions. See eqs (3.19), (3.26), and (3.27)
θ_{ij}	Thermal wake function. The effect on the adiabatic temperature of the i-th element caused by heating of the j-th element.
θ_1	Adiabatic temperature rise of the first element downstream of a heated element, K (see eq (3.29)).

11.2 Part II

A	Area, m^2
b	Channel wall-to-wall spacing m; same as H
B	Element height, m
C_p	Specific heat at constant pressure, KJ/KgC
C_d	local channel drag coefficient
C_p	pressure-loss coefficient

$\bar{C}_d$	Channel drag coefficient averaged over the length of the channel
D_h	fA_c/P_w, hydraulic diameter, m
f_d	Local wall drag per unit width of channel, N/m^2
F_d	Total channel drag, N
F_d'	Total channel drag per unit channel length, N/m
g	Gravitational acceleration, m/s^2
g_c	Conversion constant
h	Heat transfer coefficient based on inlet channel temperature, W/m^2C
$h_{ad,o}$	Element heat-transfer coefficient, W/m^2C
H	Channel wall-to-wall spacing, m; same as b
k	Thermal conductivity, W/m^2C
L	Channel length, m
$\dot{m}$	Mass flow rate, Kg/s
$\dot{m}'$	Mass flow rate, Kg/s
M	Induced channel flow rate Re_0/Gr for UWT, Re_0/Gr^* for UHF; dimensionless induced channel flow rate
p	Static pressure, N/m^2
$\bar{p}$	$Nu_{opt}/Ra_{opt}^{1/2}$, as defined by Aung (1973) for the optimally spaced UHF channel
P	Ra/b^4, as defined by Bar-Cohen and Rosenhow (1984) for the UWT channel
P_w	Wetted perimeter, m
P_h	Heated perimeter, m
q	Wall heat flux, for smooth-walled channel, W/m^2
q''	Local channel power per unit width, per unit length, W/m^2
Q	Heat, W, or dimensionless flow rate; same as M
Q'	Total channel power per unit channel width, W/m
Q''	Total channel power per unit channel length and width, W
r_H	q_2/q_1, ratio of wall heat flux; $q_1 \geq q_2$
r_T	$(T_2 - T_0)/(T_1 - T_0)$, ratio of wall temperatures; $T_1 \geq T_2$

R Ra^*/b^5, as defined by Bar-Cohen and Rosenhow (1984) for the UHF channel

S Center-to-center spacing between elements, m, or $Gr_L/Gr_{L,max'}$ power distribution factor for channel with axially varying heat flux per Ortega (1986)

T Temperature, C or K

T_m Fluid mixed mean temperature, C

T_0 Fluid inlet temperature, C

u Streamwise velocity component, m/s

u_0 Average velocity at channel inlet, m/s

v Normal velocity component, m/s

x Channel streamwise coordinate, m

y Channel normal coordinate, m

Dimensionless Groups

Gr Channel Grashof number, UWT

$Gr_{ad,o}$ Element Grashof number,

Gr^* Channel Grashof number, UHF

Gr_L^* Modified channel length Grashof number

Gr_L Channel length Grashof number

Gr_x Local plate Grashof number

Nu Channel Nusselt number

$Nu_{ad,o}$ Element Nusselt number, where h is h_{ad} or h_0, depending on subscript

Nu_H Local channel Nusselt number

Nu_L Nusselt number evaluated at channel exit for smooth channel

$Nu_{L/2}$ Nusselt number evaluated at channel mid-height for smooth channel

Nu_m Local channel Nusselt number based on mixed mean temperature and channel spacing

Nu_x Local plate Nusselt number

Pr Prandtl number

Ra Modified channel Rayleigh number for UWT

Ra^* Modified channel Rayleigh number for UHF

Re_B Element Reynolds number

Re_H Channel Reynolds number; same as Re_0

Re_0 Inlet channel Reynolds number

Re_{D_h} Hydraulic diameter Reynolds number

Greek

β Thermal expansion coefficient, K^{-1}

θ dimensionless temperature for smooth channel
as defined by Aung et al (1973), or
thermal wake function for array of elements
per Moffat et al (1985)

θ_1 θ for element one row behind heated element

μ Dynamic viscosity $Kg/m - s$

ν Kinematic viscosity $Kg/m - s$

ρ Density Kg/m^3

ϕ Induced channel flow Graetz number
as defined by Miyatake and Fujii (1974)

Subscripts

ad Element adiabatic temperature

c Flow area

e Element

L At channel exit

m Mean, as in mixed mean fluid temperature

o At channel inlet or at infinity

T Total, as in total channel heat transfer rate

w At channel wall

∞ At infinity, far from the plate or channel entrance

1   At wall 1, the hotter wall, per Aung (1973)

2   At wall 2, the cooler wall, per Aung (1973)

12 REFERENCES

Aihara, T (1973). Effects of inlet boundary condition on numerical
solutions of free convection between vertical parallel plates,
Report of Inst High-Speed Mech **28**, Tohoku Univ., Japan,
1-27.

Aihara, T (1986). Laminar free convective heat transfer in vertical
uniform heat flux ducts (numerical solutions with constant/

temperature-dependent fluid properties), *Heat Transfer- Japanese Research* **15** (3).

Aihara, T (1987). Natural convection air cooling, *Int Symp Cooling Tech for Electronic Equipment Proc Pacific Inst for Thermal Eng, ASME, JSME,* Honolulu, Hawaii.

Anderson, A M and Moffat, R J (1987). Buoyancy-induced forced convection on an isolated plate, rough and smooth, *Proc 37th Elec Components Conf, IEEE Comp, Hybrids and Manufacturing Tech Society,* Cat. No. 87CH2448-9.

Arvizu, D E (1981). PhD thesis, Mech Eng Dept., Stanford Univ., Stanford, CA.

Arvizu, D E and Moffat, R J (1981). Experimental heat transfer from an array of heated cubical elements on an adiabatic channel wall, *Report No. HMT-33,* Thermosciences Div., Dept. of Mech Eng, Stanford Univ., Stanford, CA.

Arvizu, D E and Moffat, R J (1982). The use of superposition in calculating cooling requirements for circuit board mounted electronic components, *Proc 32nd Elec Components Conf,* San Diego, CA, 133-144.

Asako, Y and Faghri, M (1987). Three-dimensional heat transfer and fluid flow analysis of arrays of rectangular blocks encountered in electronic equipment, *ASME Pre-print 87-HT-73.*

Aung, W (1972). Fully developed laminar free convection between vertical flat plates heated asymmetrically, *Int J Heat and Mass Transfer* **15**, 1577–1580.

Aung, W (1973). Heat transfer in electronic systems with emphasis on asymmetric heating, *Bell System Tech J* **52** (6).

Aung, W, Fletcher, L S and Sernas, V (1972). Developing laminar free convection between vertical flat plates with asymmetric heating, *Int J Heat and Mass Transfer* **15**, 2293–2307.

Aung, W, Kessler, T J and Beitin, K I (1972). Natural convection cooling of electronic cabinets containing arrays of vertical circuit cards, *ASME 72-WA/HT-40, ASME Winter Annual Meeting, Heat Transfer Div.,* New York.

Aung, W, Kessler, T J and Beitin, K I (1973). Free convection cooling of electronic systems, *IEEE Trans Parts, Hybrids and Packaging* **PHP-9** (2), 75-86.

Bar-Cohen, A and Rohsenow, W M (1981). Thermally optimum spacing of vertical, natural convection cooled, parallel plates, *Heat Transfer in Electronic Equipment, ASME HTD 20*, ASME, WAM, Washington D.C., 11- 18.

Bar-Cohen, A and Rosenhow, W M (1984). Thermally optimum spacing of vertical, natural convection cooled, parallel plates, *ASME J Heat Transfer* 106 (1), 116-123.

Bibier, C A and Sammakia, B G (1986). Transport from discrete heated components in a turbulent channel flow, *ASME Preprint 86- WA/HT-68*.

Birnbreier, H (1981). Experimental investigations on the temperature rise of printed circuit boards in open cabinets with natural ventilation, *Heat Transfer in Electronic Equipment, ASME HTD 20, ASME WAM*, Washington D.C., 19-23.

Bodoia, J R and Osterle, J F (1962). The development of free convection between heated vertical plates, *J Heat Transfer, Trans ASME*, Series C 84 (1), 40-44.

Buller, M L and Kilburn, R F (1981). Evaluation of surface heat transfer coefficients for electronic module packages, *Heat Transfer in Electronic Equipment*, ASME HTD 20, ASME WAM, Washington, D.C.

Carey, V P and Mollendorf, J C (1977). The temperature field above a concentrated heat source on a vertical adiabatic surface, *Int J Heat and Mass Transfer* 20, 1059-1067.

Chang, M J, Shyu, R J and Fang, L J (1987). An experimental study of heat transfer from surface mounted components to a channel airflow, *ASME Preprint 87-HT-75, Natl Heat Transfer Conf.*

Churchill, S W and Chu, H H (1975). Correlating equations for laminar and turbulent free convection from a vertical plate, *Int J Heat Mass Transfer* 18, 1323-1329.

Churchill, S W and Usagi, R (1972). A general expression for the correlation of rates of transfer and other phenomena, *AIChE J* 18 (6), 1121-1138.

Davalath, J and Bayazitoglu, Y (1987). Forced convection cooling across rectangular blocks, *ASME J Heat Transfer* 109, 321-328.

Elenbaas, W (1942). Heat dissipation of parallel plates by free convection, *Physica* IX (1), 1-28.

Ellison, G N (1984). *Thermal Computations for Electronic Equipment*, Van Nostrand Reinhold Co., New York.

Engel, R K and Mueller, W K (1967). An analytical investigation of natural convection in vertical channels, *ASME Paper 67-HT-16*.

Fujii, T, Fujii, M and Takenchi, M (1973). Influence of various surface roughness on natural convection, *Int J Heat and Mass Transfer* 16, 629.

Gebhart, B (1971). *Heat Transfer*, 2nd Edition, McGraw-Hill, New York.

Ghaddar, N K, Korczak, K Z, Mikic, B and Patera, A T (1986). Numerical investigation of incompressible flow in grooved channels: Stability and self-sustained oscillations, *J Fluid Mechanics* 163, 99.

Gosman, A D and Ideriah, F J K (1976). TEACH-T: A general purpose computer program for two-dimensional, turbulent, recirculating flows, *Calculation of Recirculating Flows*, Dept. of Mech Eng, Imperial College of London.

Heaton, H S, Reynolds, W C and Kays, W M (1964). Heat transfer in annular passages. Simultaneous development of velocity and temperature fields in laminar flow, *Int J Heat and Mass Transfer* 7, 763-781.

Horton, S F (1981). *Natural Convection from Parallel Plates with Grooved Surfaces*, MSc Thesis, Dept. of Mech Eng, Mass. Inst of Technology.

Ishizuka, M, Miyakzaki, Y and Sasaki, T (1984). On the cooling of natural air-cooled electronic equipment casings-proposal of a practical formula for thermal design, *JSME Bulletin* 29 (247), 119-123.

Ishizuka, M, Miyazaki, Y and Sasaki, T (1986). Aerodynamic resistance for perforated plates and wire nettings in natural convection, *ASME Paper 84-WA/HT-87, ASME Winter Annual Meeting*.

Jaluria, Y (1982). Thermal plume interaction with vertical surfaces, *Letters Heat and Mass Transfer* 9 (2), 207- 118.

Jaluria, Y (1982). Buoyancy-induced flow due to isolated thermal sources on a vertical surface, *ASME J Heat Transfer* **104** (2), 223-227.

Jaluria, Y (1985). Natural convective cooling of electronic equipment, *Natural Convection, Fundamentals and Applications*, S Kakac, W Aung and R Viskanta, eds., Hemisphere, New York.

Johnson, A E and Torok, D (1985). Software for fluid flow and heat transfer analyses for electronic packaging, *Computers in Mech Eng*, 41-46.

Johnson, C E (1986). Evaluation of correlations for natural convection cooling of electronic equipment, *Heat Transfer in Electronic Equipment-1986*, A Bar-Cohen, ed., *ASME HTD* **57**.

Kays, W M and Crawford, M E (1980). *Convective Heat and Mass Transfer*, 2nd Edition, McGraw-Hill, New York.

Kays, W M and London, A L (1964). *Compact Heat Exchangers*, 2nd Edition, McGraw-Hill, New York.

Kettleborough, C F (1972). Transient laminar free convection between heated vertical plates including entrance effects, *Int J Heat and Mass Transfer* **15** (5), 883-896.

Kraus, A D and Bar-Cohen, A (1983). *Thermal Analysis and Control of Electronic Equipment*, McGraw-Hill, New York.

Lehmann, G L and Wirtz, R A (1984). Convection from surface mounted repeating ribs in a channel flow, *ASME Preprint 84-WA/HT- 88*.

Liu, C K, Kline, S J and Johnston, J P (1966). An experimental study of turbulent boundary layers on rough walls, *Thermosciences Div. Report MD-15*, Stanford Univ., CA. Also available through university microfilms as PhD thesis of C K Liu.

Lundberg, R E, Reynolds, W C and Kays, W M (1963). *NASA TN D-1972*.

Miyatake, O and Fujii, T (1972). Free convective heat transfer between vertical plates - one plate isothermally heated and the other thermally insulated, *Heat Transfer - Japanese Research* **1** (3), 30-38.

Miyatake, O and Fujii, T (1973). Natural convection heat transfer between vertical parallel plates at unequal uniform temperatures, *Heat Transfer - Japanese Research* **2** (4), 79-88.

Miyatake, O and Fujii, T (1974). Natural convective heat transfer between vertical parallel plates with unequal heat fluxes, *Heat Transfer - Japanese Research* **3** (3), 29- 33.

Miyatake, O, Fujii, T, Fujii, M and Tanaka, H (1973). Natural convective heat transfer between vertical parallel plates - one plate with a uniform heat flux and the other thermally insulated, *Heat Transfer - Japanese Research* **2** (1), 25-33.

Moffat, R J (1982). Contributions to the theory of single-sample uncertainty analysis, *1980-81 AFOSR-HTTM Conf Complex, Turbulent Flows: Comparison of Computation and Experiment,* Stanford Univ., CA.

Moffat, R J, Arvizu, D E and Ortega, A (1985). Cooling electronic components: Forced convection experiments with an air-cooled array, *23rd AICHE/ASME Natl Heat Transfer Conf,* Denver, CO.

Moffat, R J and Ortega, A (1986). Buoyancy-induced forced convection, *Heat Transfer in Electronic Equipment - 1986,* A Bar-Cohen, ed., *ASME HTD* **57**.

Ortega, A (1985). Calibration of hot-wire anemometers for air velocities less than 0.5 m/s, including temperature compensation, *IL Report 82,* Thermosciences Div., Mech Eng Dept., Stanford Univ., CA.

Ortega, A and Moffat, R J (1985). Heat transfer from an array of simulated electronic components: Experimental results for free convection with and without a shrouding wall, *Heat Transfer in Electronic Equipment - 1985,* S Oktay and R J Moffat, eds., *ASME HTD* **48**.

Ortega, A and Moffat, R J (1986). Experiments on buoyancy-induced convection heat transfer from an array of cubical elements on a vertical channel wall, *Report HMT-38,* Thermosciences Div., Dept. Mech Eng, Stanford Univ., CA.

Ortega, A and Moffat, R J (1986). Buoyancy-induced convection in a nonuniformly heated array of cubical elements on a vertical channel wall, *Heat Transfer in Electronic Equipment - 1986,* A Bar-Cohen, Ed., *ASME HTD* **57**.

Piatt, J D (1986). *Heat Transfer from an Array of Heated Cylindrical Elements on an Adiabatic Channel Wall,* Masters Thesis, Naval Postgraduate School, Monterey, CA.

Quintiere, J G and Mueller, W K (1973). An analysis of laminar free and forced convection between finite vertical parallel plates, *ASME J Heat Transfer* 95, 53-59.

Raithby, G D and Hollands, K G T. A general method of obtaining approximate solutions to laminar and turbulent free convection problems, *Advances in Heat Transfer* 11, T F Irvine and J P Hartnett, eds., Academic Press, New York, 266-315.

Ravine, T L and Richards, D E (1984). Natural convection cooling of a finite-sized thermal source on the wall of a vertical channel, *ASME Paper 84-WA/HT-90*.

Schmidt, R C and Patankar, S V. A numerical study of laminar forced convection across heated rectangular blocks in two-dimensional ducts, *ASME Paper 86-WA/HT-88*.

Sellars, R J, Tribus, M and Klein, J S (1956). *Trans ASME* 78, 441-448.

Shah, R K and London, A L (1978). *Laminar Flow Forced Convection in Ducts: A Source Book for Compact Heat Exchanger Analytical Data*, Academic Press, New York.

Shakerin, S, Bohn, M and Loehrke, R I (1986). Natural convection in an enclosure with discrete roughness elements on a vertical, heated wall, *Proc 8th Int Heat Transfer Conf* 4, San Francisco, CA, 1519-1525.

Sparrow, E M and Ansari, M A (1983). A refutation of King's rule for multi-dimensional external natural convection, *Int J Heat Mass Transfer* 26, 1357-1364.

Sparrow, E M and Boessneck, D S (1983). Effect of transverse misalignment on natural convection from a pair of parallel, vertically stacked, horizontal cylinders, *ASME J Heat Transfer* 105 (2), 241-247.

Sparrow, E M and Chukaev, A (1980). Forced-convection heat transfer in a duct having spanwise periodic rectangular protuberances, *Numerical Heat Transfer* 3, 149-167.

Sparrow, E M, Cook, D S and Chrysler, G M (1982). Heat transfer by natural convection from an array of short, wall-attached horizontal cylinders, *ASME J Heat Transfer* 104 (1), 125-131.

Sparrow, E M, Niethhammer, J E and Chaboki, A (1982). Heat transfer and pressure drop characteristics of arrays of rectangular modules encountered in electronic equipment, *Int J Heat and Mass Transfer* **25** (7), 961-973.

Sparrow, E M and Pfeil, D R (1984). Enhancement of natural convection heat transfer from a horizontal cylinder due to vertical shrouding surfaces, *Trans ASME J Heat Transfer* **105** (1), 124-130.

Steinberg, D S (1980). *Cooling Techniques for Electronic Equipment*, John Wiley and Sons, New York.

Wirtz, R A and Dykshoorn, P (1984). Heat transfer and pressure drop characteristics of arrays of rectangular modules encountered in electronic equipment, *Int J Heat and Mass Transfer* **25** (7), 961-973.

Wirtz, R A and Stutzman, R J (1982). Experiments on free convection between vertical plates with symmetric heating, *ASME Trans J Heat Transfer* **104**, 501-507.

Yao, L S (1983). Free and forced convection in the entry region of a heated vertical channel, *Int J Heat and Mass Transfer* **26** (1), 65-72.

Yao, L S (1983). Buoyancy effects on a boundary layer along an infinite cylinder with a step change of surface temperature, *J Heat Transfer* **105**, 96.

Zukauskaus, A A (1972). Heat transfer for tubes in crossflow, *Advances in Heat Transfer* **8**, Academic Press, New York.

13 BIBLIOGRAPHY

Acrivos, A (1958). Combined laminar free and forced convection heat transfer in external flows. *AIChE J.* **4**, 285- 289.

Afzal, N (1980). Convective wall plume: Higher order analysis, *Int J Heat and Mass Transfer* **23**, 505-513.

Afzal, N (1983). Mixed convection in an axisymmetric buoyant plume, *Int J Heat and Mass Transfer* **26** (3), 381-388.

Aihara, T (1973). Effects of inlet boundary condition on numerical solutions of free convection between vertical parallel plates, *Report of Inst High-Speed Mech* **28**, Tohoku Univ., Japan, 1-27.

Aihara, T (1986). Laminar free convective heat transfer in vertical uniform heat flux ducts (numerical solutions with constant/

temperature-dependent fluid properties), *Heat Transfer-Japanese Research* **15** (3).

Aihara, T (1987). Natural convection air cooling, *Int Symp Cooling Tech for Electronic Equipment Proc Pacific Inst for Thermal Eng, ASME, JSME*, Honolulu, Hawaii.

Anderson, A (1987). Private communication, based on work in progress on the same apparatus.

Anderson, A M and Moffat, R J (1987). Buoyancy-induced forced convection on an isolated plate, rough and smooth, *Proc 37th Elec Components Conf, IEEE Comp, Hybrids and Manufacturing Tech Society*, Cat. No. 87CH2448-9.

Arvizu, D E (1981). PhD thesis, Mech Eng Dept., Stanford Univ., CA.

Arvizu, D E and Moffat, R J (1981). Experimental heat transfer from an array of heated cubical elements on an adiabatic channel wall, *Report No. HMT-33*, Thermosciences Div., Dept. of Mech Eng, Stanford Univ., CA.

Arvizu, D E and Moffat, R J (1982). The use of superposition in calculating cooling requirements for circuit board mounted electronic components, *Proc 32nd Elec Components Conf*, San Diego, CA, 133-144.

Asako, Y and Faghri, M (1987). Three-dimensional heat transfer and fluid flow analysis of arrays of rectangular blocks encountered in electronic equipment, *ASME Pre-print 87-HT-73*.

Aung, W (1972). Fully developed laminar free convection between vertical flat plates heated asymmetrically, *Int J Heat and Mass Transfer* **15**, 1577–1580.

Aung, W (1973). Heat transfer in electronic systems with emphasis on asymmetric heating, *Bell System Tech J* **52** (6).

Aung, W and Chimah, B (1981). Laminar heat exchange in vertical channels – application to cooling of electronic systems, *Proc Advanced Study Inst on Low Reynolds Number Forced Convection in Channels and Bundles*, Ankara, Turkey.

Aung, W, Fletcher, L S and Sernas, V (1972). Developing laminar free convection between vertical flat plates with asymmetric heating, *Int J Heat and Mass Transfer* **15**, 2293–2307.

Aung, W, Kessler, T J and Beitin, K I (1973). Free convection cooling of electronic systems, *IEEE Trans Parts, Hybrids and Packaging* **PHP-9** (2), 75-86.

Badr, R M (1983). A theoretical study of laminar mixed convection from a horizontal cylinder in a cross-stream, *Int J Heat and Mass Transfer* **26** (5), 639-653.

Bailey, F and Baker, E (1972). Liquid cooling of microelectronic devices by free and forced convection, *Microelectronic Reliability* **11**, 213.

Bar-Cohen, A (1977). Constant heating transient natural convection in partially insulated vertical enclosures, *ASME Paper 77-HT-35, ASME Heat Transfer Div., AIChE-ASME Heat Transfer Conf*, Salt Lake City, UT.

Bar-Cohen, A, Kraus, A D and Davidson, S F (1983). Thermal frontiers in the design and packaging of microelectronic equipment, *Mech Eng* **105** (6), 53-59.

Bar-Cohen, A and Rosenhow, W M (1981). Thermally optimum spacing of vertical, natural convection cooled, parallel plates, *Heat Transfer in Electronic Equipment, ASME HTD* **20**, ASME WAM, Washington D.C., 11- 18.

Bar-Cohen, A and Rosenhow, W M (1984). Thermally optimum spacing of vertical, natural convection cooled, parallel plates, *ASME J Heat Transfer* **106** (1), 116-123.

Beck, J V (1979). Effects of multiple sources in the contact conductance theory, *Trans ASME J Heat Transfer* **101C**, 132-136.

Beckett, P M and Friend, I E (1984). Combined natural and forced convection between parallel walls: Developing flow at higher Rayleigh numbers, *Int J Heat Mass Transfer* **27** (4), 611-621.

Bejan, A (1984). *Convection Heat Transfer*, John Wiley and Sons, New York.

Bergeles, A E, Chu, R C and Seeley, J H (1972). Survey of heat transfer techniques applied to electronic equipment, *ASME Paper 72-WA/HT-39, ASME WAM*, New York.

Bibier, C A and Sammakia, B G (1986). Transport from discrete heated components in a turbulent channel flow, *ASME Preprint 86- WA/HT-68*.

Bill, R G and Gebhart, B (1979). The development of turbulent transport in a vertical natural convection boundary layer, *Int J Heat Mass Transfer* **22**, 267-277.

Birnbreier, H (1981). Experimental investigations on the temperature rise of printed circuit boards in open cabinets with natural ventilation, *Heat Transfer in Electronic Equipment, ASME HTD* **20**, *ASME WAM* , Washington D.C., 19-23.

Bodoia, J R and Osterle, J F (1962). The development of free convection between heated vertical plates, *J Heat Transfer, Trans ASME*, Series C **84** (1), 40-44.

Braaten, M E and Patankar, S V (1984). Analysis of laminar mixed convection in shrouded arrays of heated rectangular blocks, *ASME HTD* **32**, *22nd Natl Heat Transfer Conf* , 77-84.

Brosh, A, Degani, D and Zalmanovich, S (1982). Conjugate heat transfer in a laminar boundary layer with heat source at the wall, *Trans ASME J Heat Transfer* **104**, 90-95.

Buller, M L and Duclos, T G (1982). Thermal characteristics of horizontally oriented electronic components in an enclosed environment, *Proc 32nd Electronic Components Conf*, San Diego, CA, 153-157.

Campo, W C, Kerjilian, G and Shaukatullah, H (1982). Prediction of component temperatures on circuit cords cooled by natural convection, *Proc 32nd Electronic Components Conf*, San Diego, CA, 150-152.

Carey, V P and Mollendorf, J C (1977). The temperature field above a concentrated heat source on a vertical adiabatic surface, *Int J Heat and Mass Transfer* **20**, 1059-1067.

Carpenter, J R, Briggs, D G and Sernas, V (1976). Combined radiation and developing laminar free convection between vertical flat plates with asymmetric heating, *ASME J Heat Transfer* **98**, 95-100.

Chang, M J, Shyu, R J and Fang, L J (1987). An experimental study of heat transfer from surface mounted components to a channel airflow, *ASME Preprint 87-HT-75, Natl Heat Transfer Conf.*

Cheesewright, R (1968). Turbulent natural convection from a vertical plane surface, *Trans ASME, J Heat Transfer* **90**, 1-8.

Chow, L C, Husain, S R and Campo, A (1984). Effects of free convection and axial conduction on forced-convection heat transfer inside a vertical channel at low Peclet numbers, *Trans ASME J Heat Transfer* **106**, 297-303.

Chu, H H, Churchill, S W and Patterson, C V S (1976). The effect of heater size, location, aspect ratio and boundary conditions on two dimensional, laminar natural convection in rectangular channels, *ASME J of Heat Transfer* **98** (2), 194-201.

Churchill, S W (1977). A comprehensive correlating equation for buoyancy induced flow in channels, *Letters Heat and Mass Transfer* **4** (3), 193-200.

Churchill, S W (1977). Comprehensive correlating equation for laminar, assisting, forced and free convection, *AIChE J* **23** (1), 10-16.

Churchill, S W and Chu, H H (1975). Correlating equations for laminar and turbulent free convection from a vertical plate, *Int J Heat Mass Transfer* **18**, 1323-1329.

Churchill, S W and Usagi, R (1972). A general expression for the correlation of rates of transfer and other phenomena, *AIChE J* **18** (6), 1121-1138.

Coyne, J C (1984). An analysis of circuit board temperatures in electronic equipment frames cooled by natural convection, *ASME HTD* **32**, *22nd Natl Heat Transfer Conf* **32**, 59-65.

Currie, I G and Newman, W A (1970). Natural convection between isothermal vertical surfaces, *Paper NC-2-7, 4th Int Heat Transfer Conf Proc*, Elsevier Publishing Co.

Dalbert, A M, Penot, F and Peube, J L (1981). Convection naturelle laminaire dans un canal vertical chauffe a flux constant, *Int J Heat Mass Transfer* **24** (9), 1463-1473.

Dalbert, A M (1982). Natural, mixed and forced convection in a vertical channel with asymmetric uniform heating, *Proc 7th Int Heat Transfer Conf* **3**, Hemisphere, New York, 431-434.

Davalath, J and Bayazitoglu, Y (1987). Forced convection cooling across rectangular blocks, *ASME J Heat Transfer* **109**, 321-328.

Davis, L P and Perona, J P (1971). Development of free convective flow of a gas in a heated vertical open tube, *Int J Heat and Mass Transfer* **14**, 889-903.

Dyer, J R (1975). The development of laminar natural convective flow in a vertical uniform heat flux duct, *Int J Heat and Mass Transfer* **18**, 1455-1465.

Dyer, J R (1978). Natural convection flow through a vertical duct with a restricted entry, *Int J Heat and Mass Transfer* **21**, 1341-1354.

Eichhorn, R and Wang, G S (1979). Natural convection from vertical plates with semi-circular leading edges, *Studies in Heat Transfer*, J P Hartnett, et al., eds., McGraw-Hill, New York, 187-198.

Elenbaas, W (1942). Heat dissipation of parallel plates by free convection, *Physica* **IX** (1), 1-28.

Ellison, G N (1976). Theoretical calculation of the thermal resistance of a conducting and convecting surface, *IEEE Trans Parts, Hybrids and Packaging* **PHP-12** (3), 265-266.

Ellison, G N (1984). *Thermal Computations for Electronic Equipment*, Van Nostrand Reinhold Co., New York.

Engel, R K (1965). *The Development of Natural Convection on Vertical Flat Plates and in Vertical Channels*, PhD thesis, School of Engineering and Science, New York Univ., New York.

Engel, R K and Mueller, W K (1967). An analytical investigation of natural convection in vertical channels, *ASME Paper 67-HT-16*.

Farouk, B and Guceri, S I (1983). Natural convection from horizontal cylinders in interacting flow fields, *Int J Heat and Mass Transfer* **26** (2), 231-243.

Flack, R D and Turner, B L (1980). Heat transfer correlations for use in naturally cooled enclosures with high-power integrated circuits, *IEEE Trans Components, Hybrids and Manufacturing Tech* **CHMT-3** (3), 449-452.

Frank, L E (1985). CINDA enhancements at Hughes for high-power density electronic packaging, *ASME HTD* **48**, *ASME Natl Heat Transfer Conf*, 61-68.

Fujii, T, Fujii, M and Takenchi, M (1973). Influence of various surface roughness on natural convection, *Int J Heat and Mass Transfer* **16**, 629.

Gebhart, B, Audunson, T and Pera, L (1970). Forced, mixed, and natural convection from long horizontal wires: Experiments at various Prandtl numbers, *Paper NC-3.2, 4th Int Heat Transfer Conf*, Paris.

Gebhart, B, Pera, L and Schorr, A W (1970). Steady laminar natural convection plumes above a horizontal line heat source, *Int J Heat and Mass Transfer* **13**, 161-177.

Gebhart, B, Shaukatullah, H and Pera, L (1976). Interaction of unequal laminar plane plumes, *Int J Heat and Mass Transfer* **19**, 751-756.

George, W K and Capp, S P (1979). A theory for natural convection turbulent boundary layers next to heated vertical surfaces, *Int J Heat and Mass Transfer* **22**, 813-826.

Ghaddar, N K, Korczak, K Z, Mikic, B and Patera, A T (1986). Numerical investigation of incompressible flow in grooved channels: Stability and self-sustained oscillations, *J Fluid Mechanics* **163**, 99.

Gosman, A D and Ideriah, F J K (1976). TEACH-T: A general purpose computer program for two-dimensional, turbulent, recirculating flows, *Calculation of Recirculating Flows*, Dept. of Mech Eng, Imperial College of London.

Greif, R (1978). An experimental and theoretical study of heat transfer in vertical tube flows, *ASME J Heat Transfer* **100**, 86-91.

Grella, J J and Faeth, G M (1975). Measurements in a two-dimensional thermal plume along a vertical adiabactic wall, *J Fluid Mechanics* **71**, 701-710.

Gtyzagordis, J (1977). Velocity profiles on a vertical plate in combined convection, *Letters Heat Mass Transfer* **4** (6), 477-480.

Haaland, S E and Sparrow, E M (1983). Mixed convection plume above a horizontal line source situated in a forced convection approach flow, *Int J Heat and Mass Transfer* **26** (3), 433-444.

Heaton, H S, Reynolds, W C and Kays, W M (1964). Heat transfer in annular passages. Simultaneous development of velocity and temperature fields in laminar flow, *Int J Heat and Mass Transfer* **7**, 763-781.

Hein, V L (1967). Convection and conduction cooling of substrates containing multiple heat sources, *Bell System Tech J* **XLVI** (8), 1659-1678.

Hetherington, H J and Patten, T D (1976). Laminar flow natural convection from the open vertical cylinder with uniform heat flux at the wall, *Int J Heat and Mass Transfer* **19**, 1121- 1125.

Hieber, C A and Gebhart, B (1969). Mixed convection from a sphere at small Reynolds and Grashof numbers, *J Fluid Mechanics* **38**, Part 1, 137-159.

Horton, S F (1981). *Natural Convection from Parallel Plates with Grooved Surfaces*, MSc Thesis, Dept. of Mech Eng, Mass. Inst of Technology.

Hsieh, C K and Coldeway, R W (1977). The natural convection of air over a heated plate with forward facing step, *ASME J Heat Transfer* **99** (3), 439.

Hunt, R and Wilks, G (1982). Mixed convection - a comparison between experiment and an exact numerical solution of the boundary layer equations, *Letters in Heat and Mass Transfer* **9**, 291-298.

Hwang, V P (1979). Effect of flow direction on mixed convection heat transfer from a vertical flat plate and a plate with square protuberances, *Letters Heat and Mass Transfer* **6** (6), 459-468.

Ishizuka, M, Miyakzaki, Y and Sasaki, T (1984). On the cooling of natural air-cooled electronic equipment casings-proposal of a practical formula for thermal design, *JSME Bulletin* **29** (247), 119-123.

Ishizuka, M, Miyazaki, Y and Sasaki, T (1986). Aerodynamic resistance for perforated plates and wire nettings in natural convection, *ASME Paper 84-WA/HT-87, ASME Winter Annual Meeting*.

Jackson, T W, Harrison, W B and Boteler, W C (1958). Combined free and forced convection in a constant-temperature vertical tube, *ASME J Heat Transfer* **80**, 739-745.

Jackson, T W and Yen, H H (1971). Combining forced and free convection equations to represent combined heat transfer coefficients for a horizontal cylinder, *ASME J Heat Transfer* **93**, 247-248.

Jaluria, Y (1976). Natural convection flow interaction above a heated body, *Letters Heat and Mass Transfer* **3** (5), 457-466.

Jaluria, Y (1980). *Natural Convection Heat and Mass Transfer*, Pergamon Press, New York.

Jaluria, Y (1981). Mixed convection in a wall plume, *Paper ASME 81-HT-37, 20th ASME/AIChE Natl Heat Transfer Conf*, Milwaukee, WI.

Jaluria, Y (1982). Thermal plume interaction with vertical surfaces, *Letters Heat and Mass Transfer* 9 (2), 207-118.

Jaluria, Y and Gebhart, B (1977). Buoyancy induced flow arising from a line thermal source on an adiabatic vertical surface, *Int J Heat and Mass Transfer* 20, 153-157.

Joffre, R J and Barron, R F (1967). Effects of roughness on free convection, *ASME Paper 67-WA/HT-38*.

Johnson, A E and Torok, D (1985). Software for fluid flow and heat transfer analyses for electronic packaging, *Computers in Mech Eng*, 41-46.

Johnson, C E (1986). Evaluation of correlations for natural convection cooling of electronic equipment, *Heat Transfer in Electronic Equipment-1986*, A Bar-Cohen, ed., *ASME HTD* 57.

Kao, T T (1975). Laminar free convection heat transfer response along a vertical flat plate with step jump in surface temperature, *Letters Heat and Mass Transfer* 2 (5), 419-428.

Karniadakis, G, Mikic, B and Patera, A T (1987). Minimum dissipation transport enhancement by flow destabilization: Reynolds analogy revisited. Accepted for publication by *J Fluid Mechanics*.

Kays, W M and Crawford, M E (1980). *Convective Heat and Mass Transfer*, 2nd Edition, McGraw-Hill, New York.

Kays, W M and London, A L (1964). *Compact Heat Exchangers*, 2nd Edition, McGraw-Hill, New York.

Kelleher, M (1971). Free convection from a vertical plate with discontinuous wall temperature, *J Heat Transfer, Trans ASME* 93 (4), 349-356.

Kennedy, K J and Zebib, A (1983). Combined free and forced convection between horizontal parallel planes: Some case studies, *Int J Heat and Mass Transfer* 26 (3), 471- 474.

Kettleborough, C F (1972). Transient laminar free convection between heated vertical plates including entrance effects, *Int J Heat and Mass Transfer* 15 (5), 883-896.

Kline, S J and McClintock, F A (1953). Describing uncertainties in single-sample experiments, *Mechanical Engineering*, 3-8.

Kozlu, H, Mikic, B and Patera, A T. Minimum dissipation heat removal by scale-matched flow destabilization, *Int J Heat and Mass Transfer.*

Kraus, A D (1965). *Cooling Electronic Equipment*, Prentice- Hall, Englewood Cliffs, New Jersey.

Kraus, A D and Bar-Cohen, A (1983). *Thermal Analysis and Control of Electronic Equipment*, McGraw-Hill, New York.

Kraus, A D, Bergles, A E and Mollendorf, J C (1977). *Directions of Heat Transfer in Electronic Equipment*, Report of Research Workshop, College of Engineering, Univ. of South Florida, Tampa, FL.

Kraus, A D, Chu, R C and Bar-Cohen, A (1982). Thermal management of microelectronics: Past, present and future, *Computers in Mechanical Engineering*, 69-79.

Kreith, F (1962). *Radiation Heat Transfer for Spacecraft and Solar Power Plant Design*, Int Textbook Co., Scranton, PA.

Krisnamurth, R and Gebhart, B (1984). Mixed convection in wall plumes, *Int J Heat and Mass Transfer* **27** (10), 1679-1689.

Lab Report, ME 268 (1981). *Heat Transfer from Two Dimensional Roughness Elements*, cited in Arvizu and Moffat, HMT-33.

Lauber, T S and Welch, A U (1966). Natural convection heat transfer between vertical flat plates with uniform heat flux, *Proc 3rd Int Heat Transfer Conf*, Chicago, IL, 126-131.

Lawrence, W T and Chato, J C (1966). Heat transfer effects on the developing laminar flow inside vertical tubes, *ASME J Heat Transfer*, 214-222.

Lee, T S (1979). *Natural Convection in a Vertical Channel with Heat and Mass Transfer and Flow Reversal*, PhD Thesis, Dept. of Aeronautics and Astronautics, Stanford Univ., CA.

Lehmann, G L and Wirtz, R A (1984). Convection from surface mounted repeating ribs in a channel flow, *ASME Preprint 84-WA/HT-88.*

Levy, E K (1971). Optimum plate spacings for laminar natural convection heat transfer from parallel vertical isothermal flat plates, *J Heat Transfer, Trans ASME*, Series C **93** (4), 463-465.

Liburdy, J A and Faeth, G M (1975). Theory of a steady laminar thermal plume along a vertical adiabatic wall, *Letters Heat and Mass Transfer* **2** (5), 407-418.

Lieberman, J and Gebhart, E (1969). Interactions in natural convection from an array of heated elements, experimental, *Int J Heat and Mass Transfer* **12**, 1385-1396.

Lin, F N (1975). Laminar free convection over two-dimensional bodies with uniform surface heat flux, *Letters Heat and Mass Transfer* **3** (1), 59-68.

Liu, C K, Kline, S J and Johnston, J P (1966). An experimental study of turbulent boundary layers on rough walls, *Thermosciences Div. Report MD-15*, Stanford Univ., CA. Also available through university microfilms as PhD thesis of C K Liu.

Lloyd, J R and Sparrow, E M (1970). Combined forced and free convection flow on vertical surfaces, *Int J Heat and Mass Transfer* **13**, 434-438.

Lundberg, R E, Reynolds, W C and Kays, W M (1963). *NASA TN D-1972*.

Marner, W J and McMillan, H K (1970). Combined free and forced laminar convection in a vertical tube with constant wall temperature, *ASME J Heat Transfer* **92**, 559-562.

Marsters, G F (1972). Arrays of heated horizontal cylinders in natural convection, *Int J Heat and Mass Transfer* **15** (5), 921-933.

Mehta, R C and Bose, T K (1983). Temperature distribution in a large circular plated heated by a disk heat source, *Int J Heat and Mass Transfer* **26** (7), 1093-1095.

Melvin, L and Ortega, A (1985). Smoke-wire flow visualization in an array of wall-mounted cubical elements at low Reynolds number, *Report IL-81*, Thermosciences Div./HTTM, Mech Eng Dept., Stanford Univ., CA.

Meric, R A (1977). An analytical study of natural convection in a vertical open tube, *Int J Heat and Mass Transfer* **20**, 429-431.

Merkin, J H (1983). Free convection boundary layers over humps and indentations, *Quarterly J Mechanics and Applied Mathematics* **36** (71).

Metais, B and Eckert, E R G (1964). Forced, mixed and free convection regimes, *ASME J Heat Transfer* **86**, 295- 296.

Miyatake, O and Fujii, T (1973). Natural convection heat transfer between vertical parallel plates at unequal uniform temperatures, *Heat Transfer - Japanese Research* **2** (4), 79-88.

Miyatake, O and Fujii, T (1974). Natural convective heat transfer between vertical parallel plates with unequal heat fluxes, *Heat Transfer - Japanese Research* **3** (3), 29- 33.

Miyatake, O, Fujii, T, Fujii, M and Tanaka, H (1973). Natural convective heat transfer between vertical parallel plates - one plate with a uniform heat flux and the other thermally insulated, *Heat Transfer - Japanese Research* **2** (1), 25-33.

Moffat, R J (1980). *Experimental Methods in the Thermosciences*, Dept of Mech Eng, Thermosciences Div., Stanford Univ., CA.

Moffat, R J (1982). Contributions to the theory of single-sample uncertainty analysis, *1980-81 AFOSR-HTTM Conf Complex, Turbulent Flows: Comparison of Computation and Experiment*, Stanford Univ., CA.

Moffat, R J, Arvizu, D E and Ortega, A (1985). Cooling electronic components: Forced convection experiments with an air-cooled array, *23rd AICHE/ASME Natl Heat Transfer Conf*, Denver, CO.

Moffat, R J and Ortega, A (1986). Buoyancy-induced forced convection, *Heat Transfer in Electronic Equipment - 1986*, A Bar-Cohen, ed., *ASME HTD* **57**.

Morton, B R (1960). Laminar convection in uniformly heated vertical pi;es, *J Fluid Mechanics* **8**, 227-240.

Nakamura, H, Asako, Y and Naitou, T (1982). Heat transfer by free convection between two parallel flat plates, *Numerical Heat Transfer* **5**, 95-106.

Nakayama, W (1986). Thermal management of electronic equipment: A review of technology and research topics, *Applied Mechanics Reviews* **39** (12).

Noronha, R I (1964-65). Free convective cooling of cabinets containing heat-dissipating components, *Proc Inst. Mech Eng* **179-1** (13), 439-450.

Ofi, O and Hetherington, H J (1977). Application of the finite element method to natural convection heat transfer from the open vertical channel, *Int J Heat and Mass Transfer* **20**, 1195-1204.

Ojalvo, M S, Anand, D K and Dunbar, R P (1967). Combined forced and free turbulent convection in a vertical circular tube with volumetric heat sources and constant wall heat addition, *J Heat Transfer*, 328-334.

Oosthuizen, P H and Madan, S (1971). The effect of flow direction on combined convective heat transfer from cylinders in air, *ASME J Heat Transfer*, 240-242.

Ormiston, S J, Raithby, G D and Hollands, K G T (1985). Numerical predictions of natural convection in a Trombe wall system, *ASME Paper 85-HT-36, ASME Natl Heat Transfer Conf*, Denver, CO.

Ortega, A (1985). Calibration of hot-wire anemometers for air velocities less than 0.5 m/s, including temperature compensation, *IL Report 82*, Thermosciences Div., Mech Eng Dept., Stanford Univ., CA.

Ortega, A and Moffat, R J (1985). Heat transfer from an array of simulated electronic components: Experimental results for free convection with and without a shrouding wall, *Heat Transfer in Electronic Equipment - 1985*, S Oktay and R J Moffat, eds., *ASME HTD* **48**.

Ortega, A and Moffat, R J (1986). Experiments on buoyancy-induced convection heat transfer from an array of cubical elements on a vertical channel wall, *Report HMT-38*, Thermosciences Div., Dept. Mech Eng, Stanford Univ., CA.

Ortega, A and Moffat, R J (1986). Buoyancy-induced convection in a nonuniformly heated array of cubical elements on a vertical channel wall, *Heat Transfer in Electronic Equipment - 1986*, A Bar-Cohen, Ed., *ASME HTD* **57**.

Ostrach, S (1952). Laminar natural convection flow and heat transfer of fluids with and without heat sources in channels with constant wall temperatures, *NASA Technical Note 2863*.

Ostrach, S (1954). Combined natural and forced convection laminar flow and heat transfer of fluids with and without heat sources in channels with linearly varying wall temperature, *NASA Technical Note 3141*.

Ostrach, S (1964). *Laminar Flows with Body Forces, Theory of Laminar Flows in High Speed Aerodynamics and Jet Propulsion*, F K Moore, ed. 4, Princeton, 528-718.

Patankar, S V (1980). *Numerical Heat Transfer and Fluid Flow*, McGraw-Hill, New York.

Patankar, S V (1981). A calculation procedure for two-dimensional elliptic situations, *Numerical Heat Transfer* **4**, 409- 425.

Pera, L and Gebhart, B (1972). Experimental observations of wake formation over cylindrical surface in natural convection flows. *Int J Heat and Mass Transfer* **15**, 175.

Pera, L and Gebhart, B (1975). Laminar plume interactions, *J Fluid Mechanics* **68** (2), 259-271.

Prakash, C and Sparrow, E M (1980). Natural convection heat transfer performance evaluations for discrete (in-line or staggered) and continuous plate arrays, *Numerical Heat Transfer* **3**, 89-105.

Praslov, R S (1971). On the effects of surface roughness on natural convection heat transfer from horizontal cylinders to air, *Inzh. Fiz. Zh.* **4** (5), 601-617.

Pratt, R G and Karaki, S (1979). Natural convection bwtween vertical plates with external frictional losses - application to Trombe walls, *Proc Natl Passive Solar Conf*, San Jose, CA. Published by Intl Solar Energy Society, Inc., American Section, Univ. of Delaware, Newark.

Quintiere, J G (1970). *An Analysis of Natural Convection Between Finite Vertical Parallel Plates*, PhD Thesis, School of Eng and Sciences, New York Univ., New York.

Quintiere, J G and Mueller, W K (1973). An analysis of laminar free and forced convection between finite vertical parallel plates, *ASME J Heat Transfer* **95**, 53-59.

Raithby, G D and Hollands, K G T. A general method of obtaining approximate solutions to laminar and turbulent free convection problems, *Advances in Heat Transfer* **11**, T F Irvine and J P Hartnett, eds., Academic Press, New York, 266-315.

Ramakrishna, K, Seetharama, K N and Sarma, P K (1978). Turbulent heat transfer by free convection from a rough surface, *ASME J Heat Transfer* **100**, 727-728.

Rao, K V, Armaly, B F and Chen, T S (1984). Analysis of laminar mixed convective plumes along vertical adiabatic surfaces, *Trans ASME J Heat Transfer* **106** 552-557.

Ravine, T L and Richards, D E (1984). Natural convection cooling of a finite-sized thermal source on the wall of a vertical channel, *ASME Paper 84-WA/HT-90.*

Sastry, C V N, Murthy, V, Narayana, and Sarma, P K (1976). Effect of discrete wall roughness on free convective heat transfer from a vertical tube, *Heat Transfer and Turbulent Buoyant Convection* 2, D B Spalding and N Afgan, eds., McGraw-Hill, New York.

Savkar, S D (1970). Developing forced and free convective flows between two semi-infinite parallel plates, *Paper NC-3.8, 4th Int Heat Transfer Conf* IV, Elsevier Publishing Co., Amsterdam.

Schmidt, R C and Patankar, S V. A numerical study of laminar forced convection across heated rectangular blocks in two-dimensional ducts, *ASME Paper 86-WA/HT-88.*

Seely, J H and Chu, R C (1972). *Heat Transfer in Microelectronic Equipment*, Marcel-Dekker, Inc., New York.

Sellars, R J, Tribus, M and Klein, J S (1956). *Trans ASME* 78, 441-448.

Shah, R K and London, A L (1978). *Laminar Flow Forced Convection in Ducts: A Source Book for Compact Heat Exchanger Analytical Data*, Academic Press, New York.

Shakerin, S, Bohn, M and Loehrke, R I (1986). Natural convection in an enclosure with discrete roughness elements on a vertical, heated wall, *Proc 8th Int Heat Transfer Conf* 4, San Francisco, CA, 1519-1525.

Siebers, D L, Schwind, R G and Moffat, R J (1983). Experimental mixed convection heat transfer from a large, vertical surface in a horizontal flow, *Report HMT-36*, Thermosciences Div., Mech Eng Dept., Stanford Univ., CA.

Sobel, N, Landis, F and Mueller, W K (1966). Natural convection heat transfer in short vertical channels including the effects of stagger, *3rd Int Heat Transfer Conf AICHE* II, New York, 121-125.

Sparrow, E M and Ansari, M A (1983). A refutation of King's rule for multi-dimensional external natural convection, *Int J Heat Mass Transfer* 26, 1357-1364.

Sparrow, E M and Azevedo, L F A (1985). Vertical channel natural convection spanning the fully-developed limit and the single-plate boundary layer limit, *Int J Heat and Mass Transfer* **28** (10), 1847-1857.

Sparrow, E M and Behrami, P A (1980). Experiments on natural convection from vertical parallel plates with either open or closed edges, *J Heat Transfer, Trans ASME* **102** (2), 221-227.

Sparrow, E M and Boessneck, D S (1983). Effect of transverse misalignment on natural convection from a pair of parallel, vertically stacked, horizontal cylinders, *ASME J Heat Transfer* **105** (2), 241-247.

Sparrow, E M and Chrysler, G M (1981). Natural convection heat transfer coefficients for a short horizontal cylinder attached to a vertical plate, *ASME J Heat Transfer* **103** (4), 630- 637.

Sparrow, E M and Chrysler, G M (1982). Natural convection fluid flow patterns resulting from the interaction of a heated vertical plate and an attached horizontal cylinder, *Trans ASME J Heat Transfer* **104** (4), 798-800.

Sparrow, E M, Chrysler, G M and Azevedo, L F (1984). Observed flow reversals and measured-predicted Nusselt numbers for natural convection in a one-sided heated vertical channel, *ASME J Heat Transfer* **106** (2), 325-332.

Sparrow, E M and Chukaev, A (1980). Forced-convection heat transfer in a duct having spanwise periodic rectangular protuberances, *Numerical Heat Transfer* **3**, 149-167.

Sparrow, E M, Cook, D S and Chrysler, G M (1982). Heat transfer by natural convection from an array of short, wall-attached horizontal cylinders, *ASME J Heat Transfer* **104** (1), 125-131.

Sparrow, E M and Faghri, M (1980). Natural convection heat transfer from the upper plate of a co-linear, separated pair of vertical plates, *ASME J Heat Transfer* **102**, 623-629.

Sparrow, E M and Gregg, J L (1958). Similar solutions for free convection from a nonisothermal vertical plate, *Trans ASME*, 379-386.

Sparrow, E M and Gregg, J L (1959). Buoyancy effects in forced convection flow and heat transfer, *J Applied Mechanics* **2**, 133-134.

Sparrow, E M, Medendes, S, Ansari, M A and Price, A T (1983). Duct-flow versus external flow natural convection at a short, wall-attached horizontal cylinder, *Int J Heat and Mass Transfer* **26** (6), 881-889.

Sparrow, E M, Niethhammer, J E and Chaboki, A (1982). Heat transfer and pressure drop characteristics of arrays of rectangular modules encountered in electronic equipment, *Int J Heat and Mass Transfer* **25** (7), 961-973.

Sparrow, E M and Pfeil, D R (1984). Enhancement of natural convection heat transfer from a horizontal cylinder due to vertical shrouding surfaces, *Trans ASME J Heat Transfer* **105** (1), 124-130.

Sparrow, E M and Prakash, C (1980). Enhancement of natural convection heat transfer by a staggered array of discrete vertical plates, *J Heat Transfer, Trans ASME* **102** (2), 215-220.

Sparrow, E M, Shah, S and Prakash, C (1980). Natural convection in a vertical channel: 1. Interacting convection and radiation; 2. The vertical plate with and without shrouding, *Numerical Heat Transfer* **3** (3), 297-314.

Tao, L N (1960). On combined free and forced convection in channels, *ASME J Heat Transfer* **82** (3), Series C, 233-238.

Tierney, J K and Koczkur, E (1971). Free convection heat transfer from a totally enclosed cabinet containing simulated electronic equipment, *IEEE Trans on Parts, Hybrids and Packaging* **PHP-7** (3), 115-123.

Tokura, I, Saito, H, Kishinami, K and Muramoto, K (1983). An experimental study of free convection heat transfer from a horizontal cylinder in a vertical array set in free space between parallel walls, *Trans ASME J Heat Transfer* **105**, 102- 107.

Torok, D (1984). Augmenting experimental studies for flow and thermal field prediction by the finite element method, *Proc 1984 Int Computers Eng Conf and Exhibit* **1**, 509-517.

Turner, B L and Flack, R D (1980). The experimental measurement of natural convective heat transfer in rectangular enclosures with concentrated energy sources, *ASME J Heat Transfer* **102**, 236-241.

Vajravelu, K and Sastri, K S (1977). Fully developed laminar free convection flow between two parallel vertical walls - I, *Int J Heat and Mass Transfer*, 655-660.

Wilks, G (1977). A mixed convection universal profile, *Letters Heat and Mass Transfer* **4** (3), 217-222.

Wirtz, R A and Dykshoorn, P (1984). Heat transfer and pressure drop characteristics of arrays of rectangular modules encountered in electronic equipment, *Int J Heat and Mass Transfer* **25** (7), 961-973.

Wirtz, R A and Stutzman, R J (1982). Experiments on free convection between vertical plates with symmetric heating, *ASME Trans J Heat Transfer* **104**, 501-507.

Yang, K T and Jergen, E W (1964). First-order perturbations of laminar free convection boundary layers in a vertical plate, *ASME J Heat Transfer* **86**, 107-115.

Yao, L S (1983). Free and forced convection in the entry region of a heated vertical channel, *Int J Heat and Mass Transfer* **26** (1), 65-72.

Yao, L S (1983). Buoyancy effects on a boundary layer along an infinite cylinder with a step change of surface temperature, *J Heat Transfer* **105**, 96.

Yuge, T (1960). Experiments on heat transfer from spheres including combined natural and forced convection, *ASME Trans J Heat Transfer*, Series C **82**, 214-220.

Zeldin, B and Schmidt, F W (1972). Developing flow with combined forced-free convection in an isothermal vertical tube, *ASME J Heat Transfer*, 211-223.

Zhang, Z and Patankar, S V (1984). Influence of Buoyancy on the vertical flow and heat transfer in a shrouded fin array, *Int J Heat and Mass Transfer* **27** (1), 137-140.

Zinnes, A E (1970). The coupling of conduction with laminar natural convection from a vertical flat plate with arbitrary surface heating, *ASME J Heat Transfer*, Series C **92** (3), 528-535.

Zinnes, A E (1969). *An Investigation of Steady Two Dimensional Laminar Natural Convection from a Vertical Plate of Finite Thickness with Plane Localized Heat Sources on its Surface*, PhD Dissertation, Lehigh Univ., Bethlehem, PA.

Zukauskaus, A A (1972). Heat transfer for tubes in crossflow, *Advances in Heat Transfer* **8**, Academic Press, New York.

Chapter 4

APPLICATION OF HEAT PIPES TO ELECTRONICS COOLING

P. J. Marto
Professor of Mechanical Engineering
Naval Postgraduate School
Monterey, California

G. P. Peterson
Assistant Professor of Mechanical Engineering
Texas A& M University
College Station, Texas

1   INTRODUCTION

The concept of a passive two-phase heat transfer device capable of transferring large quantities of heat with a minimal temperature drop was first introduced by Gaugler (1942). This device received little

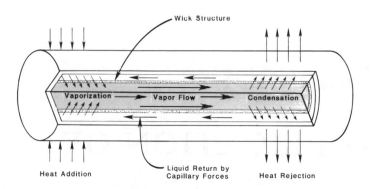

Figure 4.1: Heat pipe operation.

attention until Grover et al. (1964) published the results of an independent investigation and first applied the term "heat pipe." Since that time, heat pipes have been employed in numerous applications ranging from temperature control of the permafrost layer under the Alaska pipeline to the thermal control of optical surfaces in spacecraft.

A heat pipe typically consists of a sealed container lined with a wicking material. The container is evacuated and backfilled with just enough liquid to fully saturate the wick. Because only pure liquid and vapor are present within the container, the working fluid will remain at saturation conditions as long as the operating temperature is between the freezing point and the critical state.

As shown in Fig. 4.1, the heat pipe operates on a closed two-phase cycle and consists of three distinct regions, the evaporator or heat addition region, the condenser or heat rejection region, and the adiabatic or isothermal region. Heat added to the evaporator region of the container causes the working fluid in the evaporator wicking structure to be vaporized. The high temperature and corresponding high pressure in this region result in flow of the vapor to the other, cooler end of the container where the vapor condenses, giving up its latent heat of vaporization. The capillary forces existing in the wicking structure then pump the liquid back to the evaporator section. Other similar devices, referred to as two-phase thermosyphons have no wick, and utilize gravitational forces to provide the liquid return.

Eastman (1968) identified several characteristics of heat pipes, four of which make them particularly useful in the design of thermal control systems for electronic components and devices. First, since the heat pipe operates on a closed two-phase cycle, the heat transfer capacity may be several orders of magnitude higher than even the best solid conductors. This results in a relatively small thermal resistance and allows physical separation of the evaporator and condenser without a high penalty in overall temperature drop. Second, increases in the heat flux in the evaporator result in an increase in the rate at which the working fluid is vaporized, without significant increases in the operating temperature. Thus the heat pipe functions as a nearly isothermal device, adjusting the evaporation rate to accommodate a wide range of power inputs, while maintaining a relatively constant source temperature. Third, the evaporator and condenser portions of a heat pipe function independently, needing only common liquid and vapor streams; for this reason, the area over which heat is introduced can differ in size and shape from the area over which it is rejected, provided that the rate at which the liquid is vaporized does not exceed the rate at which it can be condensed. Hence, high heat fluxes generated over relatively small areas can be dissipated over larger areas with reduced heat fluxes. Finally, by using a two-phase cycle, the thermal response time of heat pipes is much less than that associated with solid conductors.

The high heat transfer characteristics, the ability to maintain constant evaporator temperatures under different heat flux levels, and the diversity and variability of evaporator and condenser sizes, make the heat pipe an effective device for use in the thermal control of electronic components.

As is the case with most two-phase cycles, the presence of noncondensible gases creates a problem due to the partial blockage of the condensing area. Heat pipes are no exception. During normal operation, any noncondensible gases present are carried to the condenser and remain there, reducing the effective condenser area. This characteristic although normally undesirable, can be used to control both the direction and amount of heat transferred (Tien and Chen, (1984)).

Figure 4.2a illustrates one such application of this concept referred to as a gas-loaded, variable conductance heat pipe. In this type of device the thermal conductance of the heat pipe varies as a function of the gas front position. As the heat available at the evap-

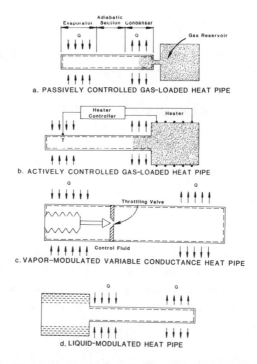

Figure 4.2: Heat pipe control schemes [Chi (1976)] (a) Passively controlled gas-loaded heat pipe. (b) Actively controlled gas-loaded heat pipe. (c) Vapor-modulated variable conductance heat pipe. (d) Liquid-modulated heat pipe.

orator varies, the vapor temperature varies and the gas contained within the gas reservoir expands or contracts, moving the gas front. This in turn results in a variation in the thermal conductance, i.e., as the heat flux increases, the gas front recedes and the thermal conductance increases due to the larger condenser surface area. In this way, the temperature drop across the evaporator and condenser can be maintained fairly constant even though the evaporator heat flux may fluctuate.

While in most applications heat pipes operate in a passive manner, adjusting the heat flow rate to compensate for the temperature difference between the evaporator and condenser (Van Buggenum and Daniels, 1987), several active control schemes have been developed (Sakuri et al., 1984). Most notable among these are: (i) gas-loaded heat pipes with some type of feedback system, (ii) excess-liquid heat pipes, (iii) vapor flow-modulated heat pipes, and (iv) liquid flow-modulated heat pipes (Chi, 1976). Figure 4.2b illustrates an example of an actively controlled, gas-loaded heat pipe in which the gas volume at the reservoir end can be controlled externally. In the example shown, a temperature sensing device at the evaporator provides a signal to the reservoir heater. This heater when activated can heat the gas contained in the reservoir, causing it to expand and thereby reducing the condenser area.

Excess-liquid heat pipes operate in much the same manner as gas-loaded heat pipes, but utilize excess working fluid to block portions of the pipe and control the condenser size or prevent reversal of heat transfer. Vapor flow-modulated heat pipes utilize a throttling valve to control the amount of vapor leaving the evaporator. Figure 4.2c illustrates an example of one such control scheme. Increased evaporator temperatures result in an expansion of the bellows chamber containing the control fluid. This in turn closes down the throttling valve and reduces the flow of vapor to the condenser. This type of device is typically applied in situations where the evaporator temperature varies and a constant condenser temperature is desired.

Figure 4.2d illustrates the principle used in liquid flow-modulated heat pipes. This type of heat pipe has two separate wicking structures, one to transport liquid from the evaporator to the condenser and the other which serves as a liquid trap. As the temperature gradient is reversed, the liquid moves into the trap and starves the evaporator of fluid (Groll et al., (1982)), (Brost and Mack, (1987)). In addition to these liquid-vapor control schemes, the quantity and direction of heat transfer can also be controlled through internal or external pumps (Furukawa, (1987)) or through actual physical contact with the heat sink (Ollendorf, (1983)).

2 TRANSPORT LIMITATIONS

During steady-state operation of a heat pipe, several important mechanisms can exist that limit the maximum amount of heat transferred. Among these are the capillary wicking limit, viscous limit, sonic limit, entrainment and boiling limits. The capillary wicking limit and viscous limits deal with the pressure drops occurring in the liquid and vapor phases respectively. The sonic limit results from the occurrence of choked flow in the vapor passage, while the entrainment limit is due to the high liquid-vapor shear forces developed when the vapor passes in counter-flow over the liquid saturated wick. The boiling limit is reached when the heat flux applied in the evaporator portion is high enough that nucleate boiling occurs in the evaporator wick, creating vapor bubbles that partially block the return of fluid.

For moderate temperature heat pipes, the most significant of these limits is usually the capillary wicking limit. However, the significance of this limit decreases somewhat for reduced gravity applications. In low temperature applications such as those using cryogenic working fluids, either the viscous limit or capillary limit occur first, while in high temperature heat pipes, such as those that use liquid metal working fluids, the sonic and entrainment limits are of increased importance. The theory and fundamental phenomena which cause each of these limitations have been the object of a considerable number of investigations and are well documented in Chi (1976), Dunn and Reay (1983), Tien (1975), and the proceedings from the six International Heat Pipe Conferences held over the past fifteen years.

Although transient modeling and start-up dynamics have been the subject of many recent investigations (Antoniuk (1987), Chang and Colwell (1985), Cullimore (1985) and Merrigan et al. (1985)) and may be of significant importance in electronics applications, due to space limitations, these subjects are not included in this chapter, and only a brief synopsis of the predictive techniques for steady-state operation is presented. For further information on the theory and fundamental operation of heat pipes, the references listed at the end of this chapter should be consulted.

2.1   Capillary Limitation

Although heat pipe performance and operation are strongly dependent on shape, working fluid, and wick structure, the fundamental phenomenon that governs the operation arises from the difference in the capillary pressure across the liquid-vapor interfaces in the evaporator and condenser.

In order to function properly, the net capillary pressure difference between the condenser and the evaporator in a heat pipe must be greater than the pressure losses throughout the liquid and vapor flow paths. This relationship can be expressed as

$$\Delta P_c \geq \Delta P_+ + \Delta P_{||} + \Delta P_\ell + \Delta P_v \qquad (4.1)$$

where

ΔP_c = net capillary pressure difference
ΔP_+ = normal hydrostatic pressure drop
$\Delta P_{||}$ = axial hydrostatic pressure drop
ΔP_ℓ = viscous pressure drop occurring in the liquid phase
ΔP_v = viscous pressure drop occurring in the vapor phase.

If the summation of the viscous pressure losses, ΔP_ℓ and ΔP_v, and the hydrostatic pressure losses, ΔP_+ and $\Delta P_{||}$, is greater than the capillary pressure difference between the evaporator and condenser, the working fluid is not supplied rapidly enough to the evaporator to compensate for the liquid lost through evaporation, and the wicking structure becomes starved of liquid and dries out. This condition, referred to as the capillary wicking limitation, varies according to the wicking structure, working fluid, evaporator heat flux and operating temperature.

Capillary Pressure

Following the development by Chi (1976), the capillary pressure difference at a liquid-vapor interface, defined as $(P_v - P_\ell)$, or ΔP_c, can be found from the LaPlace-Young equation,

$$\Delta P_c = \sigma \left(\frac{1}{r_1} + \frac{1}{r_2} \right) \qquad (4.2)$$

where r_1 and r_2 are the principal radii of curvature and σ is the surface tension. For most heat pipe wicking structures, the maximum capillary pressure may be written in terms of a single radius of curvature, r_c. The above expression then reduces to

$$\Delta P_{c,m} = \left(\frac{2\sigma}{r_{ce}} \right) - \left(\frac{2\sigma}{r_{cc}} \right). \qquad (4.3)$$

where r_{ce} and r_{cc} represent the radii of curvature in the evaporator and condenser regions respectively.

In a heat pipe operating normally, the vaporization occurring in the evaporator causes the liquid meniscus to recede into the wick,

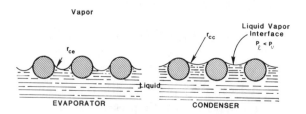

Figure 4.3: Radius of curvature of the liquid-vapor interface in the evaporator and condenser.

reducing the local capillary radius, r_{ce}, as shown in Fig. 4.3. In the condenser section, condensation causes flooding of the wick, which in turn increases the local capillary radius, r_{cc}. The resulting difference in the two radii of curvature causes a pressure difference and hence, pumping of the liquid from the condenser to the evaporator. During steady-state operation, it is generally assumed that the capillary radius in the condenser, r_{cc}, approaches infinity, so that the maximum capillary pressure for a heat pipe operating at steady-state can be expressed as a function only of the capillary radius of the evaporator wick,

$$\Delta P_{c,m} = \left(\frac{2\sigma}{r_{ce}}\right). \tag{4.4}$$

Values for the effective capillary radius, r_c, can be found theoretically for simple geometries (Chi, 1976) or experimentally for pores or structures of more complex geometry (Ferrell and Alleavitch (1969), Freggens (1969) or Eninger (1975)). Limited information on the transient behavior of capillary structures is also available (Colwell and Chang, (1984)). Table 4.1 from Chi (1976) gives values for some of the more common wicking structures.

Normal Hydrostatic Pressure Drop

The normal hydrostatic pressure drop, ΔP_+, occurs only in heat pipes in which circumferential communication of the liquid in the wick is possible. It is a result of the component of the body force acting perpendicular to the longitudinal axis of the heat pipe. The normal hydrostatic pressure drop can be expressed as

$$\Delta P_+ = \rho_\ell g d_v \cos \Psi \tag{4.5}$$

Table 4.1: Expressions for the Effective Capillary Radius for Several Wick Structures [from Chi (1976)]

Structure	r_c	Data	
Circular cylinder (artery or tunnel wick)	r	$r =$	radius of liquid flow passage
Rectangular groove	w	$w =$	groove width
Triangular groove	$w/\cos\beta$	$w =$	groove width
		$\beta =$	half-included angle
Parallel wires	w	$w =$	wire spacing
Wire screens	$(w + d_w)/2 = 1/2N$	$d =$	wire diameter
		$N =$	screen mesh number
		$w =$	wire spacing
Packed spheres	$0.41 r_s$	$r_s =$	sphere radius

where ρ_ℓ is the density of the liquid, g is the gravitational acceleration, d_v is the diameter of the vapor portion of the pipe and Ψ is the angle the heat pipe makes with respect to the horizontal.

Axial Hydrostatic Pressure Drop

The axial hydrostatic pressure drop, $\Delta P_{\parallel}$, results from the component of the body force acting along the longitudinal axis. It can be expressed as

$$\Delta P_{\parallel} = \rho_\ell g L \sin\Psi \tag{4.6}$$

where L is the overall length of the heat pipe.

In a gravitational environment, the axial hydrostatic pressure term may either assist or hinder the capillary pumping process depending upon whether the tilt of the heat pipe promotes or hinders the flow of liquid back to the evaporator (i.e., the evaporator lies either below or above the condenser). In a zero-g environment, both this term and the normal hydrostatic pressure drop term can be neglected because of the absence of body forces.

Liquid Pressure Drop

The viscous forces in the liquid result in a pressure drop, ΔP_ℓ, which resists the capillary flow through the wick. Since the liquid pressure gradient may vary along the longitudinal axis of the heat pipe, the total liquid pressure drop can be determined by integrating the pressure gradient over the length of the flow passage, or

$$\Delta P_\ell(x) = -\int_0^x \frac{dP_\ell}{dx} dx \tag{4.7}$$

where the limits of integration are from the evaporator end to the condenser end ($x = 0$) and dP_ℓ/dx is the gradient of the liquid pressure resulting from frictional drag. This frictional drag, due to the shear stress, can be written as

$$\frac{dP_\ell}{dx} = -\frac{2\tau_\ell}{(r_{h,\ell})} \tag{4.8}$$

where τ_ℓ is the frictional shear stress at the liquid-solid interface and $(r_{h,\ell})$ is the hydraulic radius, defined as twice the cross-sectional area divided by the wetted perimeter.

At this point, the Reynolds number, Re_ℓ and drag coefficient, f_ℓ, can be introduced as

$$Re_\ell = \frac{2(r_{h,\ell})\rho_\ell V_\ell}{\mu_\ell} \tag{4.9}$$

and

$$f_\ell = \frac{2\tau_\ell}{\rho_\ell V_\ell^2} \tag{4.10}$$

where V_ℓ is the local liquid velocity which is related to the local heat flow

$$V_\ell = \frac{q}{\epsilon A_w \rho_\ell \lambda}. \tag{4.11}$$

In this expression, A_w is the wick cross-sectional area, ϵ is the wick porosity, and λ is the latent heat of vaporization.

Combining equations (4.9) through (4.11) with (4.8) yields,

$$\frac{dP_\ell}{dx} = - \left(\frac{(f_\ell Re_\ell)\,\mu_\ell}{2\epsilon A_w \,(r_{h,\ell})^2 \,\lambda \rho_\ell} \right) q. \qquad (4.12)$$

This equation can be written as

$$\frac{dP_\ell}{dx} = \left(\frac{\mu_\ell}{K A_w \lambda \rho_\ell} \right) q \qquad (4.13)$$

where K is the permeability expressed as

$$K = \frac{2\epsilon \,(r_{h,\ell})^2}{(f_\ell Re_\ell)}. \qquad (4.14)$$

For laminar flow, $(f_\ell Re_\ell)$ is constant and depends only on the passage shape. Thus, the permeability, K, is independent of the flow path, x. Values of K for several common wick structures are provided in Table 4.2. For more complex wicks, the permeability will have to be determined experimentally.

For constant heat addition and removal, eq (4.13) can be substituted into eqn (4.7) and integrated over the length of the heat pipe to yield

$$\Delta P_\ell = - \left(\frac{\mu_\ell}{K A_w \lambda \rho_\ell} \right) L_{eff}\, q \qquad (4.15)$$

where L_{eff} is the effective heat pipe length defined as

$$L_{eff} = 0.5 L_e + L_a + 0.5 L_c. \qquad (4.16)$$

Vapor Pressure Drop

Determination of the vapor pressure drop in heat pipes is complicated by the mass addition and removal in the evaporator and condenser respectively, and by the compressibility of the vapor phase. Performing a mass balance on a section of the adiabatic region of the heat pipe ensures that for continued operation, the liquid mass flow rate and vapor mass flow rate must be equal. Because of the large difference in the density of these two phases, the vapor velocity must necessarily be significantly higher than the velocity of the liquid phase. For this reason, in addition to the pressure gradient resulting from frictional drag, the pressure gradient due to variations

Table 4.2: Wick Permeability for Several Wick Structures [from Chi (1976)].

Structure	K	Data	
Circular cylinder (artery or tunnel wick	$r^2/8$	$r =$	radius of liquid flow passage
Open rectangular grooves	$2\epsilon \left(r_{h,\ell} \right)^2 / \left(f_\ell Re_\ell \right) = w/s$	$\epsilon =$	wick porosity
		$w =$	groove width
		$s =$	groove pitch
		$\delta =$	groove depth
		$\left(r_{h,\ell} \right) =$	$2w\delta/(w + 2\delta)$
Circular annular wick	$2 \left(r_{h,\ell} \right)^2 / \left(f_\ell Re_\ell \right)$	$\left(r_{h,\ell} \right) =$	$r_1 - r_2$
Wrapped screen wick	$\frac{1}{122} d_w^2 \epsilon^3 / (1 - \epsilon)^2$	$d_w =$	wire diameter
		$\epsilon =$	$1 - (1.05\pi N d_w/4)$
		$N =$	mesh number
Packed sphere	$\frac{1}{37.5} r_s^2 \epsilon^3 / (1 - \epsilon)^2$	$r_s =$	sphere radius
		$\epsilon =$	porosity (dependent on packing mode)

in the dynamic pressure must also be considered. Both Chi (1976) and Dunn and Reay (1983) have addressed this problem. Chi (1976) found that upon integration of the vapor pressure gradient, the dynamic pressure effects cancel. The result is an expression, which is similar to that developed for the liquid

$$\Delta P_v = \left(\frac{C \left(f_v Re_v \right) \mu_v}{2 \left(r_{h,v} \right)^2 A_v \rho_v \lambda} \right) L_{eff} q \qquad (4.17)$$

where $\left(r_{h,v} \right)$ is the hydraulic radius of the vapor space and C is a constant, which depends on the Mach Number.

As mentioned previously, during steady-state operation, at any axial position, the liquid mass flow rate, m_ℓ, must equal the vapor mass flow rate, m_v, and while the liquid flow regime is always laminar, the vapor flow may be either laminar or turbulent. It is therefore necessary to determine the vapor flow regime as a function of the heat flux. This can be accomplished by evaluating the local axial Reynolds number in the vapor, defined as

$$Re_v = \frac{2(r_{h,v})q}{A_v\mu_v\lambda} \qquad (4.18)$$

In addition, it is necessary to determine if the flow should be treated as compressible or incompressible by evaluating the local Mach number, defined as

$$Ma_v = \frac{q}{A_v\rho_v\lambda(R_vT_v\gamma_v)}1/2 \qquad (4.19)$$

where R_v is the gas constant, T_v is the vapor temperature and γ_v is the ratio of specific heats, which is equal to 1.67, 1.4, or 1.33 for monatomic, diatomic, and polyatomic vapor, respectively (Chi, 1976).

Previous investigations summarized by Kraus and Bar-Cohen (1983) have demonstrated that the following combinations of these conditions can be used with reasonable accuracy.

$$
\begin{aligned}
Re_v &< 2300, Ma_v < 0.2 \\
(f_vRe_v) &= 16 \\
C &= 1.00
\end{aligned}
\qquad (4.20)
$$

$$
\begin{aligned}
Re_v &< 2300, Ma_v > 0.2 \\
(f_vRe_v) &= 16 \\
C &= \left[1 + \left(\frac{\gamma_v - 1}{2}\right)Ma_v^2\right]^{-1/2}
\end{aligned}
\qquad (4.21)
$$

$$
\begin{aligned}
Re_v &> 2300, Mv < 0.2 \\
(f_vRe_v) &= 0.038\left(\frac{2(r_{h,v})q}{A_v\mu_v\lambda}\right)^{3/4} \\
C &= 1.00
\end{aligned}
\qquad (4.22)
$$

$$
\begin{aligned}
Re_v &> 2300, Ma_v > 0.2 \\
(f_vRe_v) &= 0.038\left(\frac{2(r_{h,v})q}{A_v\mu_v\lambda}\right)^{3/4} \\
C &= \left[1 + \left(\frac{\gamma_v - 1}{2}\right)Ma_v^2\right]^{-1/2}.
\end{aligned}
\qquad (4.23)
$$

Since the equations used to evaluate both the Reynolds number and the Mach number are functions of the heat transport capacity, it is necessary to first assume the conditions of the vapor flow. Using these assumptions, the maximum heat capacity, $q_{c,m}$, can be determined by substituting the values of the individual pressure drops into eqn (4.1) and solving for $q_{c,m}$. Once the value of $q_{c,m}$ is known, it can then be substituted into the expressions for the vapor Reynolds number and Mach number to determine the accuracy of the original assumption. Using this iterative approach, which is covered in more detail in Chi (1976), accurate values for the capillary limitation as a function of the operating temperature can be determined in units of watt-m or watts for $(qL)_{c,m}$ and $q_{c,m}$ respectively.

2.2 Viscous Limitation

At very low operating temperatures, the vapor pressure difference between the closed end of the evaporator (the high pressure region) and the closed end of the condenser (the low pressure region) may be extremely small. Because of this small pressure difference, the viscous forces within the vapor region may prove to be dominant and hence, limit the heat pipe operation. Dunn and Reay (1983) discuss this limit in more detail and suggest the criterion

$$\frac{\Delta P_v}{P_v} < 0.1 \qquad (4.24)$$

for determining when this limit might be of a concern. Due to the operating temperature range, this limitation will normally be of little consequence in the design of heat pipes for use in the thermal control of electronic components and devices.

2.3 Sonic Limitation

Sonic limitations in heat pipes are analogous to the sonic limitations in converging-diverging nozzles (Chi, (1976)). In a converging- diverging nozzle, the mass flow rate is constant and the vapor velocity varies because of the changing cross-sectional area. In heat pipes, the reverse occurs; the area is constant and the vapor velocity varies because of the evaporation and condensation along the heat pipe. As in nozzle flow, decreased outlet pressure, or in this case condenser temperature, results in a decrease in the evaporator temperature until the sonic limitation is reached. Any further increase in the heat rejection rate does not reduce the evaporator temperature or the maximum heat transfer capability, but only reduces the condenser temperature, due to the existence of choked flow.

The sonic limitation in heat pipes can be determined as

$$q_{s,m} = A_v \rho_v \lambda \left(\frac{\gamma_v R_v T_v}{2 (\gamma_v + 1)} \right)^{1/2} \tag{4.25}$$

where T_v is the mean vapor temperature within the heat pipe.

2.4   Entrainment Limitation

Since the liquid and vapor flow in opposite directions in a heat pipe, at high enough vapor velocities, liquid droplets may be picked up or entrained in the vapor flow. This entrainment results in excess liquid accumulation in the condenser and hence, dryout of the evaporator wick. Therefore, it is necessary to evaluate when the onset of entrainment begins in a counter- current two-phase flow. The most commonly quoted criterion to determine this onset is when the Weber number, We, defined as the ratio of the viscous shear force to the force resulting from the liquid surface tension,

$$We = \frac{2 (r_{h,w}) \rho_v V_v^2}{\sigma}, \tag{4.26}$$

is equal to unity. To prevent the entrainment of liquid droplets in the vapor flow, the Weber number must therefore be less than one.

By relating the vapor velocity to the heat transport capacity

$$V_v = \frac{q}{A_v \rho_v \lambda}, \tag{4.27}$$

a value for the maximum transport capacity based on the entrainment limitation may be determined as

$$q_{e,m} = A_v \lambda \left(\frac{\sigma \rho_v}{2 (r_{h,w})} \right)^{1/2} \tag{4.28}$$

where $(r_{h,w})$ is the hydraulic radius of the wick structure, defined as twice the area of the wick pore at the wick-vapor interface divided by the wetted perimeter at the wick-vapor interface. A somewhat different approach has been proposed by Rice and Fulford (1987) who developed an expression that defines the critical dimensions for wicking structures in order to prevent entrainment.

2.5   Boiling Limitation

When the input heat flux is sufficient, nucleate boiling may occur in the wicking structure and bubbles may become trapped in the wick, blocking the liquid return and resulting in evaporator dryout. This phenomenon, referred to as the boiling limit, differs from the other limitations previously discussed in that it depends on the evaporator heat flux as opposed to the axial heat flux.

Determination of the boiling limitation is based on nucleate boiling theory and is comprised of two separate phenomena–bubble formation and the subsequent growth or collapse of the bubbles. Bubble formation is governed by the number and size of nucleation sites on a solid surface; bubble growth or collapse depends on the liquid temperature and corresponding pressure caused by the vapor pressure and surface tension of the liquid. By performing a pressure balance on any given bubble and using the Clausius-Clapeyron equation to relate the temperature and pressure, an expression for the heat flux beyond which bubble growth will occur may be developed Chi (1976). This expression, which is a function of the fluid properties, can be written as

$$q_{b,m} = \left(\frac{2\pi L_{eff} k_{eff} T_v}{\lambda \rho_v \ell n \left(r_i / r_v \right)} \right) \left(\frac{2\sigma}{r_n} - \Delta P_{c,m} \right) \qquad (4.29)$$

where k_{eff} is the effective thermal conductivity of the liquid-wick combination, given in Table 4.3, r_i is the inner radius of the heat pipe wall, and r_n is the nucleation site radius, which, according to Dunn and Reay (1983), can be assumed to be from 2.54×10^{-5} to 2.54×10^{-7} meters for conventional heat pipes.

After the power level associated with each of the four limitations is established, determination of the maximum heat transport capacity is only a matter of selecting the lowest limitation for any given operating temperature. As mentioned previously, because of the desired operating temperature range of electronic components and devices, the capillary wicking limitation is typically the first one

Table 4.3: Effective Thermal Conductivity for Liquid-Saturated Wick Stuctures [from Chi (1976)]

Wick Structure	k_{eff}
Wick and liquid in series	$\dfrac{k_\ell k_w}{\epsilon k_w + k_\ell (1-\epsilon)}$
Wick and liquid in parallel	$\epsilon k_\ell + k_w(1-\epsilon)$
Wrapped screen	$\dfrac{k_\ell \left[(k_\ell + k_w) - (1-\epsilon)(k_\ell - k_w) \right]}{(k_\ell + k_w) + (1-\epsilon)(k_\ell - k_w)}$
Packed spheres	$\dfrac{k_\ell \left[(2k_\ell + k_w) - 2(1-\epsilon)(k_\ell - k_w) \right]}{(2k_\ell + k_w) + (1-\epsilon)(k_\ell - k_w)}$
Rectangular grooves	$\dfrac{(w_f k_\ell k_w \delta) + w k_\ell (0.185 w_f k_w + \delta k_\ell)}{(w + w_f)(0.185 w_f k_f + \delta k_\ell)}$

encountered. However, with the increased heat fluxes generated by newer devices, in some instances the boiling limit may become the most significant limitation.

3   DESIGN AND MANUFACTURING CONSIDERATIONS

Several references, which present a thorough review of the problems associated with the design and manufacture of heat pipes exist. Most notable are those by Feldman (1976) and Brennan and Kroliczek (1979). In addition to such factors as cost, size, weight, reliability, fluid inventory, and construction and sealing techniques, the design and manufacture of heat pipes are governed by three operational considerations: the effective operating temperature range, which is determined by the selection of the working fluid; the maximum power the heat pipe is capable of transporting, which is determined by the ultimate pumping capacity of the wick structure (for the capillary wicking limit); and the maximum evaporator heat flux, which is determined by the point at which nucleate boiling occurs.

3.1   Working Fluid

Since the basis for operation of a heat pipe is the vaporization and condensation of the working fluid, selection of a suitable working fluid is an important factor in the design and manufacture of heat pipes. Care must be taken to ensure that the operating temperature range is adequate for the application. As discussed by Kraus and Bar-Cohen (1983), most applications of heat pipes in the thermal control of electronic components require the selection of a working fluid with a boiling temperature between 250 and 375 K. This includes fluids such as ammonia, Freon 11 or 113, acetone, methanol and water. For a capillary wick limited heat pipe, the characteristics of a good working fluid are a high latent heat of vaporization, a high surface tension, a high liquid density and a low liquid viscosity. Chi (1976) combined these properties into a parameter referred to as the liquid transport factor or figure of merit, which is defined as

$$N_\ell = \frac{\rho_\ell \sigma \lambda}{\mu_\ell}. \qquad (4.30)$$

This grouping of properties can be used to evaluate various working fluids at specific operating temperatures. The concept of a parameter for evaluating working fluids has been extended by Gosse (1987), where it was demonstrated that the thermophysical properties of the liquid-vapor equilibrium state could be reduced to three independent parameters.

In addition to the thermophysical properties of the working fluid, consideration must be given to the ability of the working fluid to wet the wick and wall materials. Further criteria for the selection of the working fluids have been presented by Heine and Groll, (1984) in which a number of other factors including the liquid and vapor pressure, and the compatibility of the materials are considered.

3.2   Wicking Structures

The wicking structure has two functions in heat pipe operation: it is both the vehicle and the mechanism through which the working fluid is returned from the condenser to the evaporator, and it ensures that the working fluid is evenly distributed over the evaporator surface. Figure 4.4 illustrates several common wicking structures presently in use, along with several more advanced concepts under development. In order to provide a flow path with low flow resistance through which the liquid can be returned from the condenser to the evaporator, an open porous structure with a high permeability is desirable. However, to increase the capillary pumping pressure, a small pore size is necessary. Solutions to this apparent dichotomy can be achieved through the use of a nonhomogeneous wick made of several different materials or through a composite wicking structure. Udell and Jennings (1984) proposed and formulated a model for a heat pipe with a wick consisting of porous media of two different permeabilities oriented parallel to the direction of the heat flux. This wick structure provided a large pore size in the center of the wick for liquid flow and a smaller pore size for capillary pressure.

Composite wicking structures accomplish the same type of effect. As shown in Fig. 4.4, the capillary pumping and axial fluid transport

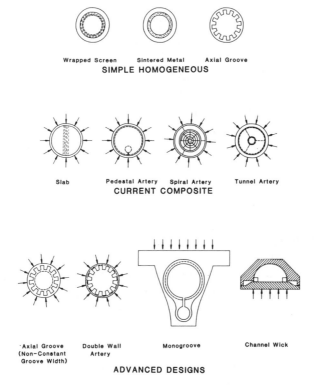

Figure 4.4: Heat pipe wick configurations.

are handled independently. In addition to fulfilling this dual purpose, several of the wicks illustrated in Fig. 4.4 separate the liquid and vapor flow. This results from an attempt to eliminate the viscous shear force that occurs during counter-current, liquid-vapor flow.

3.3 Materials Compatibility

Chemical reactions between the working fluid and the wall or wicking structure, or decomposition of the working fluid may lead to the formation of noncondensible gases or problems associated with corrosion. Because of the detrimental effect of noncondensible gases on heat pipe performance, careful consideration must be given to the selection of working fluids, and wicking and wall materials in order to prevent the occurrence of these problems over the operational life of the heat pipe. The effect of noncondensible gas formation has been discussed earlier and may result in either decreased performance or total failure. Corrosion problems can lead to physical degradation of the wicking structure since solid particles carried to the evaporator wick and deposited there will eventually reduce the wick permeability (Barantsevich et al. 1987).

Basiulis et al. (1976) conducted extensive compatibility tests with several combinations of working fluids and wicking structures, the results of which are summarized in Table 4.4, along with other investigations by Basiulis and Filler (1971). Other more recent investigations such as those performed by Zaho (1987) in which the compatibility of water and mild steel heat pipes were evaluated, Roesler (1987) in which stainless steel, aluminum and ammonia combinations were evaluated, and Murakami et al. (1987) in which a statistical predictive technique for evaluating the long term reliability of copper water heat pipes was developed, provide additional insight into the compatibility of various liquid-material combinations which might be used in the thermal control of electronics equipment. Most of the data available are the result of accelerated life tests. Although a majority of it is based upon actual test results, care should be taken to ensure that the tests in which the data were obtained are similar to the application under consideration. Such factors as thermal cycling, and mean operating temperature must be considered.

Noncondensible gas generation and corrosion problems are only two of the factors to be considered when selecting heat pipe wicks and

Table 4.4: Working Fluid, Wick and Container Compatibility Data [from Dunn and Reay (1983)]

Material	Water	Acetone	Ammonia	Methanol
Copper	RU	RU	NU	RU
Aluminum	GNC	RL	RU	NR
Stainless Steel	GNT	PC	RU	GNT
Nickel	PC	PC	RU	RL
Refrasil	RU	RU	RU	RU

Material	Dow-A	Dow-E	Freon 11	Freon 113
Copper	RU	RU	RU	RU
Aluminum	UK	NR	RU	RU
Stainless Steel	RU	RU	RU	RU
Nickel	RU	RL	UK	UK
Refrasil	RU		UK	UK

RU	Recommended by past successful usage.
RL	Recommended by literature.
PC	Probably compatible.
NR	Not recommended.
NU	Not Used.
UK	Unknown.
GNC	Generation of gas at all temperatures.
GNT	Generation of gas at elevated temperatures when oxide present.

working fluids. Others include wettability of the fluid/wick combination, strength to weight ratio, thermal conductivity and stability, and cost of fabrication.

3.4 Heat Pipe Shapes and Sizes

The sizes and shapes of heat pipes are almost as varied as the applications. Sizes range from a 15 m long monogroove heat pipe developed by Alario et al. (1984) for spacecraft heat rejection to a 10 mm long expandable bellows type heat pipe developed by Peterson (1986) for the thermal control of semiconductor devices. Vapor and liquid flow cross-sectional areas also vary significantly from those encountered in flat plate heat pipes which have very large flow areas to a commercially available heat pipe manufactured by ITOH Research and Development Company, with a cross-sectional area of less than 0.30 mm^2 or one developed by Haug et al. (1986) which has a diameter of less than 3 mm. Heat pipes may be fixed or variable in length and

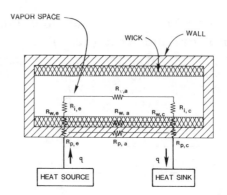

Figure 4.5: Electrothermal analog for a heat pipe.

either rigid or flexible, such as those developed by Bliss (1970) for situations where relative motion or vibration poses a problem.

4 REVIEW OF HEAT PIPE ADVANTAGES COMPARED TO CONDUCTION

Since heat pipes utilize the latent heat of vaporization of the working fluid as opposed to sensible heat, the required temperature difference is small, and hence the effective thermal conductivity may be several orders of magnitude greater than that of even the best solid conductors. For example, assume that it is necessary to transfer 20 watts of thermal energy over a distance of 0.5 m in a device 1.27 cm in diameter. Use of a pure aluminum rod would result in a temperature difference of approximately 460°C. A pure copper rod would result in a temperature difference of 206°C, while use of a simple copper-water heat pipe with a screen wick would result in a temperature difference between the evaporator and condenser surfaces of only about 6°C.

4.1 Heat Pipe Thermal Resistance

The temperature drop between the evaporator and condenser of a heat pipe is of particular interest to the designer of heat pipe thermal control systems and is most readily found by utilizing an analogous electrothermal network. Figure 4.5 illustrates the electrothermal analog for the heat pipe illustrated in Fig. 4.1. As shown, the overall thermal resistance is comprised of nine different resistances arranged in a series/parallel combination. These nine resistances can be summarized as follows:

R_{pe} - The radial resistance of the pipe wall at the evaporator

R_{we} - The resistance of the liquid/wick combination at the evaporator

R_{ie} - The resistance of the liquid/vapor interface at the evaporator

R_{va} - The resistance of the adiabatic vapor section

R_{pa} - The axial resistance of the pipe wall

R_{wa} - The axial resistance of the liquid/wick combination

R_{ic} - The resistance of the liquid/vapor interface at the condenser

R_{wc} - The resistance of the liquid/wick combination at the condenser

R_{pc} - The radial resistance of the pipe wall at the condenser

Asselman and Green (1973) give estimates of the order of magnitude for each of these resistances (Table 5). In comparing these values, it is apparent that several simplifications can be made. First, because of the comparative magnitudes of the resistance of the vapor space and the axial resistances of the pipe wall and liquid/wick combinations, the axial resistance of both the pipe wall and the liquid/wick combination may be treated as open circuits and neglected. Second, again because of the comparative resistances, the liquid/vapor interface resistances and the axial vapor resistance can, in most situations, be assumed to be negligible. This leaves only the pipe wall radial resistances and the liquid/wick resistances at both the evaporator and condenser.

The radial resistances at the pipe wall can be computed from Fourier's law as

$$R_{pe} = \frac{\delta}{k_p A_e} \tag{4.31}$$

Table 4.5: Comparative Values for Heat Pipe Resistances [from Asselman and Green (1973)].

Resistance	°C/Watt
R_{pe} and R_{pc}	10^{-1}
R_{we} and R_{wc}	10^{+1}
R_{ie} and R_{ic}	10^{-5}
R_{va}	10^{-8}
R_{pa}	10^{+2}
R_{wa}	10^{+4}

for flat plates, where δ is the plate thickness and A_e is the evaporator area, or

$$R_{pe} = \frac{\ell n \, (d_o/d_i)}{2\pi L_e k_p} \tag{4.32}$$

for cylindrical pipes, where L_e is the evaporator length. An expression for the equivalent thermal resistance of the liquid/wick combination in circular pipes is

$$R_{we} = \frac{\ell n \, (d_o/d_i)}{2\pi L_e k_{eff}} \tag{4.33}$$

where values for the effective conductivity, k_{eff}, can be found in Table 4.3. For sintered metals, an investigation performed by Peterson and Fletcher (1987) found that the method presented by Alexander (1972), and later by Ferrell et al. (1972), for determining the effective thermal conductivity of saturated sintered materials was the most accurate over the widest range of conditions.

The adiabatic vapor resistance, although usually negligible, can be found as

$$R_{va} = \frac{T_v \, (P_{v,e} - P_{v,c})}{\rho_v \lambda q} \tag{4.34}$$

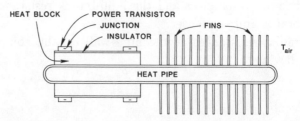

Figure 4.6: Schematic drawing of a heat pipe heat sink.

where $P_{v,e}$ and $P_{v,c}$ are the vapor pressures at the evaporator and condenser.

Combining these individual resistances provides a mechanism by which the overall thermal resistance can be computed and hence the temperature drop associated with various axial heat fluxes can be computed.

4.2 Relative Weight

In addition to the substantial reduction in the temperature difference, the use of heat pipes normally results in significant decreases in weight when compared to solid conductors. In the example in section 4, the solid copper and aluminum rods would have weights which would be 13.7 and 4.2 times as heavy as the simple copper-water heat pipe described.

5 HEAT PIPE HEAT SINKS

5.1 Cooling Components and Hot Spots: Earth Environment

Since the thermal resistance of a heat pipe is very small compared to conduction in a solid metal rod, it is not surprising to find heat pipes used as heat sinks to cool electronic components. The simplest heat sinks are in the shape of cylindrical heat pipes that are available commercially in a wide range of sizes and materials. The most common type of moderate-temperature heat pipe today is constructed of copper with water as the working fluid and with either a screen wick structure or internal axial grooves. These pipes have diameters

Figure 4.7: Cross-section profile of spiral rectangular groove wick [Murase et al. (1982)].

as small as 4 *mm* and can be bent into various shapes to accommodate irregular geometries. A typical heat pipe heat sink is shown
in Fig. 4.6. Generally, the heat source (e.g., a power transistor, a
thyristor, or a thermoelectric generator) is attached mechanically to
the heat pipe, forming the evaporator end. A substantial thermal
resistance can exist at this mechanical junction unless proper care
is taken by clamping the device with sufficient pressure and/or by
applying thermal grease to the contacting surfaces. A series of fins
are attached (either by soldering, brazing or press fitting) to the
condenser end to permit adequate heat rejection to the ambient air.
The air may be blown over the fins with a fan or convected away
naturally. The fin design depends on whether convection is forced or
natural.

One such heat sink, manufactured by Furukawa Electric Co.,
Ltd., is the HEAT KICKER (Murase et al., (1982) and Yoshida
et al., (1984)). Depending on the application, the heat source can
be located at one end of the heat pipe, as shown in Fig. 4.6, or in
the center of the heat pipe with heat-rejecting fins at either end.
There are, however, different wicks used in these two configurations.
The configuration shown in Fig. 4.6, has a 150-mesh copper screen
wick, whereas the centrally-heated version has a spirally-grooved (either triangular or rectangular groove) wick. Figure 4.7 shows a cross
section of the spiral, rectangular groove wick.

The thermal performance of the screen-wick heat pipe is shown
in Fig. 4.8 where the sensitivity of the heat pipe to inclination angle
is readily observed. For example, in a horizontal orientation at a
vapor temperature of 80°C, the heat pipe can transport about 500
watts. If the heat pipe is tilted 5 degrees with the evaporator end
down (i.e., gravity assisted), the heat transport capacity increases
to about 600 watts. When the heat pipe is tilted 5 degrees in the
opposite direction, however, with the condenser end down, the heat
transport capacity drops to about 300 watts. Similar curves for the
centrally-heated, grooved heat pipe are presented in Fig. 4.9. With
this wick, the heat pipe thermal transport capacity at the same vapor
temperature of 80°C is about 1400 watts when horizontal and about
1100 watts when inclined ±5 degrees. Clearly, the performance of
this heat pipe is far superior to the screen-wick heat pipe, which ex-
hibits a larger liquid flow resistance than the axial grooves. Murase

et al. (1982) showed that, as a heat sink in an audio amplifier system, the grooved heat pipe has a lower thermal resistance than an extruded aluminum heat sink. This improved performance can lead to a 50 percent savings in weight, as shown in Fig. 4.10.

The use of multiple, parallel heat pipes to cool high power thyristors has been demonstrated by Murase et al. (1987). Several heat pipes are mounted in parallel into a common heater block with common, heat rejecting fins. The heat pipes are made of copper/groove-wick/water, with an outside diameter of 15.9 *mm* and a length of 1200 *mm*. The aluminum fins are placed at a pitch of 8 *mm* to be suitable for natural convection. A black coating enhances radiation effects. Kolb (1987) has shown that a 500 watt, gas-fired, thermoelectric generator can be effectively cooled with 13 gravity-assisted, copper/water heat pipes containing aluminum fins. Aakalu and Carlen (1984) demonstrated the use of cylindrical heat pipes to transfer heat from circuit modules of a large computer to water-cooled, heat exchanger units.

In all of these applications, the heat pipes must be chosen properly to ensure good system integration with the fins and the air cooling flow available. The overall thermal resistance of the heat pipe must be balanced against the other thermal resistances such as the contact resistance and the convective resistance of the air. Some potential users still express concern over the reliability of heat pipes.

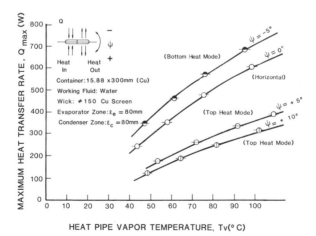

Figure 4.8: Maximum heat transfer characteristics (Screen heat pipe) [Murase et al. (1982)].

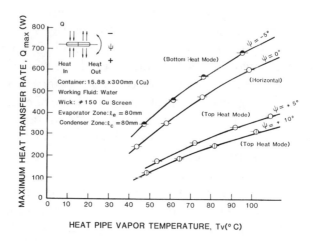

HEAT PIPE VAPOR TEMPERATURE, Tv(° C)

Figure 4.9:   Maximum   heat   transfer   characteristics (Grooved heat pipe) [Murase et al. (1982)].

As mentioned earlier, if heat pipe materials are not selected properly, or if heat pipes are not cleaned or sealed properly, they may suffer a gradual deterioration in thermal performance as noncondensible gases collect in the condenser. If there is any uncertainty in the heat transport capacity of a heat pipe, the designer may derate the heat pipe and use several of them in parallel to ensure sufficient cooling capacity. Properly manufactured heat pipes achieve significant lifetimes. However, lifetime tests similar to those performed by Basiulis et al. (1976), Basiulis and Filler (1971), and others, must be continually updated to convince potential users that heat pipes can be built reliably for long, effective, operational lifetimes.

5.2   Heat Sinks: Space Environment

As larger spacecraft are developed, there is an increasing need for heat pipes to transfer heat from hot components to suitable heat rejection devices. A heat sink concept called the capillary pumped loop (CPL) has a unique two-phase thermosyphon together with a capillary wick to transfer heat in a zero-g environment (Chalmers et al. (1986) and Ku et al. (1987)). A schematic of the capillary pumped loop is shown in Fig. 4.11. Multiple, parallel evaporators and condensers allow fabrication of rectangular cold plates. Only the evaporator section contains a capillary wick structure. The condenser is made of smooth tubing, but internally-finned tubing could be used to enhance the heat transfer. A cross section of an evaporator tube is shown in Fig. 4.12. Extruded aluminum tubing with internal, longitudinal, trapezoidally-shaped fins is press fit with a porous tubular wick (pore size $\simeq 10 \ \mu m$). Liquid from the condenser flows to the evaporator in the center liquid return channel and is then distributed radially to the tips of the trapezoidal fins. Heat added to the evaporator wall flows to the fin-wick interface and evaporates the working fluid. The vapor flows into the axial vapor flow channels and proceeds toward the condenser. With this design using ammonia, evaporator heat fluxes in excess of 15 W/cm^2 have been achieved. As the vapor enters the condenser tubes and begins to condense, a thin liquid film collects on the wall. Further down the condenser, large intermittent slugs of liquid and vapor collect. Eventually, all the vapor is condensed and sub-cooled liquid flows back to the evaporator to complete the loop. An isolator decouples each liquid flow

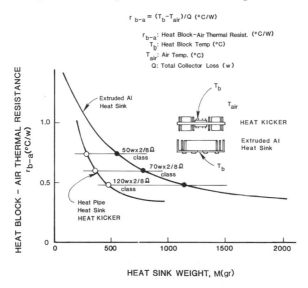

Figure 4.10: Comparison of thermal resistances vs weight for two types of heat sink [Murase et al. (1982)].

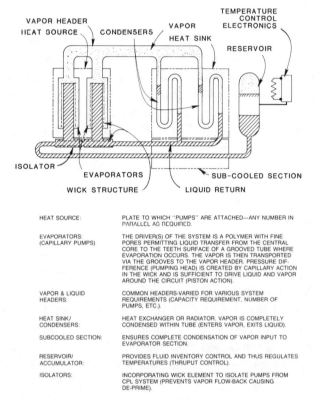

HEAT SOURCE:	PLATE TO WHICH "PUMPS" ARE ATTACHED—ANY NUMBER IN PARALLEL AS REQUIRED.
EVAPORATORS: (CAPILLARY PUMPS)	THE DRIVER(S) OF THE SYSTEM IS A POLYMER WITH FINE PORES PERMITTING LIQUID TRANSFER FROM THE CENTRAL CORE TO THE TEETH SURFACE OF A GROOVED TUBE WHERE EVAPORATION OCCURS. THE VAPOR IS THEN TRANSPORTED VIA THE GROOVES TO THE VAPOR HEADER. PRESSURE DIFFERENCE (PUMPING HEAD) IS CREATED BY CAPILLARY ACTION IN THE WICK AND IS SUFFICIENT TO DRIVE LIQUID AND VAPOR AROUND THE CIRCUIT (PISTON ACTION).
VAPOR & LIQUID HEADERS:	COMMON HEADERS-VARIED FOR VARIOUS SYSTEM REQUIREMENTS (CAPACITY REQUIREMENT, NUMBER OF PUMPS, ETC.).
HEAT SINK/ CONDENSERS:	HEAT EXCHANGER OR RADIATOR. VAPOR IS COMPLETELY CONDENSED WITHIN TUBE (ENTERS VAPOR, EXITS LIQUID).
SUBCOOLED SECTION:	ENSURES COMPLETE CONDENSATION OF VAPOR INPUT TO EVAPORATOR SECTION.
RESERVOIR/ ACCUMULATOR:	PROVIDES FLUID INVENTORY CONTROL AND THUS REGULATES TEMPERATURES (THRUPUT CONTROL).
ISOLATORS:	INCORPORATING WICK ELEMENT TO ISOLATE PUMPS FROM CPL SYSTEM (PREVENTS VAPOR FLOW-BACK CAUSING DE-PRIME).

Figure 4.11: Basic CPL schematic and functional description [Chalmers et al. (1986)].

path and prevents vapor backflow into the individual evaporators. A two-phase reservoir can achieve variable conductance by using excess liquid flooding in the condenser rather than noncondensible gases.

The concept of the capillary pumped loop has been proposed by Chalmers et al. (1986) to design an advanced two-phase cold plate to cool spacecraft electronic equipment. The evaporator heat sink contains six parallel evaporators bonded into a common aluminum plate. The specific design requirements of the prototype capillary cold plate (PCCP) are shown in Table 4.6. It is manufactured from aluminum (except for the stainless steel isolators) and uses ammonia as the working fluid. The first prototype cold plate is under construction. Proposed advanced designs include composite plate materials, superior bonding techniques, and an improved wick design to increase the capillary pumping pressure.

6 COOLING CIRCUIT BOARDS

As the power density of printed wiring boards increases, metal conductors become less attractive because of unacceptably large temperature variations that can occur on the board. This is especially critical for military avionic applications with very high speed integrated circuits (VHSIC). In this situation, the heat pipe is appealing because it can receive the heat at one heat flux and reject it at another with little temperature change. Also, the heat pipe is ideally suited for isothermal applications.

A technique to cool hot components on wiring boards is placement of the hot components in direct contact with small, flat heat pipes that can transfer the heat out to the edges of the board to be dissipated through suitable liquid-cooled or air-cooled side walls (Token (1986)). This technique is shown schematically in Fig. 4.13 where a series of dual-in-line packages are in direct contact with heat pipes aligned in parallel. The required heat pipe must be flat and very thin, and requires a novel wick design. As depicted in Fig. 4.14, the heat pipe should have dimensions of about 1-1.5 *mm* in thickness, 4-6 *mm* in width and a length of approximately 150-300 *mm*, depending on the wiring board dimensions. The heat pipe case and wick are made of stainless steel and the working fluid is methanol. Notice that one face of the heat pipe is crimped to insure that the sintered wick material is firmly pressed against the inner walls of the case.

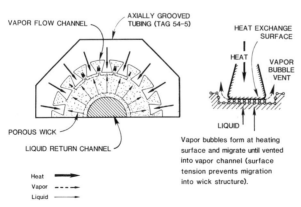

Figure 4.12: CPL evaporator pump and isolator details [Chalmers et al. (1986)].

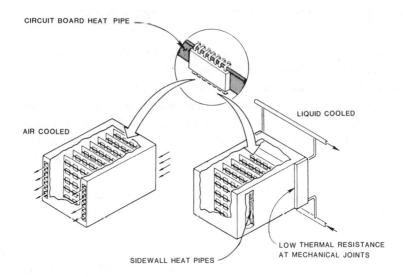

Figure 4.13: Heat pipe cooling of circuit cards [Token (1986)].

Figure 4.15 compares the thermal performance of a flat heat pipe to that of solid aluminum and copper strips of the same dimensions. The heat pipe shows nearly isothermal conditions over its entire length, with a temperature rise above the heat sink of only 7°C. The copper and aluminum strips exhibit a temperature rise above the heat sink of approximately 40 to 95°C, respectively. The heat pipe therefore has two advantages over metal conduction. Cooler and more uniform temperatures can be maintained than by pure conduction. Notice in this application that, because the heat pipe evaporator is much larger than the condenser, the heat flux in the condenser is much higher than in the evaporator.

An alternative to using individual heat pipes to cool wiring board components is to design the wiring board itself as a flat plate heat pipe, as shown in Fig. 4.16. Basiulis et al. (1987) proposed two designs: a series of flat heat pipes embedded within the walls of the wiring board or the wiring board itself as a vapor chamber, Fig. 4.17. The advantages and disadvantages of each of these techniques can lead to various design tradeoffs. For example, the embedded heat pipes enable different materials to be used for the printed wiring

Table 4.6: Summary of Main PCCP Design Requirements [from Chalmers et al. (1986)].

General Requirements/Objectives

- Compatible for inclusion in NASA two-phase thermal test beds

- Demonstrate high heat load capability of capillary pumped two-phase instrument heat acquisition device

- Demonstrate prototypical Space Station hardware

PCCP-Specific Requirements

Construction	Square plate and flat < 0.05 cm over 30.5×30.5 cm
Heat Flux (top surface)	Nominal: 3.23 W/cm^2 (3 kW over 930 cm^2) Max: 10 W/cm^2 (3 kW over 300 cm^2 area)
Heat Input/Removal	3 kW/300 W
Maximum Operating Pressure	1.41 MN/m^2 (Saturation pressure of ammonia at $T = 80^\circ C$)
Temperature Gradient	$\pm 5^\circ C$ (Point to point across and/or through the plate) at 3.23 W/cm^2
Horizontal Tilt	1.27 cm (minimum)
Capillary Static Lift	3200 N/m^2 at 20°C
Pressure Drop Through Plate	700 N/m^2 (maximum)
Evaporator Film Coefficient	8.5 $kW/m^2 C$
Priming	Reprime after dryout by reducing power by 30%

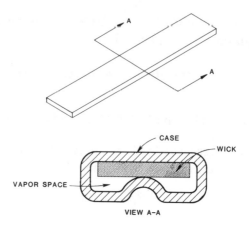

VIEW A-A

PHYSICAL DATA

- Working Fluid – Methanol
- Case Material – Type 304 Stainless Steel
- Wick Material – Sintered Type 304 Stainless Steel
  Wire Rovings

- Length – 6 in. to 14 in.
- Cross-Sectional Dimensions
  - 0.076 in. x 0 220 in.
  - 0.060 in. x 0.185 in.
  - 0.040 in. x 0.160 in.

Figure 4.14: Avionic flat heat pipe design [Token (1986)].

board and for the heat pipe. Also, heat pipes with different working
fluids may be used in different sections of the board to accommodate
non-uniform power dissipation requirements. This is not possible in
the vapor chamber design, but the vapor chamber should have more
uniform surface temperatures with fewer hot spots. With embedded
heat pipes, the wiring board skin can be constructed of materials with
low coefficients of thermal expansion in order to be compatible with
VHSIC and leadless chip carrier technology (Basiulis et al., (1987)).
Figure 4.18 is a photograph of the embedded heat pipe components
before assembly; Figure 4.19 shows a completed, heat-pipe-cooled
wiring board. A similar approach to solving hot spot problems on a
flat plate heat sink was proposed by Ciekurs (1986).

He points out that, with this type of device, the evaporative cooling that occurs near a hot spot causes a three-dimensional vapor flow,
which diffuses the heat uniformly over the heat sink face, aiding in
the effectiveness of the heat rejection process. Additional advantages
are a weight savings and an overall reduction in package size.

Nelson et al. (1978a) proposed a more intricate circuit card cooling scheme with a composite wick structure as shown in Fig. 4.20. The circuit card heat pipe case is made of beryllium copper. The wick is stainless steel screen in the evaporator and condenser. With a sintered fiber, stainless steel artery, a high capillary pumping pressure is accompanied by a low liquid flow resistance. Methanol is the working fluid.

The designs described are, as yet, costly to manufacture compared to solid metal conductors, but in some applications the heat pipe may be the only technically feasible solution. Ingenious, inexpensive fabrication techniques will serve to make these flat plate heat pipes even more attractive.

7 DIRECT COOLING OF SEMICONDUCTOR CHIPS

When the power dissipation of a semiconductor is very high, it may be necessary to cool the device directly by immersion in a dielectric liquid. Direct immersion cooling generally involves nucleate pool boiling from the chip(s). If the liquid pool is saturated, a vapor

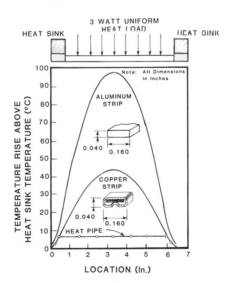

Figure 4.15: Comparison of temperature rise of heat pipe to metal conductors [Token (1986)].

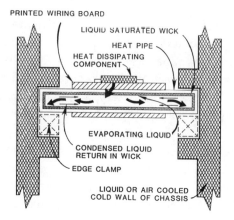

Figure 4.16: Edge-cooled heat pipe printed wiring board [Basiulis et al. (1987)].

space condenser is required, and this two-phase loop (i.e., the boiling of the liquid, the condensation of the vapor, and the return of the condensate) may be viewed as one form of a two-phase, closed loop, thermosyphon. If the pool is subcooled, a submerged condenser is required (Kraus and Bar-Cohen, 1983). Recent developments in direct liquid cooling of electronic components are reviewed by Bergles and Bar-Cohen (1988).

When nucleate boiling occurs from semiconductor chips, with the generation and release (or perhaps collapse) of vapor bubbles, there

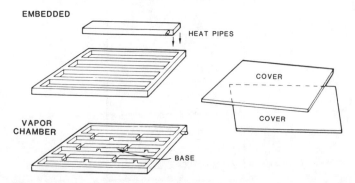

Figure 4.17: Embedded heat pipe and vapor chamber concepts for cooling printed wiring boards [Basiulis et al. (1987)].

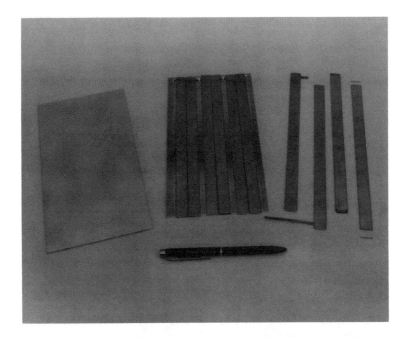

Figure 4.18: Photograph of pre-assembled embedded heat pipe board [courtesy of Hughes Aircraft Company].

are several important limitations: (1) Above the nucleate boiling critical heat flux, the vapor blankets the heat source, leading to dry out and overheating (Kraus and Bar-Cohen, (1983)), (2) the bubbles may generate dynamic forces on the chips and leads, creating high frequency mechanical vibration and subsequent failure, and (3) the electric breakdown voltage of the dielectric fluid may decrease in the presence of vapor bubbles.

An alternative to nucleate boiling is direct evaporation (with no bubble nucleation) of a very thin liquid film; a wicking structure can ensure that fresh liquid is in contact with the heat source. Thus the principle of the heat pipe, on a very small size scale, may be ideally suited to remove heat from particular semiconductor devices. Dean (1976) first proposed an integral heat pipe to cool microelectronics packages. A few years later, Nelson et al. (1978) proposed direct heat pipe cooling of semiconductor devices. They were concerned with ways of reducing the junction-to-case thermal resistance for various power transistors. Figure 4.21 shows a cross section of their proposed

Figure 4.19: Photograph of printed wiring board with embedded heat pipes [courtesy of Hughes Aircraft Company].

heat-pipe-cooled transistor. In this application, it is imperative that neither the wick material nor working fluid chemically or electrically interact with the semiconductor device. They explored various wick concepts, including glass fiber bundles (fiber glass strands 0.1 mm in diameter, spaced 0.5 mm apart to permit a high heat flux density). Difficulties were encountered in applying the fiber bundle wick to microcircuits because of the numerous lead wires. This required tedious hand placement of the fiber bundles. Figure 4.22 shows a photograph of a fiber bundle wick applied to a high frequency power transistor. As an alternative, a high performance powder wick was

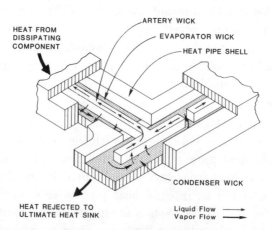

Figure 4.20: Flow paths in the Hughes circuit card heat pipe design [Basiulis et al. (1978)].

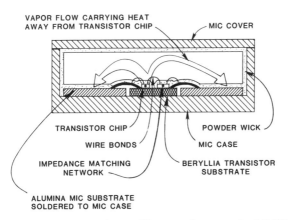

Figure 4.21: Cross section of heat pipe cooled MIC RF transistor [Nelson et al. (1978a)].

developed and patented (Sekhon et al. (1977)), which could be applied by mass production techniques. Figure 4.23 is a photograph of the powder wick applied to a high frequency transistor. Comparison of the thermal performance of these wicks to normal pool boiling reveals a substantial reduction in the thermal resistance. Figure 4.24 shows the reduction in junction temperature that occurs when a conventionally-bonded transistor and heat sink are exposed to two-phase cooling. Clearly, the powder wick heat pipe is far superior to either pool boiling or to the fiber wick heat pipe, thus demonstrating the effectiveness of this cooling technique. In addition, it would be reasonable to utilize this technique to cool various semiconductor chips within a given case.

Other direct heat pipe cooling techniques have been proposed. Eldridge and Peterson (1983) proposed the scheme shown in Fig. 4.25. Here, the reverse side of the integrated circuit chip is bonded to a silicon cooling wafer that has a top surface etched with grooves. A heat pipe is fixed to the top of this cooling wafer. Wicking material lines the heat pipe, making contact with the etched grooves to insure good distribution of liquid. The grooves result in a larger heat transfer surface area and thin film evaporation (or nucleate boiling) without disturbing the IC chip. Figure 4.26 shows a cross section of a device proposed by Kromann et al. (1986). Heat generated in the flip chip die is transferred to the liquid in the wick. The liquid evaporates and the vapor travels to the condenser which is a series

of parallel channels within the lid of the device. The liquid condensate returns by gravity to the porous wick. There are eight layers
of polyester cloth in the wick: The top layer in contact with the lid
grooves is 208 mesh; layers 2 through 7 are 302 mesh; and layer 8 is
421 mesh. Pentane is the working fluid. Figure 4.27 compares the
internal thermal resistance of this device at various die heat fluxes.
For wickless operation, as the percent fill (i.e., the ratio of working
fluid volume to the internal void volume) was increased, the internal
thermal resistance decreased. For the same percent fill of 52%, however, the presence of the wick reduced the thermal resistance over the
heat flux range of 10-18 W/cm^2. This was attributed to the ability
of the wick to provide fresh liquid in contact with the hot die surface.

Figure 4.22: Fiber bundle wick appied to a high frequency power transistor [Nelson et al. (1978a)], [courtesy
of Hughes Aircraft Company].

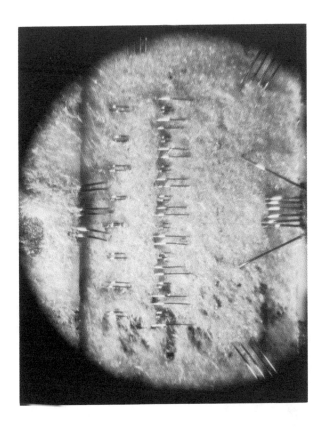

Figure 4.23: Powder wick applied to 45-watt RF power transistor [Nelson et al. (1978a)], [courtesy of Hughes Aircraft Company].

Although these results are promising, concern about the long-term reliability of the device must be investigated before commercial use is feasible. Concerns expressed by the authors include material compatibility between the working fluid and the materials it contacts (wick, lid, die, etc.), generation of noncondensible gases, degradation of the wick material by contaminants, and evaporation/boiling induced pitting in the die surface. Other two-phase cooling schemes have been proposed by Moran and Simons (1979) and Andros and Shay (1980).

Kiewra and Wayner (1986) investigated evaporation from a thin liquid film in a wickless, two-phase thermosyphon in order to cool a disc-shaped heat source. Fluid flow in ultra-thin films in the presence of evaporation was investigated in an effort to optimize the basic mechanisms of heat, mass, and momentum transfer in situations where both body and interfacial forces were present.

Cotter (1984) described very small, micro heat pipes for cooling microelectronic devices. He defines a micro heat pipe as "one so small that the mean curvature of the vapor-liquid interface is necessarily comparable in magnitude to the reciprocal of the hydraulic radius of the total flow channel." In practical terms, a micro heat pipe is a wickless, non-circular channel with a diameter of 10-500 μm and a length of about 10-20 mm. Figure 4.28 is a schematic of a possible micro heat pipe shaped in an equilateral triangle. Cotter proposes micro heat pipes for cooling the volumetric heating produced in a parallel-processing microelectronic component. The evaporator would be an integral part of the device. Heat would be removed from within the device by an array of micro heat pipes. An approximate theoretical result for the maximum heat transport capability, q_m, of a micro heat

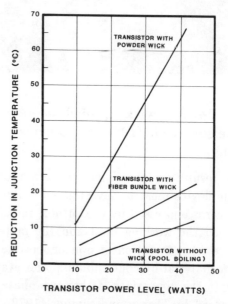

Figure 4.24: Comparison of reduction in junction temperature achieved by fiber bundle and powder wick and non-wicked transistors immersed in heat pipe fluid [Nelson et al. (1978a)].

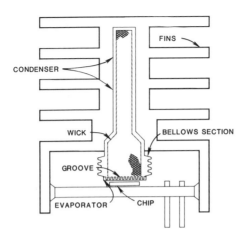

Figure 4.25: Heat pipe cooling of semiconductor chip [Eldridge and Peterson (1983)].

pipe was derived. For a micro heat pipe with an equilateral triangular shape with sides 0.2 *mm*, length 10 *mm*, and an optimum amount of methanol as the working fluid, Cotter predicts a q_m of 0.03 watts. He reasoned that, when used in a solid in an array of parallel pipes (about 10% by volume), these micro heat pipes could provide a few tens of watts per cubic centimeter of cooling. Besides the difficulties of manufacturing, cleaning and filling these micro heat pipes (Cotter believes that existing micro-mechanical technology could overcome these problems), the micro heat pipe is very sensitive to percent fill of the working fluid. The liquid-vapor interface changes continually along the pipe, as shown in Fig. 4.29, and care must be exercised to ensure proper wetted conditions without flooding the micro channels. In summary, these unique devices may play an important role in future microelectronic devices where thermal loads are modest and where isothermal conditions throughout the devices are highly desirable.

8 CABINET COOLING USING HEAT PIPE HEAT EXCHANGERS

In many applications, electronic equipment is contained in a cabinet. Frequently, the ambient air may be contaminated with dust, moisture or an oil mist, and direct contact between the air and the electronic components in this situation is very undesirable. A heat pipe heat exchanger provides a suitable solution to this dilemma. Figure 4.30 shows a schematic representation of a heat pipe heat exchanger. It consists of many parallel finned cylindrical heat pipes arranged in either a staggered or an in-line array. The hot, clean inlet air from inside the cabinet is blown over one end of the heat exchanger (the evaporator); the cold contaminated air from the ambient surroundings is blown over the other end (the condenser). Shah and Giovannelli (1987) provided detailed design information for a heat pipe heat exchanger, including the important coupling of the heat pipe thermal performance to the heat exchange capability of the finned surfaces. The selection of the heat pipe size and materials, as well as fin material and dimensions, will depend upon the amount of energy to be exchanged, the desired temperatures, and the available air flow.

Heat pipe heat exchangers for the specific purpose of cooling electronic equipment mounted in sealed cabinets are available in several countries around the world. These heat exchangers may be mounted either in the ceiling or the sides of the cabinet. Figure 4.31(a) shows

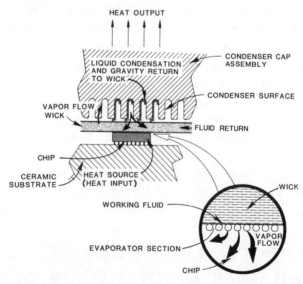

Figure 4.26: Integral heat pipe showing the fluid/vapor transport and the resulting heat transfer [Kromann et al. (1986)].

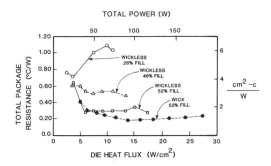

Figure 4.27: Total package thermal resistance [Kromann et al. (1986)].

one such design by Gerak et al. (1987). The heat exchanger has 8 rows, each row with 17 finned heat pipes 16 *mm* in diameter. They are made of copper and contain R-12 as the working fluid. They are mounted vertically with the condenser end up (i.e., gravity assisted). The air side fins may be either copper or aluminum. The overall dimensions of the heat exchanger are: 645 *mm* in width, 264 *mm* in depth and 700 *mm* in height. Figure 4.31(b) demonstrates the dependence of the heat transfer capability of the heat exchanger on the inlet air temperature difference between the hot and cold sides. At a temperature difference of 10 K, over 2000 watts can be transferred.

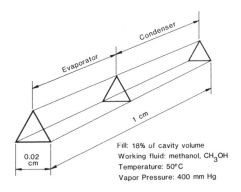

Figure 4.28: Example of a micro heat pipe [Cotter (1984)].

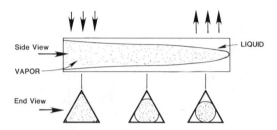

Figure 4.29: Steady state operation of a micro heat pipe [Cotter (1984)].

9 SUMMARY AND OUTLOOK

Heat pipe technology has been developing since the mid-sixties. Cylindrical heat pipes are available commercially in a wide range of sizes. They can be utilized as heat sinks to cool electronic components and packages or as heat pipe heat exchangers to cool sealed cabinets. Cylindrical heat pipes offer a low weight, long life solution to cooling electronics systems. Little additional research is needed on these simple devices provided that materials are chosen with proven compatibility and proper cleaning, filling, and sealing procedures are followed.

Flat-plate-type heat pipes or vapor chambers which are an integral part of cold plates and/or printed wiring boards can provide a surface with minimal non-uniformities in temperature. They require additional developmental work to arrive at the best choice of wick material, evaporator/condenser dimensions, working fluid percent fill, etc. They can be manufactured for one-of-a-kind applications, and are therefore costly however, their expense will decrease in the future as designs become more standardized and production volumes increase. The use of heat pipes (or two-phase thermosyphons) to directly cool semiconductor devices is particularly promising. As semiconductor devices perform more and more operations over a shorter period of time, increases in power density will require the use of phase change heat transfer for temperature control. More research is therefore needed on small-scale, two-phase flow experiments, including film evaporation and surface-tension-driven condensation in order to arrive at the best designs. The use of micro heat pipes should be explored thoroughly, including their theoretical performance and their

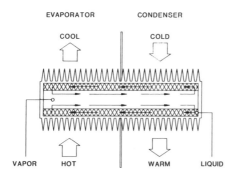

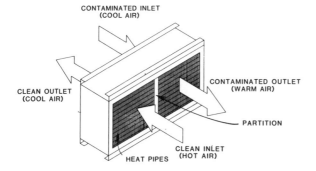

Figure 4.30: Heat pipe heat exchanger.

practical size limitations. Heat pipes in space, in a zero-gravity environment, will continue to expand as larger spacecraft are developed. More effort, perhaps, should be directed to cryogenic applications.

10 NOMENCLATURE

A area, m^2
d diameter, m
f drag coefficient, dimensionless
g gravitational constant, 9.807 m/s^2
k thermal conductivity, $W/m\text{-}K$
K wick permeability, m^2
L length, m

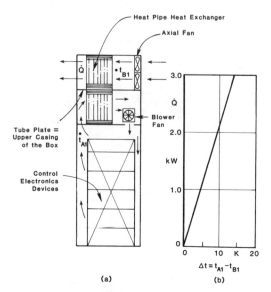

(a)   (b)

Figure 4.31: Heat pipe heat exchanger to cool electronic equipment in a sealed cabinet (Gerak et al. (1987)).

Ma Mach number, dimensionless

N screen mesh number, m^{-1}

N_ℓ liquid Figure of Merit, W/m^2

P pressure, N/m^2

q heat flow rate, W

R thermal resistance, K/W; or universal gas constant, J/kg-

Re Reynolds number, dimensionless

r radius, m

s groove pitch, m

S crimping factor, dimensionless

T temperature, K

V velocity, m/s

w groove width, m; or wire spacing, m

We Weber number, dimensionless

Greek

β half of included angle, degrees or radians

γ specific heat ratio, dimensionless

δ groove depth, m

ϵ wick porosity, dimensionless

λ latent heat of vaporization, J/kg

μ dynamic viscosity, kg/m-s

ρ density, kg/m^3
σ surface tension, N/m
τ shear stress, N/m^2
Ψ angle of inclination, degrees or radians

Subscripts

a adiabatic section, air
b boiling
c capillary, capillary limitation, condenser
e entrainment, evaporator section
eff effective
f fin
h hydraulic
i inner
ℓ liquid
m maximum
n nucleation
o outer
p pipe
s sonic or sphere
v vapor
w wire spacing, wick
$\parallel$ axial hydrostatic pressure
$+$ normal hydrostatic pressure

11 REFERENCES

Aakalu, N G and Carlen, R A (1984). Heat pipe links for water-cooled large computer circuit assemblies, IBM Technical Disclosure Bulletin **27**, 3551-3552.

Alario, J, Brown, R and Otterstadt, P (1984). Space constructable radiator prototype test program, *AIAA Paper 84-1793*.

Alexander, E G Jr (1972). Structure-property relationships in heat pipe wicking materials, Ph.D. Thesis, North Carolina State University, Dept. of Chemical Engineering.

Andros, F E and Shay, R J (1980). Micro bellows thermo-capsule, U. S. Patent 4,212,349.

Antoniuk, D (1987). Generalized modeling of steady-state and transient behavior of variable conductance heat pipes, *AIAA Paper 87-1615*.

Asselman, G A A and Green, D B (1973), Heat pipes, *Phillips Technical Review* **16**, 169-186.

Barantsevich, V L, Barakove, L V and Tribunskaja, I A (1987). Investigation of the heat pipe corrosion resistance and service characteristics. *Proc 6th Int Heat Pipe Conf*, Grenoble, France, 188-193.

Basiulis, A and Filler, M (1971). Operating characteristics and long life capabilities of organic fluid heat pipes, *AIAA Paper 71-408*.

Basiulis, A, Prager, R C and Lamp, T R (1976). Compatibility and reliability of heat pipe materials, *Proc 2nd Int Heat Pipe Conf*, Bologna, Italy, 357-372.

Basiulis, A, Tanzer, H, and McCabe, S (1987). Heat pipes for cooling of high density printed wiring boards, *Proc 6th Int Heat Pipe Conf*, Grenoble, France, 531-536.

Bergles, A E and Bar-Cohen, A (1988). Recent developments in direct liquid cooling of electronic components, *Research and Modeling of Thermal Phenomena in Electronic Systems* 1, Hemisphere Publishing Corporation, New York.

Bliss, F E (1970). Construction and test of a flexible heat pipe, *ASME Paper 70-HT/SpT-13*.

Brennan, P J and Kroliczek, E J (1979). *Heat Pipe Design Handbook*, B&K Engineering, Inc., Towson, MD.

Brost, O and Mack, H (1987). A liquid controlled variable conductance heat pipe, *Proc 6th Int Heat Pipe Conf*, Grenoble, France, 549-555.

Chalmers, D R, Kroliczek, E J and Ku, J (1986). Design of an advanced two-phase capillary cold plate, *Proc SAE 16th Intersoc Conf on Env Systems*, San Diego, CA.

Chang, W S and Colwell, G T (1985). Mathematical modeling of the transient operating characteristics of a low-temperature heat pipe, *Numerical Heat Transfer* **8**, 169-185.

Chi, S W (1976). *Heat Pipe Theory and Practice*, McGraw-Hill Publishing Company, New York.

Ciekurs, P V (1986). An approach to solving hot spot cooling problems in electronic packaging, *ASME* **86-WA/EEP-1**.

Colwell, G T and Chang, W S (1984). Measurements of the transient behavior of a capillary structure under heavy thermal loading, *Int J of Heat and Mass Transfer* **27**, (4), 541-551.

Cotter, T P (1984). Principles and prospects of micro heat pipes, *Proc 5th Int Heat Pipe Conf*, Tsukuba, Japan, 328-335.

Cullimore, B (1985). Modeling of transient heat pipe effects using a generalized thermal analysis program, *AIAA Paper 85-0938*.

Dean, D S (1976). An integral heat pipe package for microelectronic circuits, *Proc 2nd Int Heat Pipe Conf*, Bologna, Italy, 481-502.

Dunn, P D and Reay, D A (1982). *Heat pipes*, 3rd edition, Pergamon Press, New York.

Eastman, G Y (1968). The heat pipe, *Scientific American* **218**, (12), 38-46.

Eldridge, J M and Peterson, K E (1983). Heat pipe vapor cooling etched silicon structure, IBM Technical Disclosure Bulletin **25**, (8), 4118-4119.

Eninger, J E (1975). Capillary flow through heat pipe wicks, *AIAA Paper 75-661*.

Feldman, K T (1976). *The heat pipe: theory, design and applications*, Technology Application Center, Univ. of New Mexico, Albuquerque, NM.

Ferrell, J K and Alleavitch, J (1969). Vaporization heat transfer in capillary wick structures, Preprint **6**, *ASME-AIChE Heat Transfer Conf*, Minneapolis, MN.

Ferrell, J K, Alexander, E G and Piver, W T (1972). Vaporization heat transfer in heat pipe wick materials, *AIAA Paper 72-256*.

Freggens, R A (1969). Experimental determination of wick properties for heat pipe applications, *Proc 4th Intersoc Energy Conversion Eng Conf*, Washington, D.C., 888-897.

Furukawa, M, Imai, R, Miyazaki, Y and Oshima, S (1987). Pump assisted Heat Pipe, *Proc 6th Int Heat Pipe Conf*, Grenoble, France, pp. 315-320.

Gaugler, R S (1944). Heat transfer devices, U. S. Patent 2,350,348.

Gerak, A, Horvath, L, Jelinek, F, Stulc, P and Zboril, V (1987). Examples of heat pipe application in chemical, electrical and other industries, *Proc 6th Int Heat Pipe Conf*, Grenoble, France, 522-530.

Gosse, J (1987). The thermo-physical properties of fluids on liquid-vapor equilibrium: An aid to the choice of working fluids for heat pipes, *Proc 6th Int Heat Pipe Conf*, Grenoble, France, 17-21.

Groll M, Supper, W and Savage, C J (1982). Shutdown characteristics of an axial-grooved liquid-trap heat pipe thermal diode, *J of Spacecraft* **19**, (2), 173-178.

Grover, G M, Cotter, T P and Erikson, G F (1964). Structures of very high thermal conductivity, *J Appl Phys* **35**, 1190-1191.

Haug, F, Prenger, F C and Chrisman, R H (1986). Measurement of performance limits in cryogenic heat pipes, *AIAA Paper 86-1255*.

Heine, D and Groll, M (1984). Compatibility of organic fluids with commercial structure materials for use in heat pipes, *Proc of the 5th Int Heat Pipe Conf*, Tsukuba, Japan, 170-174.

Kiewra, E W and Wayner, P C (1986). A small sale thermosyphon for the immersion cooling of a disc heat source, Heat Transfer in Electronic Equipment, ASME Symposium **HTD** (57), Bar-Cohen (Ed), 77-82.

Kolb, H (1987). Heat pipe cooling for thermoelectric generators, *Proc Int Symp on Cooling Technology for Electronic Equipment*, Honolulu, HI 161-172.

Kraus, A D and Bar-Cohen, A (1983). *Thermal Analysis and Control of Electronic Equipment*, McGraw Hill Publishing Company, New York.

Kromann, G B, Hannemann, R J and Fox, L R (1986). Two-phase internal cooling technique for electronic packages, Heat Transfer in Electronic Equipment, ASME Symposium **HTD** (57), Bar-Cohen (Ed), 61-66.

Ku, J, Kroliczek, E J and McIntosh, R (1987). Capillary pumped loop technology development, *Proc 6th Int Heat Pipe Conf*, Grenoble, France, 288-298.

Merrigan, M A, Keddy, E S and Sena, J T (1985). Transient heat pipe investigations for space power systems, Los Alamos Technical Report **LA-UR-85-3341**.

Moran, K P and Simons, R E (1979). Self-regulating evaporative/conductive thermal link, IBM Technical Disclosure Bulletin **21**, 3281-3282.

Murakami, M and Arai, K (1987). Statistical prediction of long-term reliability of copper water heat pipes from acceleration test data, *Proc 6th Int Heat Pipe Conf*, Grenoble, France, 2194-199.

Murase, T, Yoshida, K, Fujikake, J, Koizumi, T and Ishida, N (1982). Heat pipe heat sink HEAT KICKER for cooling of semiconductors, *Furukawa Review* **2**, 24-33.

Murase, T, Tanaka, S and Ishida, S (1987). Natural convection type long heat pipe heat sink "POWERKICKER-N" for the cooling of GTO thyristor, *Proc 6th Int Heat Pipe Conf*, Grenoble, France, 537-542.

Nelson, L A, Sekhon, K S and Fritz, J E (1978). Direct heat pipe cooling of semiconductor devices, *Proc 3rd Int Heat Pipe Conf*, Palo Alto, CA, 373-376.

Nelson, L A, Sekhon, K S and Ruttner, L E (1978). Application of heat pipes in electronic modules, *Proc 3rd Int Heat Pipe Conf*, Palo Alto, CA, 367-372.

Ollendorf, S (1983) Heat pipe thermal switch, *NASA Technical Briefs* **8**, (1), 84.

Peterson, G P (1986) Analytical development and computer modeling of a bellows type heat pipe for the cooling of electronic components, ASME **86-WA/HT-69**.

Peterson, G P and Fletcher, L S (1987). Effective thermal conductivity of sintered heat pipe wicks, *AIAA J of Thermophysics and Heat Transfer* **1**, (3), 36-42.

Rice, G and Fulford, D (1987) Influence of a fine mesh screen on entrainment in heat pipes, *Proc 6th Int Heat Pipe Conf*, Grenoble, France, 168-172.

Roesler, S, Heine, D and Groll, M (1987). Life testing with stainless steel/ammonia and aluminum/ammonia heat pipe, *Proc 6th Int Heat Pipe Conf*, Grenoble, France, 211-216.

Sakuri, Y, Masumoto, H, Kimura, H, Furukawa, M and Edwards, D K (1984). Flight experiments for gas-loaded variable conductance heat pipe on ETS-III active control package, *Proc 5th Int Heat Pipe Conf*, Tsukuba, Japan, 26-32.

Sekhon, K S, Nelson, L A and Fritz, J E (1977). Transistor cooling by heat pipes having a wick of dielectric powder, U. S. Patent 4,047,198.

Shah, R K and Giovannelli, A D (1987). Heat pipe heat exchanger design theory, *Heat Transfer Equipment Design*, Hemisphere Publishing Corporation, Washington, D. C.

Tien, C L (1975). Fluid mechanics of heat pipes, *Annual Review of Fluid Mechanics*, 167-186.

Tien, C L and Chen, S J (1984) Noncondensible gases in heat pipes, *Proc 5th Int Heat Pipe Conf*, Tsukuba, Japan, 97-101.

Token, K (1986). Trends in aircraft thermal management, *Proc Printed Wiring Board Heat Pipe Workshop*, Hughes Aircraft Company, Electron Dynamics Division, Torrance, CA.

Udell, K S and Jennings, J D (1984). A composite porous heat pipe, *Proc 5th Int Heat Pipe Conf*, Tskuba, Japan, 41-47.

Van Buggenum, R I J and Daniels, D H V (1987). Development, manufacturing and testing of a gas loaded variable conductance heat pipe, *Proc 6th Int Heat Pipe Conf*, Grenoble, France, 242-249.

Yoshida, K, Ogiwara, S, Murase, T and Ishida, S (1984). Flat plate heat pipes for cooling devices, *Proc 5th Int Heat Pipe Conf*, Tsukuba, Japan, 174-178.

Zaho, R D, Zhu, Y H and Liu D C (1987). Experimental investigation of the compatibility of mild carbon steel and water heat pipes, *Proc 6th Int Heat Pipe Conf*, Grenoble, France, 200-204.

Chapter 5

THERMAL STRESS FAILURES IN MICROELECTRONIC COMPONENTS – REVIEW AND EXTENSION

E. Suhir
AT&T Bell Laboratories
Murray Hill, New Jersey

Mechanical failure can result in a significant loss of components during fabrication and testing as well as during the normal operating life of the system. This problem has become especially important in the recent decade, mostly in connection with the rapidly developing VLSI technology. As is known, mechanical failure in VLSI devices is one of the major concerns of engineers and designers, since the high degree of integration, high speed, high power level and the high production cost of advanced electronic products have imposed more stringent reliability requirements on individual components.

In this chapter, consideration is given to thermal stress failures – the most typical and, perhaps, the most critical mechanical mode of failure in microelectronics. The emphasis is on the thermal stresses arising in bi- and multimaterial structures due to the thermal expansion (contraction) mismatch of the constituent materials. Accordingly the chapter opens with a brief review of the general thermal stress literature and proceeds to define the state-of-the-art in thermal stress failures of microelectronic components. It concludes with the theoretical evaluation of stresses in adhesively bonded and soldered assemblies used in the electronic industry.

1 THERMAL STRESSES IN NON-ELECTRONIC STRUCTURES

Thermal stress failures are a serious concern in the design and operation of gas turbines, internal combustion engines, aerospace structures, and power stations, and, as a consequence, an enormous number of publications has been devoted to thermoelasticity, thermoplasticity and thermal creep in engineering structures. Many of the methods and solutions discussed in this literature can be successfully applied to microelectronics components. Thermomechanical design problems are usually defined in terms of one or more of the following categories:

- Thermoelastic deformation associated with a comparatively small and uniform change in temperature,
- Stress fields resulting from large variations in temperature and displacements, including nonuniform temperature distribution effects and thermoplasticity,
- Elastic or elasto-plastic deformation, considering time dependent effects, such as thermal shock and other dynamic problems, creep, stress relaxation, stress rupture and thermal fatigue and
- Problems associated with brittle material.

The problem of thermoelastic stresses was first addressed by Poisson in 1829. A more detailed investigation of this problem was carried out in 1838 by Duhamel who presented the equations of thermoelasticity in their modern form and proved the uniqueness of the solution to these equations and the acceptability of the superposition of the effects due to external and thermal loads. Duhamel also examined many specific problems of thermoelasticity. The next important steps were taken by Neuman in 1841 and by Kelvin in 1878 who applied methods of thermodynamics to evaluate thermal stresses. General methods for the solution of two- and three-dimensional problems of thermoelasticity were developed by Borchardt in 1873 and Rayleigh in 1901. Thus, from the point of view of the principle itself, the linear theory of thermoelasticity was completed in the nineteenth century. The fundamentals of thermoelasticity and numerous solutions to various practical problems can be found in monographs by Biot (1956), Gatewood (1957), Parkus (1958, 1959, 1968), Boley and Weiner (1960), Nowacki (1962), and Timoshenko and Goodier (1970). Dynamic problems of thermoelasticity were examined by Lessen (1956), Smith and Robinson (1958), Chadwick (1960) and other investigators.

The problems of thermoplasticity and creep are essentially nonlinear and therefore significantly more complicated. Some fundamental results in this area were obtained only a few decades ago by Prager (1959), Hill (1950), Hoff (1956) and Rabotnov (1969) and the current status of knowledge in the area of thermoplasticity and creep is reflected in the monographs by Finnie and Heller (1959), Manson (1961), Dorn, (1961), Goldenblat and Nikolaenko (1964), Johns (1965), Hilton (1967), and Tien (1985). These books also contain solutions to many thermoelastic problems.

Thermal fatigue mechanisms were studied by Coffin (1954, 1962, 1968), Manson (1961, 1981), Howes (1976), and Raynor and Skelton (1985). The Coffin-Manson formula modified by Engelmaier (1984) for the fatigue life of ductile materials is currently in wide use for the evaluation of the reliability of solder joints in electronic components subjected to thermal cycling.

The behavior of "sandwich" structures, such as bi-metallic thermostats or composite materials, are of especially great importance from the standpoint of microelectronics applications. This behavior was examined by Timoshenko (1925), Aleck (1949), Kuraniski (1959), Dillon (1967), Grimado (1978), Chen et al (1982), and Suhir (1986). Of course, numerous papers dealing with the mechanics of adhesive bonding, can also be of great help in understanding the behavior of "sandwich" structures, subjected to heating or cooling. These include the works of Eley, ed., (1969), Benson (1961), Goland and Reissner (1944), Keer and Chantaramungkorn (1975), Delale et al (1981), Du Chen and Shun Cheng (1983).

A large number of important problems, involving thermal stresses in beams, plates, shells and various solid structures has been solved in recent years. Many useful solutions and a detailed bibliography, in the area of thermal stress and related issues in reliability, can be found in the Journal of Thermal Stresses, published by Hemisphere Publishing.

2 THERMAL STRESS FAILURES IN MICROELECTRONICS

2.1 Failure Modes

The mechanical behavior of materials under elevated or low temperatures is an important and an extremely complicated area of the mechanics of materials. Very few, if any, practical engineering considerations in microelectronics can be isolated from temperature effects. The thermal expansion (contraction) mismatch of materials in multimaterial structures and/or nonuniform temperature distributions result in thermal stresses which are often large enough to cause unacceptable distortions or mechanical damage. The diversity of materials and structures, as well as the wide range of processing, testing and operating conditions encountered in microelectronic devices result in a wide variety of complex thermal stress failure mechanisms.

Mechanical failure is generally defined as any change in the size, shape, material properties or integrity of a structure that renders it incapable of performing its intended function. Mechanical failure of an electronic component might be brought about by one or a combination of many different responses to loads and environments, while in manufacturing, testing, handling or operation. Obviously, success in creating competitive products, while averting premature failures, can be achieved consistently only by recognizing and evaluating all potential failure modes early in the development process.

The major thermal-stress-induced failure modes, which can occur in microelectronic components are as follows:

- Transistor chip junction failure,
- Excessive elastic or plastic deformation,
- Ductile rupture failure,
- Brittle fracture failure,
- Fatigue failure,
- Creep failure,
- Thermal relaxation failure,
- Thermal shock failure and
- Stress corrosion failure.

Transistor junction failure occurs because of the elevated thermal stresses in the IC circuit if the heat produced by a chip cannot readily escape. This results in an irreversible thermal break-down of a p-n junction.

Temperature-induced elastic or plastic deformation failure occurs whenever this deformation, brought about by appreciable temperature difference and/or temperature gradients, exceeds geometric tolerances at the relevant packaging level. Thus, lead-frame expansion may snap delicate wire-bonds and film deposition can result in a processing failure.

Ductile rupture occurs when the plastic deformation in a material results in a progressive local reduction in cross-sectional area. Failures of this type could occur in solder joints of printed wiring boards during manufacturing or testing.

Brittle fracture is encountered when the elasto-plastic deformation results in breaking of the primary interatomic bonds. Pre-existing flaws or fatigue cracks can serve as initiation sites for the rapidly growing granular, multifaceted fracture surfaces often associated with this failure mode. Brittle fracture is the dominant failure mode for silicon and ceramics and often occurs as a consequence of the thermal expansion mismatch between these materials and bonded layers of metals, plastics and polyimides.

Fatigue failure is a general term given to the sudden and catastrophic rupture of a material as a result of the application of fluctuating loads or deformations over a period of time (fatigue life). The loads and deformations that result in a fatigue failure are, as a rule, significantly smaller than the static failure level. When loads or deformations are of such magnitude that more than about ten thousand cycles are required to produce failure, the phenomenon is usually termed "high-cycle fatigue." Otherwise, the term "low-cycle fatigue" is used. Low-cycle fatigue at high stress levels, or at low stress levels as in the case of very ductile materials, is usually characterized by elasto- plastic behavior. Fatigue is, probably, the most typical failure mode for microelectronic components and can well occur in solder joints, wire bonds, various metallizations and nonmetallic interconnections.

Creep failures result whenever the plastic deformation in the material, accruing over time under the influence of stress and temperature, leads to unacceptably large dimensional changes. The following three stages of creep are often observed: (1)"primary" or transient creep, during which time the rate of strain decreases; (2) "secondary" or steady-state creep, during which time the rate of strain is virtually constant, and (3) "tertiary" creep, during which time the rate of strain rapidly increases until rupture occurs. The occurrence of this last stage depends on the stress-temperature history of the component. Creep behavior is typical for solders at elevated temperatures.

Thermal relaxation failure occurs when the dimensional changes, caused by creep, result in excessive relaxation of a prestrained or prestressed structural member. Such stress relaxation in a prestressed flange bolt, in a "sandwich" structure of the type often used in electronic packages, can lead to catastrophic deterioration in the load-carrying capability of this member.

Thermal shock failures occur when extremely high temperature gradients lead to stresses which exceed the yield or fracture limit of the material. Such a failure could occur in a component undergoing rapid temperature cycling tests, in liquids maintained at different temperatures or during immersion in liquid nitrogen. Stress-corrosion failure occurs when the thermal stress in a corrosive environment generates a field of localized surface cracks, usually along grain boundaries, which trigger or facilitate the propagation of major flaws and/or cracks. While these last two failure modes may be encountered in metals used in microelectronic components, such failures are relatively rare.

2.2 Effect of Thermal Stress on the Transistor Junction

An increase in the reverse voltage applied to a transistor junction results in greater dissipated power. This causes an increase in temperature both in the junction and in the adjacent areas of the semiconductor. The temperature increase could be especially great in VLSI devices, where superposition of several IC layers on a chip effectively "buries" the transistor junction. The temperature increase, in its turn, results in a further increase in the reverse voltage and in dissipated power. If the heat cannot readily escape, then the avalanche increase in temperature whose distribution is essentially nonuniform, induces elevated thermal stresses. These can result in an irreversible thermal break-down of the p-n junction. Such a break-down often causes total degradation of the IC device reliability. Obviously, this could happen only in relatively powerful transistors.

The effect of thermal stress on transistor junction performance and reliability was addressed by Lang et al (1970), Ristič and Cvekič (1978), Moeschke and Wlodarsky (1979), Mikoshiba (1981), Anderson et al (1985), and other investigators. Almost all the developed thermal models consider the junction temperature to be in a steady-state condition. A few papers (Anderson et al) consider heat dissipation and thermally induced stresses under transient conditions. This is of importance, for example, for semiconductor power switching devices experiencing dynamic and shock thermal loading.

2.3 Thermal Stresses in Adhesively Bonded Joints

2.3.1 Strength of Bonded Joints Under External Loading

The problems of the mechanical behavior of bonded joints under external and thermal loading are closely related and, therefore, publications addressing the strength of adhesively bonded joints subjected to external loading are of great help in understanding thermally-induced failures. The following failure modes can take place in bonded joints:

- Adherend failure,
- Cohesive failure of the bonding material, i.e., failure of the adhesive material itself and

• Adhesive failure of the bonding material, i.e., failure at the adherend/adhesive interface

The last failure mode is generally encountered at a relatively low load and should not occur in a properly fabricated joint. It is therefore often regarded as a quality control problem rather than a structural problem.

Most of the basic work in the field of bonded joint strength has been reviewed by a number of authors such as Liu (1976), Thongcharoen (1977) and Cooper and Sawyer (1979). The first and the simplest analysis of adhesively bonded joints was performed by Volkersen (1938). It was a "shear lag" analysis, where the only factors considered were the stresses in the adherends and the shear deformation of the adhesive. Obviously, the relationships obtained are more representative of a double, than a single lap joint, since in the latter, overall bending of the adherends inevitably occurs. An approximate analysis of single lap joints with consideration of bending effects is due to Goland and Reissner (1944). In their classical paper transverse normal ("peeling") stresses are determined, in addition to shearing stresses in the interface. A typical distribution of interfacial shearing and peeling stresses, predicted by Goland-Reissner theory, is shown in Fig. 5.1. Note, that the shear stress is non-zero at the ends. This violates the stress-free boundary condition at the edge and is a consequence of ignoring the finite compliance of the adhesive and the variation of the peeling stress in the through-thickness direction.

More accurate and substantially more complicated theories accounting for the above variation were suggested by several researchers such as Kelsey and Benson (1966), Pahoja (1972), Pirvics (1974) and Allman (1977). The main effect of these refined analyses is to move the position of the peak shear stress to about one adherend thickness from the ends. Experimental confirmation of this result was presented by Ojalvo and Eidinoff (1978). The results generated by these theories have also indicated that the simplified approach used by Goland and Reissner (1944) provides acceptable values for the maximum shearing stress.

The linear approach, utilized in the above studies, is appropriate for joints under relatively low loadings when the stresses in the adhesive are not too close to those resulting in its failure. Evidently,

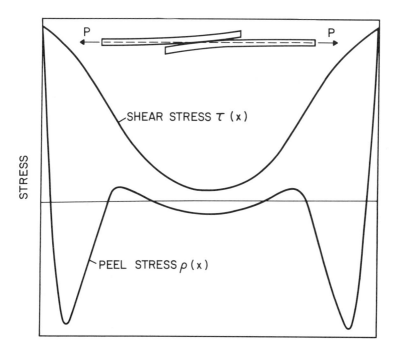

Figure 5.1: Typical variations of shear and peel stress in a single lap joint.

a linear analysis is conservative since it underestimates the ultimate static load for the joint, especially for "ductile" adhesives. Experimental evidence of the importance of adhesive nonlinearity is given, among others, by Grant (1976). A theoretical model, taking into account the elasto-plastic behavior of the attachment material and, in particular, of solder interconnections, was suggested by Suhir (1986).

Considerable experimental work has been performed to verify and extend the theoretical results. Many experiments were carried out using rubber or photoelastic models. The results obtained in this area prior to 1970 were reviewed by Niranjan (1970). By the mid-1960s many researchers started to apply numerical, mainly finite element, methods to the solution of this problem. While a number of authors utilized constant strain quadrilateral elements, others such as Thongcharoen (1977) used elements in which the strain was assumed to vary linearly. Most of the investigations were linear and used mesh sizes that were too coarse to adequately predict the adhesive shear-

stress behavior near free edges where the interfacial stresses were
the greatest. Several investigations such as the one performed by
Liu (1976) were conducted using a nonlinear finite-element procedure to determine the distribution of stresses in the plastic range of
the adhesive material but did not consider the geometric nonlinear
behavior.

The effect of geometric nonlinearity on the distribution of the
elastic stresses in single lap joints was analyzed first by Cooper
and Sawyer (1979). Their results showed good agreement with the
Goland-Reissner theory despite the fact that this theory did not consider the variation of stress through the adhesive thickness. Many
authors used finite element techniques to account for the nonlinear
elastic behavior of the adhesive. A number of researchers investigated the stress concentration and represented the adhesive as an
elastic/ideally-plastic material. It is noteworthy that the cost and
time required to apply the finite element method may limit its utility as a design tool.

Although the state-of-the-art in testing adhesives is currently almost totally, based on static strength properties, there are several
papers dealing with fatigue damage of adhesively bonded joints primarily from the standpoint of the relationship between the static
and fatigue strength. Sawyer and Cooper (1980) conducted static
and fatigue tests using epoxy with aluminum adherends. According
to the Johnson and Mall review (1984), the available experimental
results indicate that the ratio of static strength to fatigue strength
varies from 2.3 to 4.7, depending on the adhesive and specimen configuration. This range is similar to that of metals.

Analyses aimed at minimizing the stresses and maximizing the
fatigue life of adhesively bonded joints, subjected to external loading
have led to the following recommendations:

- equalize the in-plane and bending stiffness of the adherends,
  and use identical adherends, if possible,
- use as high an adherend in-plane stiffness as possible,
- use low modulus adhesives (as an alternative to using a low
  modulus adhesive throughout the joint, Strinivas (1975) demonstrated that the maximum stresses in the interface can be
  reduced by using such an adhesive only at the ends of the

overlap, in the region) of high interfacial stresses, while a higher modulus adhesive is used in the central region,

- vary, if possible, the adherend thickness along the joint in a proper way (Ojalvo and Eidinoff, (1978)) and
- keep the stresses within the elastic range and minimize peeling stresses.

2.3.2 Thermal Stresses In Bonded Joints

Heterogeneous structures, whose materials are cemented, soldered, welded or brazed together are very common in microelectronic applications. Once the attachment material hardens, all the constituents of the assembly, including the attachment material itself, form a single entity. Usually, the coefficients of thermal expansion and/or the temperatures of the constituent materials are different. If so, the thermal expansion or contraction mismatch results in stresses that could exceed the ultimate or fatigue strength of the structure. This situation is often referred to as the "thermostat-effect" or "bi-metallic strip effect."

The problem of thermal stresses arising in bi-metal thermostats subjected to uniform heating or cooling, was first addressed by Timoshenko (1925) The Timoshenko theory is widely utilized in various engineering applications, including the area of microelectronics (see, for instance, Glascock and Webster (1984)). Valuable insight into the thermally-induced stresses in bi-material structures has been provided by Aleck (1949) who obtained an approximate solution for the stresses induced by a uniform change in temperature of a thin rectangular plate clamped along the edge. Aleck's work also extended the analysis to long-edge, clamped plates with an aspect ratio greater than 4. While Timoshenko's approach uses elementary beam theory, Aleck's solution is based on methods using two-dimensional theory-of-elasticity. The distribution of stresses in Aleck's theory is shown in Fig. 5.2.

Thermal stresses in bonded joints within microelectronics devices were apparently first studied by Dash (1955) who evaluated residual thermoelastic stresses in a silicon wafer due to the thermal contraction mismatch between silicon and other materials bonded to it. The evaluation and control of these stresses has always been an important problem in the design and fabrication of semiconductor devices.

Thermal stresses and fracture in germanium were studied by Morton and Forgue (1959). A survey of the literature devoted to this type of failure has been given by Taylor (1959).

Stresses in a silicon chip, eutectically bonded to a gold-plated molybdenum block, were studied photoelastically by Riney (1961). The distribution of the interfacial thermal stresses in semiconductor device structures due to differential thermal contraction was, evidently, first considered by Taylor and Yuan (1962). These authors adopted the approach used previously in the lap joint theory by Volkersen (1938) and Goland and Reissner (1944). Zeyfang (1970), using the energy minimization method developed in Aleck's paper (1949), has shown that, for long plates bonded to substrates, it is possible to divide the stress, strain and displacement distributions into two parts: a uniform part, which is valid for the inner portion of the plate, and a nonuniform part, applicable to the portions located in the vicinity of the short edges.

Chin and Nelson (1979) developed several easy-to-use analytical models of thermal stresses in bonded joints which highlight the relative importance of different geometrical and material parameters and which provide some insight into different failure modes in electronic device structures. The authors examined structures typical of microelectronic applications: three elastic layers with two bonded joints (adhesive, solder); two elastic layers joined by one bonded joint; two circular sheets bonded together, and two elastic layers with one joint allowing free flexure.

As has been mentioned above, the classical Timoshenko theory of bi-metal thermostats is widely used for thermal stress analysis in engineering. Unfortunately, however, the Timoshenko theory, which deals only with the stresses in the attached strips, cannot be used to evaluate the effect of the attachment material itself. Extensions of this theory have been recently suggested by Chang (1981) and Suhir (1986).

The magnitude and the distribution of stresses in adhesively bonded bi-material assemblies, with consideration of the attachment compliance,was studied by Suhir (1986). Compliant attachments, providing a strain buffer between thermally-mismatched materials, can offer substantial stress relief and, for this reason, are considered in many cases as an acceptable solution to the thermal mismatch

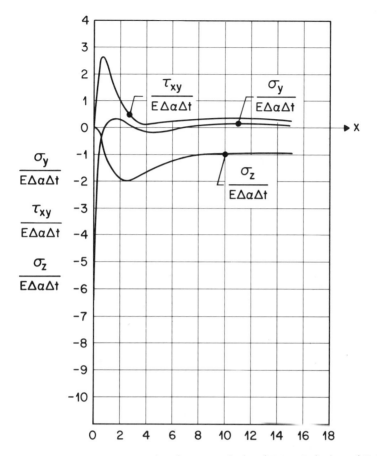

Figure 5.2: Stresses $(\sigma_z/E\Delta\alpha\Delta t), (\sigma_y/E\Delta\alpha\Delta t), (\tau_{xy}/E\Delta\alpha\Delta t)$ along clamped edge.

problem. Suhir (1986) concluded that, for relatively small assemblies, compliant attachment could, indeed, result in substantial stress relief in both the adherends and the adhesives. For large structures, however, the increased compliance of the attachment has a small effect on the stresses in the adherends and on the assembly bowing, but still has an essentially favorable effect on the interfacial stresses. Suhir (1986, 1987) has also suggested a simple engineering theory of thermally-induced stresses in adhesively bonded and soldered assemblies. In the case of soldered assemblies, the author assesses the effect of the elasto-plastic behavior of solders on the stress level.

2.2.3   Thermal Fatigue

Fatigue failure of materials and structures is due to the initiation and propagation of fatigue fractures under the action of the repeated removal or reversal of the applied load. If this load is produced by thermally-induced stresses, then thermal fatigue occurs.

Although randomly-oriented small fractures always exist in almost every material for low level cyclic loads, they do not affect the material strength. Under elevated cyclic loads, however, irreversible mechanical changes accumulate in the material. Fatigue fractures usually start at some discontinuity, stress-raiser, inclusion, or the like. Once the fracture is initiated, it propagates until the reduced cross-section is no longer able to sustain the load and this remaining section cracks suddenly. Although the precise mechanism of fatigue is still not known, it is believed that, at the first stage, plastic strains accumulate at the most highly stressed grains, while the microscopic stresses still remain within the elastic range. These strains initiate "sliding lines" in the grains. At the second stage, the "sliding lines" increase in number and gradually merge, forming "sliding strips" and creating the first submicroscopic fractures. At the third stage, these fractures combine. This creates favorable conditions for the development of a progressing, macroscopic fracture. At the fourth stage, an accelerated development of the visible fractures takes place and, suddenly, in the fifth stage the fatigue failure occurs.

The first three stages of fatigue failure are usually associated with 60% – 90% of the total number of cycles-to-failure. Fractures propagate rather slowly and generally do not threaten the integrity of the structure. However, undetected microcracks, subjected to dynamic loads, and especially in combination with low temperatures, can lead to a brittle fracture resulting in a sudden, disastrous failure. Since microcrack detection in microelectronic devices is most difficult, the material selection and the design loads must insure that the duration of the first three stages of fatigue failure exceed the projected life-time of the design.

The evaluation of microelectronic package reliability often involves power or thermal cycling. Fatigue damage of multimaterial assemblies, subjected to such evaluation, may occur as a result of:

- repeated thermal expansions and contractions,

- thermal stresses caused by the nonuniform distribution of temperature, and/or
- repeated reversal of loadings caused by the mismatch of the thermal expansion coefficients of the constituents.

Although in the last decade there has been a marked increase in research efforts aimed at a better understanding of mechanical fatigue at elevated temperatures, the majority of papers in the non-electronic area deals with high-temperature low-cycle fatigue of metals. There are a limited number of publications related to thermal fatigue, even for metal structures. This may be partially attributed to the perception that thermal fatigue is simply a special case of high-temperature, low-cycle fatigue. Although there is, of course, a great similarity between thermal and mechanical fatigue, some essential differences do exist. These are:

- During thermal fatigue, plastic deformation tends to concentrate in the regions of the highest temperature and, therefore, the distribution of this deformation can be different from the case of mechanical loading,
- In thermal fatigue, the alteration from tension to compression does not produce the "work hardening" result as is sometimes the case in mechanical fatigue at constant temperature and
- Temperature variation can result in nonhomogeneous structural effects and thus cause anisotropy of mechanical properties. This would not take place during mechanical fatigue.

In analyzing microelectronic devices, attention must also be devoted to the fatigue strength of the adhesive material. It may be anticipated that fatigue fractures will occur at the assembly ends, where the interfacial stresses are maximum. These fractures will then propagate to the inner portions of the assembly, as the bearing ability of the end portions becomes exhausted, initiating new fatigue fractures and enhancing existing ones. The fatigue failure of the bonded joining occurs when the entire bonded area is covered by fatigue fractures. Total time-to-failure for the entire joint is significantly greater than the time-to-failure for the outer portions where the initial shearing and peeling stress concentration takes place.

The reader is reminded that the analysis of fatigue strength is rather difficult even for metals whose fatigue behavior has been studied for years, and at room temperature where there are no additional complications due to thermal effects. Clearly, structures with non-metals, at elevated temperatures, are far less amenable to analytic evaluation. Thermal cycling fatigue, under nonsteady loading, due to variable temperatures, is significantly more complicated. In micro-electronics, in particular, the problem of the fatigue life evaluation is discouragingly difficult since very often there is not much information available for the mechanical properties of the adhesive (non-metal) materials such as epoxies and silicone materials. Therefore, the overwhelming majority of the studies in this area is experimental and investigators heavily rely on the results of thermal cycling tests to determine the thermal fatigue limits of materials and structures.

2.4   Multimaterial and Heteroepitaxial Structures

Multimaterial and heteroepitaxial (stacks of thin-layers) structures are widely used in microelectronics. With appropriate modifications the methods employed in the prediction of stresses in composite materials can be applied to these heteroepitaxial structures. Fortunately, the development of new materials and processes in recent years has created great interest and stimulated extensive research in the area of composite materials, including their thermoelastic behavior.

A simplified engineering theory for the evaluation of interlaminar stresses in layered or laminated beam composites was first suggested by Boley and Testa (1969). Obviously, the formulation and analysis of layered structures becomes more complicated and laborious when more than one bond surface is present. Rigorous solution of such a structure, requires the introduction of many more unknowns and the simultaneous solution of a potentially large number of differential equations. To rectify this situation, Boley and Testa (1969) considered the individual beam layers, above and below the bond plane of interest, to be rigidly attached. This approach enables them to treat the actual nonhomogenous beam as a series of two-layered beams. This is, of course, a conservative approach, since when all except one bonded element are rigid (and thus constrain the movement of the beam layers), the calculated interlaminar stresses would be greater than the true values. A similar approach was later taken by Grimado (1978) who evaluated interlaminar stresses in layered beam composites subjected to differential expansion or contraction of individual beam components. In Grimado's study, shear deformations in the beam elements were neglected and all of the materials were assumed to transmit only longitudinal shear and out-of-plane normal stresses, the adhesive layers were assumed thin, and the bonded structure was considered free of initial stresses.

From the standpoint of structural analysis, multilayered heteroepitaxial structures can be considered as a special case of mulitmaterial assemblies when the thicknesses of the layers are very small compared to the thickness of the substrate. Interestingly, however, the first theoretical formula for stresses, arising in a thin film prepared on a thick substrate was obtained by Stoney in 1909 without any connection to the layered structures used in microelectronics. Stoney's formula, originally suggested for electroplated films, is still widely used for the determination of stress in any film structure when the deformation of the underlying substrate is known. This can be expressed by

$$\sigma_f = E_s^0 \frac{h_s^2}{6\rho h_f} \tag{5.1}$$

where σ_f is the stress in the film, $E_s^0 = E_s/(1 - \nu_s)$ is the generalized Young's modulus for the substrate material, E_s and ν_s are the Young's modulus and Poisson's ratio for this material, h_s and h_f are thicknesses of the substrate and the film and ρ is the radius of curvature.

Stoney's formula, eq (5.1), can be easily obtained on the basis of the following elementary considerations. The equation of equilibrium for a thermally mismatched bi-material assembly, shown in Fig. 5.3, is

$$M_1 + M_2 - \frac{h_1 + h_2}{2} T = 0 \tag{5.2}$$

where T is the force due to this mismatch, h_1 and h_2 are the thicknesses of the components, and M_1 and M_2 are bending moments acting over the component cross-sections. These moments are related to the moments of inertia $I_1 = h_1^3/12$ and $I_2 = h_2^3/12$ of the cross-sectional areas (for components of unit width) by

$$M_1 = E_1^0 \frac{I_1}{\rho} = \frac{E_1^0 h_1^3}{12\rho} \tag{5.3a}$$

and

$$M_2 = E_2^0 \frac{I_2}{\rho} = \frac{E_2^0 h_2^3}{12\rho} \qquad (5.3b)$$

where the generalized Young's moduli E_1^0 and E_2^0 are used to account for the two-dimensional stress condition. From eqs (5.2) and (5.3) one finds that

$$T = \frac{E_1^0 h_1^3 + E_2^0 h_2^3}{6\rho (h_1 + h_2)} \qquad (5.4)$$

In the case where the component #1 is a thin film, and component #2 is a thick substrate, eq (5.4) can be simplified to

$$T = E_s^0 \frac{h_s^2}{6\rho} \qquad (5.5)$$

This results in eq (5.1).

It should be emphasized, however, that Stoney's formula, proceeding from substrate bowing must be used when the mechanical characteristics of the film material are unavailable. Otherwise, the relation

$$\sigma_f = E_f^0 (\alpha_f - \alpha_s) \Delta t \qquad (5.6)$$

should be utilized. Here α_f and α_s are coefficients of thermal expansion for the film and the substrate, respectively, and Δt is the

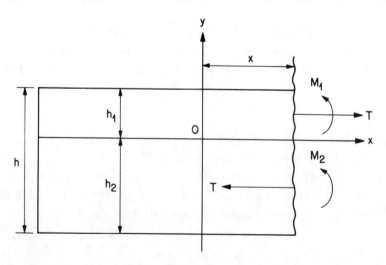

Figure 5.3: Forces and moments at the X cross section.

temperature difference. Equation (5.6) follows from the strain compatibility condition which requires that the strain in the substrate,

$$\epsilon_s = \alpha_s \Delta t + \frac{T}{E_s^0 h_s} \tag{5.7}$$

be equal to the strain in the film

$$\epsilon_f = \alpha_f \Delta t - \frac{T}{E_f^0 h_f} \tag{5.8}$$

This equality yields the following relationship for the thermal mismatch force, T, in a thin film structure

$$T = \frac{\Delta \alpha \Delta t}{1/E_s^0 h_s + 1/E_f^0 h_f} \cong E_f^0 h_f \Delta \alpha \Delta t \tag{5.9}$$

where $\Delta \alpha$ is $\alpha_f - \alpha_s$. This relationship leads to eq. (5.1).

When the material properties and temperature difference are known, the curvature of the structure can be easily obtained from eqs (5.1) and (5.6) as

$$\frac{1}{\rho} = 6 \frac{E_f^0}{E_s^0} \frac{h_f}{h_s^2} \Delta \alpha \Delta t \tag{5.10}$$

In order to find the deflection function $w(x)$, the function's second derivative, $w''(x)$, is substituted for the curvature, $1/\rho$ in eq (5.10). Then one obtains

$$w(x) = w_0 \left[1 - \left(\frac{x}{\ell} \right)^2 \right] \tag{5.11}$$

where w_0, the maximum bow, is given by

$$w_0 = 3 \frac{E_f^0}{E_s^0} \left(\frac{\ell}{h_s} \right)^2 h_f \Delta \alpha \Delta t \tag{5.12}$$

and ℓ is the half-length of the structure.

Most of the methods for the calculation of stresses in multilayered heteroepitaxial structures are based on eqs (5.1) to (5.12). Reinhart and Logan (1973) developed a planar stress model for double-heterostructure devices. Röll (1976) used eq (5.1), as a basis for interpreting experimental stress/strain distribution. Olsen and Ettenberg (1977) extended eqs (5.6) and (5.10) to the case of a multilayered structure and Vilms and Kerps (1982) suggested a simplified version of the Olsen-Ettenberg theory. Note that all of these relations are aimed at the evaluation of the internal stresses in the films and, therefore, can be used only for an indirect judgement of the level of the interfacial stresses which are responsible for film blistering and peeling. To overcome this limitation, prediction of the stresses in single- and multilayered heteroepitaxial structures, including interfacial stresses, was recently suggested by Suhir (1987).

In addition to the need for detailed thermal stress distribution in microelectronic packages, there exists a tremendous interest in the thermostructural aspects of growing high quality epitaxial layers of foreign materials on lattice-mismatched semiconductor substrates. In particular, the desirability of expitaxially depositing GaAs on silicon substrates has been recognized for many years. The state-of-the-art in this area was reviewed by Jordan et al (1986) and Shaw (1987).

Interestingly, the theoretical methods developed for thermally-induced stresses in heteroepitaxial structures could also be applied to the case of lattice mismatch. In such analyses, lattice-mismatch strain replaces the differential thermal expansion strain in the relevant formulas. Furthermore, it should be emphasized that the thermal stresses themselves are also a potentially serious problem in such hybrid semiconductor devices, primarily because of wafer bowing resulting from the different thermoelastic properties of the two materials. In severe cases, the thermal stresses can stimulate the generation of dislocations and even cause fractures in the GaAs film.

Stresses due to the thermal contraction mismatch in the film structures are also of great importance from the point of view of their effect on the diffusion of impurities. The effect of window edge stress on the shape of the junction profile was studied both theoretically and experimentally by Jaccodine and Schlegel (1966), Lawrence (1966, 1967), Hu (1979), Isomae (1981, 1985), Huang et al (1986) and others.

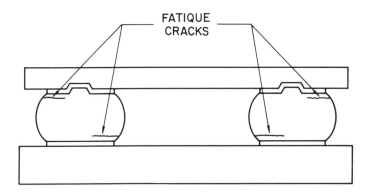

Figure 5.4: Schematics showing locations of thermal fatigue cracks.

2.5 Some Other Thermal Stress Related Failures

2.5.1 Thermal Fatigue In Solder Joints

Solder joints are widely used in flip-chip designs to electrically interconnect the chip circuits with the underlying substrate metallization and to provide mechanical connections. Because of differences in the expansion coefficients and temperatures of the silicon chip and the substrate material, a shear displacement is produced in these joints each time the device is turned on or off. If the substrate is sufficiently compliant with respect to the bending loads, the solder joints will also experience tension and compression. The strain amplitude in solder joints may approach a few tenths of a percent. The mechanical integrity of these joints is often affected by fatigue crack propagation under such strain cycling exposure (Fig. 5.4).

Various aspects of solder technology are discussed in monographs by Manko (1967) and Thwaites (1982). Thermal cycling fatigue of solder interconnections has been studied by many investigators including Norris and Landzberg (1969), Merrell (1971), Wild (1972, 1975), Coombs (1976), Zommer et al (1976), Boah and DeVore (1978, 1982), Becker (1979), Levine and Ordonez (1981), Engelmaier (1982, 1983, 1984, 1985), Waine et al (1982), Hall (1983, 1984, 1987), Burges (1984), Sherry et al (1985), Stone et al (1985, 1986) and Vaynman and Fine (1987).

Much of this attention has been focused on the development of an empirical model, which could be used to predict the fatigue life of solder joints. A model developed by Engelmaier (1983, 1985), for 63/37 and 69/49 In-Pb solders, proceeds from Manson-Coffin's relation and provides an estimate of the number of cycles-to-failure

$$N_f = \frac{1}{2}\left(\frac{2\epsilon'_f}{\Delta\gamma}\right)^{-1/c} \tag{5.13}$$

where $\Delta\gamma$ is the range of the shear (angular) strain (after total stress relaxation) and ϵ'_f and c are fatigue ductility parameters. For these solders the ϵ'_f value is approximately equal to 0.32 and the parameter c, in the exponent, can be determined by eq (5.14) derived from data given by Wilde (1971).

$$c = -0.441 - 6 \times 10^{-4}\bar{T}_s - 1.74 \times 10^{-2}\ln\left(1 + \frac{360}{t_d}\right) \tag{5.14}$$

Here $\bar{T}_s$ is the mean cycling temperature, (°C), and t_d is the dwell time in one cycle ($f = 360/t_d < 1000$). It is to be noted that for very soft solders, which do not constrain the free thermal expansion or contraction of the soldered components, the shear strain, $\Delta\gamma$, can be evaluated by the relationship

$$\Delta\gamma \cong \epsilon\frac{\ell}{h} \tag{5.15}$$

where ϵ is the thermal expansion mismatch strain of these components, ℓ is the distance of the given solder joint from the middle of the assembly and h is the solder joint height. The strain ϵ can be easily evaluated for the given temperatures and coefficients of thermal expansions of the components.

Another thermal fatigue-life model was suggested recently by Agarwala (1985) on the basis of tests conducted for 50/50 Pb-In solders. Both Engelmaier's and Agarwala's models consider shear stresses only.

Geometric optimization of solder bumps was addressed by Goldman (1969), Wilson (1981), Kreibel and Lochmann (1981), Hall (1983), Wilson and Anderson (1983) and Engel (1984). Time and temperature effects were examined by Chen et al (1972) and Baker (1979). Several authors such as Sherry et al, (1985) and LoVasco and Britman, (1987) have used finite element modeling to investigate the distribution of stresses in solder bumps. Nevertheless, despite the remarkable progress in understanding and optimizing solder joint metallurgy, many basic mechanical and structural problems of solder joint behavior remain unsolved.

2.5.2 Die and Die-Attached Reliability

Die-attachment materials provide the mechanical, thermal and often also the electrical connection between the die and the substrate. If the coefficients of thermal expansion and/or the temperatures of the materials are different, the resulting thermal stresses could induce die fracture or failure of the interconnection. Until recently, structural analysis was not usually performed for a die-attach design since experience suggested that, for small dies (typically smaller than 5 mm), a wide variety of bonding materials could be successfully utilized as reliable interconnections. The introduction of relatively large dies (1 cm and above) has, however, changed the situation and available data indicates that large dies, soldered to printed wire boards, often fracture due to the elevated stresses caused by the thermal mismatch between the die and the substrate. These failures are thought to reflect the high stress level in the die but also in the bonding material and have led to concern about both die fracture and interfacial rupture.

Die fracture is associated with large VLSI chips and has been addressed by Taylor (1959), Johnson (1974), Hund and Burchett (1983), Chiang and Shukla (1984) and others. Three key features in failure mechanism have been established by Shukla and Mencinger, (1985):

- The cracks usually show up after temperature excursions (thermal shocks and thermal cycling),
- The crack typically emanates from the bottom of the die near the die edges and corners and
- Both vertical and horizontal cracks are observed.

The occurrence of vertical cracks, forming at the rear surface of the chip, is attributed primarily to the presence of voids in the die bond region (Chiang and Shukla (1984)), while horizontal cracks are due to the elevated interfacial stresses near the die edges (Hund and Burchett (1983)).

Disbonding failures (interfacial rupture) could be caused by solder fatigue or creep rupture in the case of solder attachment (Lang, Feder and Williams (1970)), or by the adhesive failure in the case of a nonmetal adhesive (Shukla and Mencinger (1985)).

Chiang and Shukla (1984) employed a finite element stress analysis to show that the presence of voids at the die bond interface creates significant secondary stresses in silicon. In particular, the presence of voids due to unbonded die edges exerts a tensile stress field in the bonded areas of silicon near the bonded/unbonded discontinuity, causing a crack to propagate upwards away from the free edges. The stress intensification mechanism due to the voids at the die bond interface has been described qualitatively by Bolger and Mooney (1983). A quantitative analysis of the influence of the die-attachment design on thermal stresses in the die and the attachment, including some recommendations for the choice of die-attach material and thickness, was performed by Suhir (1987).

2.5.3   Thermal Failures in Wire-Bonds

Thermocompression and thermosonic wire bonding are two commonly used techniques for the interconnection of microelectronic devices. In thermocompression bonding the joining is achieved through the use of high temperatures and compressive loading, while the method of thermosonic bonding uses much lower temperatures and employs ultrasonic energy accompanied by relatively small compressive loads.

Gold or aluminum wires, typically 1 *mil* in diameter, are commonly utilized in microelectronic packages to electrically connect the chip with the package I/O terminal. The wire bond may experience elevated stresses as the temperature changes. Due to the difference in the coefficients of thermal expansion of the wires, chips and package materials, these stresses can shift or dislodge a poorly adhering bond and cause an open circuit. Yet, thermally-induced displacements of loose wires can result in intermittent short circuits. The most important failure modes for wire bonds are given by Schafft (1973):

- a gap in the wire (generally at midspan),
- a rupture of the wire at the heel of the bond and
- a lift-off of the deformed wire.

A gap in the wire usually results from melting, due to excessive current flow and is dependent on the wire length and diameter. A wire rupture is due to thermomechanical stresses and usually occurs at the heel of the bond where the cross-sectional area of the bond is the smallest. Failure of the bond through lift-off occurs as a result of thermal stresses, applied to an initially or in-life weakened bond. Another reason for wire lift-off is a fracturing of the silicon under the bond, which can be caused, in particular, by excessive thermal stresses.

The reliability of wire bonds, including thermal failures, was analyzed by Gaffney (1968), Nowakowski and Villella (1971), Schafft (1972), Ravi and Philofsky (1972) and Ebel et al (1982).

3 ANALYTICAL DETERMINATION OF THERMALLY-INDUCED STRESSES

3.1 Introductory Remarks

The analysis described herein is an extension of existing approaches which attempt to predict thermally-induced stresses in adhesively-bonded or soldered assemblies. In this development, temperature is assumed to be uniform and the methods of structural mechanics are used to determine the normal stresses, acting in the bonded layers and responsible for their ultimate and fatigue failures as well as the shearing and transverse-normal (peeling) stresses, arising in the interface and responsible for the adhesive and cohesive failure of the attachment material.

Since compliant attachments are often used as strain-buffers between thermally mismatched materials in micro-electronic devices, the analysis addresses the effect of the attachment compliance on the magnitude and distribution of stresses. The effect of the non-linear compliance of the attachment material is also assessed. This is a significant factor for many materials, and especially for solder interconnections, which exhibit elasto- plastic behavior even at low stress level.

The overwhelming majority of studies dealing with such stresses in electronic components is experimental and most investigators rely heavily on the results of thermal cycling tests to evaluate the integrity of microelectronic structures. This is, of course, a reasonable approach which is often anchored in product development. It is to be emphasized, however, that utilization of experimental techniques and numerical procedures (based, for instance, on the finite-element method) to obtain a solution to the problem in question may be associated with a considerable investment in time and resources. But even more importantly, the empirical and/or numerical testing alone, if not supported by theoretical analysis, often cannot add much to understanding of the behavior of materials and structures. This is due to the fact that the experimental data reflects the combined effect of a wide variety of factors, while it is the knowledge of the role of each particular parameter affecting structural behavior, which is usually needed for the prediction of stresses and rational structural design. The lack of such insight often results in "trial-and-error" experimental procedures and, in many cases, makes an engineer unable to extend the obtained empirical data to new materials and structures. For this reason, it is essential that a theoretical approach, as presented in this section, be considered an important constituent part of any microelectronic package reliability analysis.

A bi-material assembly, whose components are cemented or soldered together, is, strictly speaking, a tri-material body in which the attachment material layer is a "full and equal member" of the assembly. For this reason, the stress analysis set forth in subsequent sections for such a structure begins with a model for a tri-material assembly, fabricated at elevated temperature and subsequently cooled. The main purpose of this model is to determine which mechanical properties of the attachment material are crucial, from the standpoint of thermally-induced stresses and, in particular, the extent to which the thermal expansion of the attachment material affects the stresses in the bonded materials. A later section considers the simplification arising from the fact that, when an adhesive or solder is used, the interstitial component is substantially thinner and of lower Young's modulus than the adherends. This makes it possible to develop a simplified bi- material analytical model, in which only the significant factors of the attachment material are taken into account.

3.2 Tri-Material Assembly

In developing the tri-material model, it will be assumed that the coefficients of linear thermal expansion of the assembly components are related as follows: $\alpha_1 < \alpha_2 < \alpha_3$. Shearing stresses in the interfaces, when the assembly is cooled down by the temperature Δt are schematically shown in Fig. 5.5. Note, that the analytical model is quite flexible since it could be applied to assemblies with both continuous ($a = \ell$) and non-continuous ($a < \ell$) attachments. Thus, in the latter case it could be used for the analysis of stresses in soldered joints.

In order to determine the shearing stresses arising as a result of the mismatch in the thermal expansion coefficient of the materials, the conditions of the compatibility of displacements are used. These conditions require that the displacements $u_1^-(x)$ of the lower extreme fibre of component #1 be equal to the displacements $u_2^+(x)$ of the upper extreme fibre of component #2 (attachment), and that the displacements $u_2^-(x)$ of the lower extreme fibre of component #2 be equal to the displacements $u_3^+(x)$ of the upper extreme fibre of component #3:

$$u_1^-(x) = u_2^+(x), \quad u_2^-(x) = u_3^+(x). \tag{5.16}$$

If the shearing stresses $\tau_1(x)$ and $\tau_2(x)$ were known, the above displacements could be calculated by

$$u_1^-(x) = \alpha_1 \Delta t x - \lambda_1 \int_0^x T_1(\xi) d\xi + \kappa_1 \tau_1(x)$$

$$+ \frac{h_1}{2} \int_0^x \frac{d\xi}{\rho(\xi)} \tag{5.17a}$$

$$u_2^+(x) = \alpha_2 \Delta t x + \lambda_2 \int_0^x [T_1(\xi) - T_2(\xi)] d\xi - \kappa_2 \tau_1(x)$$

$$- \frac{h_2}{2} \int_0^x \frac{d\xi}{\rho(\xi)} \tag{5.17b}$$

$$u_2^-(x) = \alpha_2 \Delta t x + \lambda_2 \int_0^x [T_1(\xi) - T_2(\xi)] d\xi + \kappa_2 \tau_2(x)$$

$$+\frac{h_2}{2}\int_0^x \frac{d\xi}{\rho(\xi)} \tag{5.17c}$$

and

$$u_3^+(x) = \alpha_3\Delta tx + \lambda_3\int_0^x T_2(\xi)d\xi - \kappa_3\tau_2(x)$$

$$-\frac{h_3}{2}\int_0^x \frac{d\xi}{\rho(\xi)} \tag{5.17d}$$

In these relations, Δt is the temperature difference and λ_i and κ_i are respectively

$$\lambda_i = \frac{1-\nu_i}{E_ih_i}, \qquad \kappa_i = \frac{h_i}{3G_i}, \; i = 1,2,3, \tag{5.18}$$

the axial and interfacial compliances for the i-th component as found by Suhir (1986). In addition, E_i is the Young's modulus, ν_i is the Poisson ratio, $G_i = E_i/2(1 + \nu_i)$ is the shear modulus, h_i is the thickness of the i-th component, $\tau_i(x), i = 1, 2$, is the shear stress in the i-th interface (i.e., in the interface between the i-th and the $(i + 1)$-th components). $T_i(x)$ is defined by eq (5.19)

$$T_i(x) \equiv \int_{-\ell}^x \tau_i(\xi)d\xi \tag{5.19}$$

and is the force per unit assembly width for the i-th interface, ℓ is half the assembly length, and $\rho(x)$ is the radius of curvature. The origin of the rectangular coordinates x and y is in the middle of the assembly, on the centerline of the attachment. It is assumed that the attachment is absolutely stiff in the direction of its thickness, and, therefore, all the components have the same radius of curvature. This is supposedly a conservative assumption resulting in an overestimation of the peeling stresses.

The first terms in the relationships of eqs (5.17) are unrestricted thermal contractions. The second terms are due to the forces, $T_i(x)$, and are calculated under an assumption that these forces are uniformly distributed over the component thicknesses. The third terms account for the actual nonuniform distribution of the forces, $T_i(x)$, and are calculated under an assumption that the corresponding corrections are directly proportional to the shearing stress in the given cross-section and are not affected by the stresses in other cross-sections. The last terms are due to bending.

After substituting the relationships eqs (5.17) into the compatibility conditions of eqs (5.16) the following two equations are obtained:

$$\kappa_{12}\tau_1(x) - \lambda'_{12}\int_0^x T_1(\xi)d\xi + \lambda_2\int_0^x T_2(\xi)d\xi$$

$$+\frac{h_1 + h_2}{2}\int_0^x \frac{d\xi}{\rho(\xi)} = (\alpha_2 - \alpha_1)\Delta tx \qquad (5.20a)$$

and

$$\kappa_{23}\tau_2(x) - \lambda'_{23}\int_0^x T_2(\xi)d\xi + \lambda_2\int_0^x T_1(\xi)d\xi + \frac{h_2 + h_3}{2}\int_0^x \frac{d\xi}{\rho(\xi)}$$

$$= (\alpha_3 - \alpha_2)\Delta tx \qquad (5.20b)$$

Here $\kappa_{12} = \kappa_1 + \kappa_2$ and $\kappa_{23} = \kappa_2 + \kappa_3$ are interfacial compliances and $\lambda'_{12} = \lambda_1 + \lambda_2, \lambda'_{23} = \lambda_2 + \lambda_3$ are axial compliances of the assembly.

The relationship between the forces $T_i(x)$ and the radius of curvature $\rho(x)$ can be found on the basis of the rotational equilibrium condition for a portion of the assembly (See Fig. 5.5):

$$\frac{h_1 + h_2}{2}T_1(x) + \frac{h_2 + h_3}{2}T_2(x) = M_1(x) + M_2(x) + M_3(x) \qquad (5.21)$$

The bending moments on the left side of this relation are evaluated with regard to the center line of component #2. The bending moments $M_i(x), i = 1, 2, 3$, on the right side are related to the flexual rigidities

$$D_i = \frac{E_i h_i^3}{12(1 - \nu_i^2)} \qquad i = 1, 2, 3 \qquad (5.22)$$

of the components by

$$M_i(x) = -\frac{D_i}{\rho(x)} \qquad i = 1, 2, 3 \qquad (5.23)$$

Inserting these relations for the bending moments into eq (5.21) yields the curvature

$$\frac{1}{\rho(x)} = -T_1(x)\sqrt{\frac{\lambda_{12}}{D}} - T_2(x)\sqrt{\frac{\lambda_{23}}{D}} \qquad (5.24)$$

where $D = D_1 + D_2 + D_3$ is the total flexural rigidity of the assembly, and where

$$\lambda_{i,i+1} = \left(\frac{h_i + h_{i+1}}{2\sqrt{D}}\right)^2 \qquad i = 1, 2 \tag{5.25}$$

are additional compliances associated with bowing. Using eq (5.24), one may rewrite the eqs (5.20)

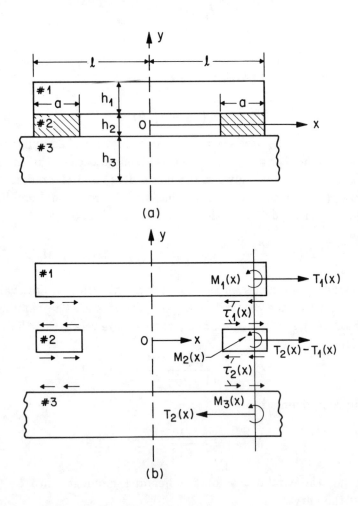

Figure 5.5: Stress analysis model.

$$\kappa_{12}\tau_1(x) - \lambda_{12}^0 \int_0^x T_1(\xi)d\xi + \lambda_2^0 \int_0^x T_2(\xi)d\xi = (\alpha_2 - \alpha_1)\Delta tx \quad (5.26a)$$

and

$$\kappa_{23}\tau_2(x) - \lambda_{23}^0 \int_0^x T_2(\xi)d\xi + \lambda_2^0 \int_0^x T_1(\xi)d\xi = (\alpha_3 - \alpha_2)\Delta tx \quad (5.26b)$$

where

$$
\begin{aligned}
\lambda_{12}^0 &= \lambda'_{12} + \lambda_{12} = \lambda_1 + \lambda_2 + \lambda_{12} \\
\lambda_{23}^0 &= \lambda'_{23} + \lambda_{23} = \lambda_2 + \lambda_3 + \lambda_{23} \\
\lambda_2^0 &= \lambda_2 - \sqrt{\lambda_{12}\lambda_{23}}
\end{aligned}
\quad (5.27)
$$

are axial compliances of the assembly with consideration of bowing.

Because the shearing stresses must by antisymmetric with respect to the origin, the solutions to the eqs (5.26) are sought in the form

$$\tau_i(x) = C_i \sinh kx \quad i = 1,2 \quad (5.28)$$

where k is the eigenvalue of the problem. Note that the solution in this form presumes that the attachment is continuous and/or the thermal expansion mismatch between the assembly components is great. If, however, components with equal thermal expansion coefficients are used and, in addition, the attachment is not continuous, then the origin must be placed in the middle of such an attachment. After substituting eq (5.28) into (5.26), the following two systems of linear algebraic equations are obtained

$$\left(\kappa_{12}k^2 - \lambda_{12}^0\right)C_1 + \lambda_2^0 C_2 = 0 \quad (5.29a)$$

$$\lambda_2^0 C_1 + \left(\kappa_{23}k^2 - \lambda_{23}^0\right)C_2 = 0 \quad (5.29b)$$

and

$$\lambda_{12}^0 C_1 - \lambda_2^0 C_2 = k\frac{\alpha_2 - \alpha_1}{\cosh k\ell}\Delta t \quad (5.30a)$$

$$-\lambda_2^0 C_1 + \lambda_{23}^0 C_2 = k\frac{\alpha_3 - \alpha_2}{\cosh k\ell}\Delta t \quad (5.30b)$$

Equations (5.29) is a system of homogeneous equations which has a nontrivial (nonzero) solution for the unknowns C_1 and C_2 if the determinant of this system is zero. Evaluation of this determinant results in a biquadratic equation for k

$$\kappa_{12}\kappa_{23}k^4 - \left(\kappa_{12}\lambda_{23}^0 + \kappa_{23}\lambda_{12}^0\right) k^2 + \delta = 0 \qquad (5.31)$$

where

$$\delta = \lambda_1\lambda_2 + \lambda_2\lambda_3 + \lambda_3\lambda_1 + \lambda_3\lambda_{12} + \lambda_2 \left(\sqrt{\lambda_{12}} + \sqrt{\lambda_{23}}\right)^2 \qquad (5.32)$$

After the value of k is found, the constants C_1 and C_2 can be determined from eqs (5.30)

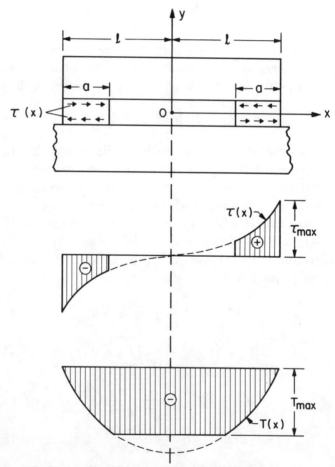

Figure 5.6: Shearing stress distribution.

$$C_1 = c\left[\lambda_2(\alpha_3 - \alpha_1) + (\lambda_3 + \lambda_{23})(\alpha_2 - \alpha_1) - \sqrt{\lambda_{12}\lambda_{23}}(\alpha_3 - \alpha_2)\right]$$
$$(5.33a)$$

and

$$C_2 = c\left[\lambda_2(\alpha_3 - \alpha_1) + (\lambda_1 + \lambda_{12})(\alpha_3 - \alpha_2) - \sqrt{\lambda_{12}\lambda_{23}}(\alpha_2 - \alpha_1)\right]$$
$$(5.33b)$$

where

$$c = \frac{k\Delta t}{\delta\cosh k\ell}$$

The solution obtained is applicable to any tri-material assembly without any additional limitations on the thickness or material properties of the components. If, however, the thickness and/or the Young's modulus of the interstitial component #2 are small, which is the case when it is an adhesive, then the λ_2 compliance becomes large and the terms containing λ_2 dominate. In this case, eqs (5.31) and (5.33) can be substantially simplified to yield simplified relationships for the eigenvalue, k, and the constants of integration

$$k = \sqrt{\frac{\lambda}{\kappa}} \qquad C_1 = C_2 = \frac{k\Delta\alpha\Delta t}{\lambda\cosh k\ell} \qquad (5.34)$$

where the parameters

$$\lambda = \lambda_{23}^0 = \lambda_1 + \lambda_3 + \lambda_{13}, \quad \kappa = \kappa_1 + 2\kappa_2 + \kappa_3 \qquad (5.35)$$

are axial and interfacial compliances, respectively, and where $\Delta\alpha = \alpha_3 - \alpha_1$ is the thermal expansion coefficient mismatch between the assembly components. In this case, the functions $\tau_1(x)$ and $\tau_2(x)$ coincide and are expressible by a simple relationship for the shearing stress

$$\tau(x) = k\frac{\Delta\alpha\Delta t}{\lambda\cosh\ell}\sinh kx \qquad (5.36)$$

The distribution of the shearing stress, calculated on this basis, is shown qualitatively in Fig. 5.6.

The foregoing analysis leads to the following important conclusions concerning the preceding relations.

1. The coefficient of thermal expansion of the attachment material, α_2, does not enter the shear stress relationships. Thus, if

the case when it is an adhesive, then the λ_2 compliance becomes large and the terms containing λ_2 dominate. In this case, eqs (5.31) and (5.33) can be substantially simplified to yield simplified relationships for the eigenvalue, k, and the constants of integration

$$k = \sqrt{\frac{\lambda}{\kappa}} \qquad C_1 = C_2 = \frac{k\Delta\alpha\Delta t}{\lambda\cosh k\ell} \qquad (5.34)$$

where the parameters

$$\lambda = \lambda_{23}^0 = \lambda_1 + \lambda_3 + \lambda_{13}, \quad \kappa = \kappa_1 + 2\kappa_2 + \kappa_3 \qquad (5.35)$$

are axial and interfacial compliances, respectively, and where $\Delta\alpha = \alpha_3 - \alpha_1$ is the thermal expansion coefficient mismatch between the assembly components. In this case, the functions $\tau_1(x)$ and $\tau_2(x)$ coincide and are expressible by a simple relationship for the shearing stress

$$\tau(x) = k\frac{\Delta\alpha\Delta t}{\lambda\cosh\ell} \sinh kx \qquad (5.36)$$

The distribution of the shearing stress, calculated on this basis, is shown qualitatively in Fig. 5.6.

The foregoing analysis leads to the following important conclusions concerning the preceding relations.

1. The coefficient of thermal expansion of the attachment material, α_2, does not enter the shear stress relationships. Thus, if the thermal expansion mismatch between the assembly components is significant, and the thickness and/or the Young's modulus of the attachment material are small (compared to the thicknesses and the Young's moduli of the assembly components), the induced stresses will be independent of α_2. The situation is different, of course, when the component materials have similar thermal expansion coefficients. However, for $\Delta\alpha \rightarrow 0$, the induced stresses in the components are small in any event.

2. The compliance κ_2 of the attachment material plays an important role in establishing the magnitude of the thermally induced stresses.

3. The constants C_1 and C_2 of integration are identical. This means that the attachment material is subjected only to shear forces. No axial forces act in the attachment material in this

case and, therefore, no normal stresses arise in its cross-sections.

4. The shearing stress in the interface depends on the thermal mismatch strain $(\Delta\alpha\Delta t)$, the axial (λ) and the interfacial (κ) compliances of the assembly and, in the general case, on the assembly size (2ℓ).

3.3   Bi-material Assembly: Simplified Theory

Basic Relationships: The results of the previous section enable one to use a simplified stress analysis model for an adhesively bonded bi-material assembly. In this model the displacement compatibility conditions formulated in eq (5.16) are replaced by the condition

$$u_1(x) = u_2(x) - 2\kappa_2\tau(x) \tag{5.37}$$

and the integral eqs (5.26) reduce to one equation for the shear stress function $\tau(x)$

$$\tau(x) - k^2 \int_0^x T(\xi)d\xi = \frac{\Delta\alpha\Delta t}{\kappa}x \tag{5.38}$$

where

$$T(x) = \int_{-\ell}^x \tau(\xi)d\xi \tag{5.39}$$

is the shearing force, $\Delta\alpha = \alpha_3 - \alpha_1$ is the thermal coefficient of expansion mismatch between the adherends, Δt is the temperature differential. Here, k, the eigenvalue of the problem (parameter of assembly stiffness), is given by

$$k = \sqrt{\lambda/\kappa} \tag{5.40}$$

where

$$\lambda = \lambda_1 + \lambda_3 + \lambda_{13} = \frac{1-\nu_1}{E_1h_1} + \frac{1-\nu_3}{E_3h_3} + \frac{h^2}{4D} \tag{5.41a}$$

$$\kappa = \kappa_1 + 2\kappa_2 + \kappa_3 = \frac{h_1}{3G_1} + \frac{2h_2}{3G_2} + \frac{h_3}{3G_3} \tag{5.41b}$$

are the axial and the interfacial compliances, respectively. The total thickness of the assembly is given by $h = h_1 + h_2 + h_3 \cong h_1 + h_3$ and

$$D \cong D_1 + D_3 = \frac{E_1h_1^3}{12(1-\nu_1^2)} + \frac{E_3h_3^3}{12(1-\nu_3^2)} \tag{5.42}$$

is its total flexural rigidity. Thus, eq (5.24) reduces to a rather simple relationship for the curvature

$$\frac{1}{\rho(x)} = -\frac{h}{2D}T(x) \tag{5.43}$$

The integral equation (5.38) has the following solution for the contact areas:

$$\tau(x) = k\frac{\Delta\alpha\Delta t}{\lambda\cosh k\ell}\sinh kx \tag{5.44}$$

Here, as shown in Fig 5.5, $-\ell < x < -(\ell - a)$ and $\ell - a < x < \ell$. Obviously, in the inner part of the assembly beyond the contact areas, when $-(\ell - a) < x < \ell - a$, the shearing stress is zero.

Equation (5.44) must satisfy the boundary conditions

$$\tau(0) = 0, \quad T(\ell) = 0 \tag{5.45}$$

The first of these conditions indicates that the shear stress in the middle of the assembly is zero, since, due to the symmetry of thermal loading, there is no displacement in the mid-cross- section where $x = 0$. The second condition reflects the fact that at the edge, where there are no external forces acting on the assembly, the force $T(x)$ must turn to zero.

Substituting the shear-stress relation, eq (5.44), into eq (5.39) allows this force to be evaluated by

$$T(x) = -\frac{\Delta\alpha\Delta t}{\lambda}X(x) \tag{5.46}$$

where the function

$$X(x) = 1 - \frac{\cosh kx}{\cosh k\ell} \tag{5.47}$$

characterizes the longitudinal distribution of these forces and the resulting normal stresses. In the inner part of the assembly, the force $T(x)$ is constant

$$T_{max} = -\frac{\Delta\alpha\Delta t}{\lambda}X_{max} \tag{5.48}$$

where the factor

$$X_{max} = X(\ell - a) = 1 - \frac{\cosh k(\ell - a)}{\cosh k\ell} \tag{5.49}$$

reflects the effects of the size of the assembly and the length of the contact area on the maximum force (and the resulting normal stress) arising in the mid-cross-section of the adherends.

The distribution of the interfacial shear stresses, as evident from eq (5.44), is characterized by the function

$$X'(x) = \frac{\sinh kx}{\cosh k\ell} \tag{5.50}$$

The maximum values of these stresses occur at the assembly end $(x = \ell)$

$$\tau_{max} = k\frac{\Delta\alpha\Delta t}{\lambda}X'_{max} \tag{5.51}$$

where the factor

$$X'_{max} = \tanh k\ell \tag{5.52}$$

accounts for the effect of the assembly size. The average value of the shear stress is given by

$$\bar{\tau} = \frac{1}{a}\int_{\ell-a}^{\ell} \tau(\xi)d\xi = \frac{\Delta\alpha\Delta t}{\lambda a}X_{max} \tag{5.53}$$

In the case of a continuous attachment, the parameter a in this relationship should be replaced by ℓ.

The distributions of the shear stress $\tau(x)$ and the resulting forces $T(x)$ are schematically shown in Fig. 5.6. The dotted lines refer to the case $a = \ell$, i.e., to the case of a continuous attachment. Note, that the increase in the length of the bonded areas results in greater forces and greater stresses in the adherends while the maximum shear stress in the interface is not affected by the length of the bonded areas.

The geometric factor defined in eq (5.49) attains its maximum value in the case of a continuous attachment $(\ell = a)$:

$$X_{max} = 1 - \frac{1}{\cosh k\ell} \tag{5.54}$$

The factors X_{max} and X'_{max} for a continuous attachment are plotted in Fig. 5.7. As is evident from this figure, for sufficiently large assemblies (large ℓ values) and/or stiff attachments (large k values) the stresses, both in the adherends ($k\ell > 4$) and in the attachment material ($k\ell > 2.5$), become independent of the assembly size.

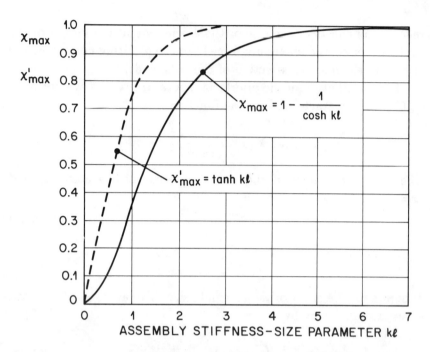

Figure 5.7: Factors reflecting the effect of the assembly size on maximum normal stresses in the components (X_{max}) and maximum shearing stress in the interface (X'_{max}).

Large $k\ell$ values: For large $k\ell$ values the stress distribution functions, $X(x)$ and $X'(x)$, can be simplified to

$$X(x) \cong 1 - e^{-k(\ell-x)}, \quad X'(x) \cong e^{-k(\ell-x)} \qquad (5.55)$$

Thus, for small x values, i.e., for the cross-sections sufficiently remote from the assembly edges, the function $X(x)$ is close to unity, and the function $X'(x)$ is close to zero. Hence, in the inner part of the assembly, the stresses in the assembly components are independent of the location of the given cross-section and can be calculated under the assumption that the assembly is infinitely large. It is noteworthy that owing to this fact, the stresses in the assembly components can be found directly and in an elementary way, as was done by Timoshenko. However, the shear stresses in the inner part of the assembly are, on the contrary, very small. Near the ends, where the coordinate x is of the same order as the assembly half length, the interfacial shear stresses increase exponentially and reach their maximum value at the end cross-sections. Alternately, the normal stresses in the assembly components vanish at the edges. Such a distribution is similar to the one predicted by Goland-Reissner theory.

Normal Stresses: The bending moments for this condition can be calculated via eq (5.23)

$$M_i(x) = -\frac{D_i}{\rho(x)} = -\frac{h\Delta\alpha\Delta t}{2\lambda}\frac{D_i}{D}X(x), \quad i = 1,2 \tag{5.56}$$

These moments result in normal stresses, linearly distributed over the component thicknesses, while the axial forces, $T(x)$, cause stresses, which are uniformly distributed over the cross-sections. Obviously, the total normal stresses are due to the combined action of the axial forces and bending moments, and the maximum values of these stresses are as follows:

$$\sigma_{1,max} = \frac{T_{max}}{h_1} + \frac{6M_{1,max}}{h_1^2} = -\frac{\Delta\alpha\Delta t}{\lambda h_1}\left(1 + 3\frac{h}{h_1}\frac{D_1}{D}\right)X(\ell - a) \tag{5.57a}$$

and

$$\sigma_{3,max} = -\frac{T_{max}}{h_3} - \frac{6M_{3,max}}{h_3^2} = \frac{\Delta\alpha\Delta t}{\lambda h_3}\left(1 + 3\frac{h}{h_3}\frac{D_3}{D}\right)X(\ell - a) \tag{5.57b}$$

where the "+" is used for tensile stresses and the "−" for compressive stresses. In the case when component #3 is very thick, no bending occurs and eqs (5.57) can be simplified to

$$\sigma_{1,max} = -\frac{E_1}{1 - \nu_1}\Delta\alpha\Delta tX(\ell - a) \tag{5.58a}$$

and

$$\sigma_{3,max} = \frac{E_1}{1 - \nu_1}\frac{h_1}{h_3}\Delta\alpha\Delta tX(\ell - a) \tag{5.58b}$$

Such a situation is encountered when component #3 is a substrate and component #1 is a thin film. In this case, the elastic constants of component #1 are of primary importance, while the elastic constants of component #3 affect the stresses only through the X factor and if the assembly is sufficiently large, do not affect the stresses at all. In all such cases, the stresses in component #1 are significantly greater than those in component #3.

External Load: The basic integral equation for the shear stress, eq (5.38), is equally applicable to the case when the stress is due to an external load P (Fig. 5.8). Such a load could be applied, for instance, during laboratory Instron testing. However, the boundary conditions in this case are essentially different from eq (5.45) and are

$$T(-\ell) = 0, \quad T(0) = \frac{P}{2}, \quad T(\ell) = P \qquad (5.59)$$

With these conditions the basic equation (5.38) has the solution

$$\tau(x) = \frac{kP}{2\sinh k\ell}\cosh kx \qquad (5.60)$$

and the maximum and the average values of the shear stress are

$$\tau_{max} = \tau(\ell) = \frac{kP}{2}\coth k\ell, \quad \bar{\tau} = \frac{P}{2\ell} \qquad (5.61)$$

Comparing these equations with eqs (5.51) and (5.53), it can be concluded that the maximum stresses for the two conditions, i.e., thermally and externally loaded structures, will be equal, if

$$\Delta\alpha\Delta t = \frac{P\lambda}{2}\coth^2 k\ell \qquad (5.62)$$

For the average stresses, the corresponding condition is

$$\Delta\alpha\Delta t = \frac{P\lambda}{2}\frac{\cosh k\ell}{\cosh k\ell - 1} \qquad (5.63)$$

For stiff and/or long assemblies, however, eqs (5.62) and (5.63) approach the identical limit, i.e.,

$$\Delta\alpha\Delta t = \frac{P\lambda}{2} \qquad (5.64)$$

This expression could be of help when comparing the stress caused by differential thermal expansion to the stress associated with the application of an external load.

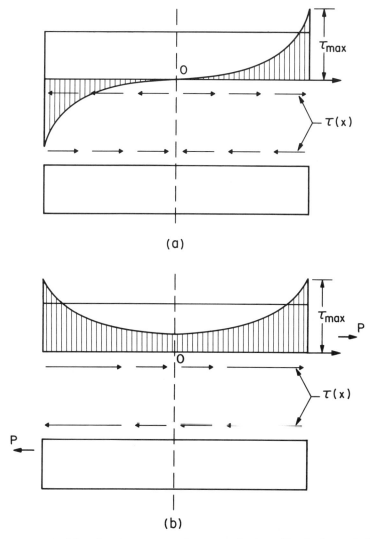

(a)

(b)

Figure 5.8: Shearing stresses due to thermally induced (a), and external (b) loading

Peeling Stress: The transverse normal ("peeling") stresses are due primarily to the fact that the assembly components are forced to bend jointly, i.e., to have the same curvature at all the cross-sections, despite the difference in their flexural rigidity. These stresses are schematically shown in Fig. 5.9. However, to simplify the analysis, we replace the distributed peeling stresses, which are directed upward and occupy relatively small areas near the edges, with concentrated forces, N_0, applied at the edges. The equation of equilibrium takes the form:

$$(x + \ell) N_0 \quad - \quad \int_{-\ell}^{x}\!\!\int_{-\ell}^{\xi} p(\xi')d\xi'd\xi' - \int_{\ell-a}^{x}\!\!\int_{-\ell}^{\xi} p(\xi')d\xi'd\xi'$$

$$= M_1(x) \quad - \quad \frac{h_1}{2}T(x) = -M_2(x) + \frac{h_2}{2}T(x) = \mu T(x) \quad (5.65)$$

where the factor

$$\mu = \frac{h_3 D_1 - h_1 D_3}{2D} \quad (5.66)$$

reflects the effect of the difference in the adherend thicknesses and flexural rigidities. By differentiating eq (5.65), a relationship for the transverse shearing force $N(x)$ is obtained

$$N(x) = \int_{-\ell}^{x} p(\xi)d\xi = N_0 - \mu\tau(x) \quad (5.67)$$

A second differentiation results in an expression for the distributed "peeling" stress

$$p(x) = -\mu\frac{d\tau(x)}{dx} = -\frac{\mu}{\kappa}\Delta\alpha\Delta t\frac{\cosh kx}{\cosh k\ell} \quad (5.68)$$

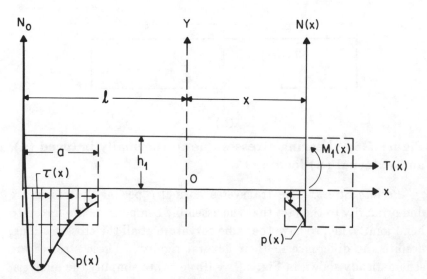

Figure 5.9: Peeling stress distribution.

Because the equilibrium condition requires that $N(\ell) = 0$, the concentrated forces, N_0, at the assembly edges can be evaluated by

$$N_0 = \mu\tau(\ell) = \mu\tau_{max} \tag{5.69}$$

This indicates that this force is proportional to the maximum shearing stress. Note that a more accurate solution for the interfacial stresses can be obtained by considering the through-thickness compliance.

Deflections: The deflection function $w(x)$ can be found by using eq (5.43) for the curvature. In the case of a continuous attachment, the maximum bow is given by

$$w_{max} = w(\ell) = \frac{h\ell^2}{4\lambda D}\Delta\alpha\Delta t\left(1 - 2\frac{\cosh k\ell - 1}{(k\ell)^2\cosh k\ell}\right) \tag{5.70}$$

For large assemblies and/or stiff attachments, $k\ell \to \infty$, this expression reduces to the simple relationship first obtained by Timoshenko

$$w_{max} = \frac{h\ell^2}{4\lambda D}\Delta\alpha\Delta t \tag{5.71}$$

3.4   Nonlinear Effects

Some attachment materials display nonlinear compliance of the attachment. In the case of solder interconnections, for instance, this nonlinearity is due to the elastoplastic behavior of the solder material, even at low stress levels. In this section, a simple analytical model for the assessment of the effect of the attachment compliance nonlinearity is proposed. It should be emphasized that other important phenomena which could affect the stresses in soldered joints, such as creep, and temperature dependence of the mechanical properties are not covered in this discussion.

An idealized stress/strain curve in shear for solders is shown in Fig. 5.10 (Goldman (1969)). This curve can be approximated by the expression

$$\tau = G_e\gamma, \text{ for } \tau \leq \tau_s; \; \tau = \tau_s\left(\frac{\gamma}{\gamma_s}\right)^n \quad \tau \geq \tau_s \tag{5.72}$$

where τ_s is the yield stress (assumed to coincide with the limit of elasticity), γ_s is the corresponding shear strain and $G_e = \tau_s/\gamma_s$ is the elastic shear modulus. The exponent n is related to the fullness coefficient $\beta = \omega/\tau_u\gamma_u$ of the stress/strain curve (here ω is the area, restricted by the γ axis and the vertical line $\gamma = \gamma_u$, τ_u is the ultimate shear stress, and γ_u is the corresponding strain) by

$$n = \frac{1-\beta}{\beta} \qquad (5.73)$$

Although, generally speaking, the n value can be any positive number, for convex curves of the type shown in Fig. **5.10**, this value is smaller than unity.

The shear modulus in the elasto-plastic zone can be found from eq (5.72) by differentiation

$$G_a = \frac{d\tau}{d\gamma} = nG_e\left(\frac{\tau_s}{\tau}\right)^\mu \qquad (5.74a)$$

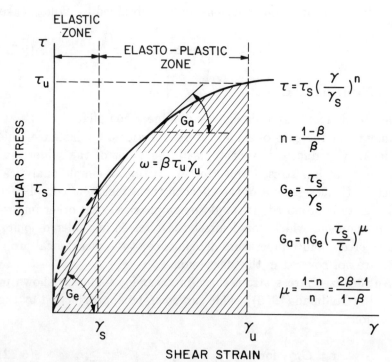

Figure **5.10**: Idealized stress/strain curve in shear for solders.

with

$$\mu = \frac{1-n}{n} = \frac{2\beta - 1}{1 - \beta} \qquad (5.74b)$$

Next, the coefficient of the attachment compliance can be calculated from

$$\kappa_a = \frac{h_a}{3G_a} = \frac{\kappa_e}{n}\left(\frac{\tau}{\tau_s}\right)^{\mu} \qquad (5.75)$$

where $\kappa_e = h_a/3G_e$ is the elastic coefficient of compliance.

Using eq (5.72) it is possible to obtain the basic integral equation (similar to eq (5.38) in the elastic case) for the shear stress function, $\tau(x)$, in the form

$$2\kappa_n \tau^{1+\mu}(x) = -(\kappa_1 + \kappa_2)\tau(x) + \lambda \int_0^x T(\xi)d\xi + \Delta\alpha\Delta tx \qquad (5.76)$$

where

$$\kappa_n = (1 + \mu)\kappa_e \tau_s^{-\mu} \qquad (5.77)$$

is the coefficient of the attachment compliance for the entire stress range. Obviously, $\kappa_n = \kappa_e$ in the linear (elastic) case where $\mu = 0, n = 1$ and $\beta = 1/2$.

The nonlinear relation for the shear stress, eq (5.76), may be solved by the Picard method, using eq (5.44) as a "zero-approximation" solution. Substituting eq (5.44) into the right side of eq (5.76), one obtains

$$\tau(x) = \tau_s\left(\frac{\sinh kx}{\eta_\tau \cosh k\ell}\right)^n, \quad \eta_\tau = \frac{\tau_s\sqrt{\lambda\kappa}}{n\Delta\alpha\Delta t} \qquad (5.78)$$

where the nonlinear effect is accounted for by the parameter n. The maximum value of the shearing stress is

$$\tau_{max} = \tau(\ell) = \tau_s\left(\frac{\tanh k\ell}{\eta_\tau}\right)^n \qquad (5.79)$$

The half length of the elastic zone can be found from eq (5.78) by setting $\tau(x)$ equal to τ_s. The location of the cross-section where the shearing stress reaches its yield point value can then be expressed by

$$x_e = \frac{1}{k}\ln\left(\eta + \sqrt{\eta^2 + 1}\right), \eta = \eta_\tau \cosh k\ell \qquad (5.80)$$

Solving for the shear stress, in the presence of non-linearities, yields the following conclusions:

- The nonlinear shear stress depends on the yield-stress level and increases with an increase in the yield stress.
- The shape of the stress/strain curve affects the shear stress through the n^n factor. This effect is strongest for n values close to 0.368, which corresponds to $\beta = 0.731$, and offers a reasonable representation of an actual stress/strain curve. The shape of the stress/strain curve also influences the shear stress indirectly.
- The distribution of the nonlinear shear stress is described by the function $\sinh^n kx$, which indicates that the nonlinearity results in a more uniform distribution of stresses over the assembly length.

For sufficiently long and stiff assemblies, eqs (5.78)–(5.80) can be simplified as follows:

$$\tau(x) = \tau_{max} e^{-kn(\ell-x)} \tag{5.81a}$$

$$\tau_{max} = \tau_s \eta_\tau^{-n} \tag{5.81b}$$

$$x_e = \ell + \sqrt{\frac{\kappa}{\lambda}} \ln\eta_\tau \tag{5.81c}$$

As evident from this last expression, the length $\ell_p = -\sqrt{\kappa/\lambda} \ln \eta_\tau$ of the plastic zone (on one side of the assembly) increases with an increase in the interfacial compliance, in the thermal-expansion-mismatch strain and in the value of the exponent n. This length reduces with the increase in the axial compliance and the yield stress.

The shear force in the x cross-section is

$$T(x) = -\int_x^\ell \tau(\xi)d\xi = -\frac{\tau_{max}}{kn}\left[1 - e^{-kn(\ell-x)}\right]$$

or

$$T(x) = -\eta_\tau^{1-n}\frac{\Delta\alpha\Delta t}{\lambda}\left[1 - e^{-kn(\ell-x)}\right] \tag{5.82}$$

Here, the factor

$$\eta_\tau^{1-n} = \left(\frac{\tau_s}{\tau_{max}}\right)^\mu \tag{5.83}$$

reflects the effect of the nonlinearity of the attachment compliance on the shear force and the resulting normal stresses in the assembly components. Note, that the reduction factor for the maximum shear stress differs from eq (5.83) by the factor of n. This indicates that the maximum shearing stress at the interface is affected by the nonlinearity of the attachment compliance more strongly than the normal stresses in the components.

3.5 Strength Requirements

The results obtained in Section 3.3 can be used to derive relations for the required material characteristics in a microelectronic structure. As an illustration, the case of a continuous die attachment is considered.

The desired interfacial compliance of the attachment material can be determined, on the basis of eq (5.57), to be

$$\kappa \geq \frac{\lambda \ell^2}{\ln^2 \left(\eta + \sqrt{\eta^2 - 1}\right)} \tag{5.84}$$

where the parameter η is related to the maximum normal stress and the allowable (design) stress $[\sigma]$ by

$$\eta = \frac{1}{1 - ([\sigma]/\sigma_\infty)} \tag{5.85}$$

where the normal stress in an infinitely large die is expressed as

$$\sigma_\infty = \frac{\Delta \alpha \Delta t}{\lambda h_1} \left[1 + 3 \frac{h}{h_1} \frac{D_1}{D}\right] \tag{5.86}$$

Assuming that the attachment material compliance is substantially greater than the interfacial compliance of the die and the substrate (which is usually the case when such an attachment is employed to lower high stresses), the compliance required to prevent die fracture is determined to be

$$\frac{h_a}{G_a} \geq \frac{3\lambda \ell^2}{2\ln^2 \left(\eta + \sqrt{\eta^2 - 1}\right)} \tag{5.87}$$

The allowable stress σ is generally chosen to be a fraction of the ultimate stress for silicon. Equation (5.87), plotted in Fig. 5.11, indicates that the required compliance of the attachment material increases rapidly with an increase in the die size.

Alternatively, the necessary attachment material compliance can be related to the shear stress, via eq (5.51) in the form

$$\frac{h_a}{G_a} \geq \frac{3\lambda\ell^2}{2u^2}$$ (5.88)

where the parameter $u = k\ell$ can be found from the equation

$$u \tanh u = \frac{[\tau]\lambda\ell}{\Delta\alpha\Delta t} = \epsilon$$ (5.89)

For sufficiently large dies and/or stiff attachments ($u > 2.5$) eq (5.88) can be simplified to yield:

$$\frac{h_a}{G_a} \geq \frac{3\lambda\ell^2}{2\epsilon^2}$$ (5.90)

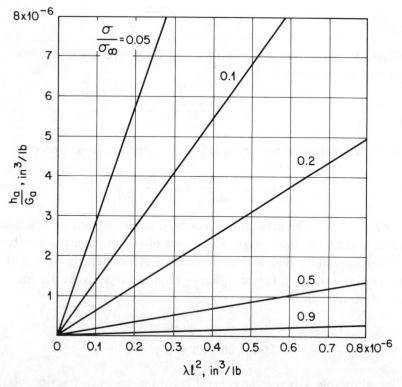

Figure 5.11: Required minimum compliance of the attachment from the standpoint of maximum normal stresses in the die.

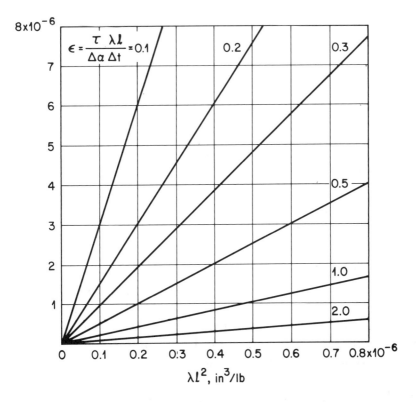

Figure 5.12: Required minimum compliance of the attachment from the standpoint of maximum shearing stresses in the attachment material.

This relationship is plotted in Fig. 5.12. The allowable stress τ can be chosen as a certain portion of the ultimate shearing stress for the bonding material when the ultimate strength condition is considered, or as a portion of the fatigue shearing stress when the fatigue strength conditions are addressed.

3.6 Numerical Examples

To aid the reader in interpreting the preceding developments, numerical values for the key parameter associated with several distinct microelectronic package configurations were determined and are shown in Table 5.1. These examples involve a 1.23 in long silicon substrate attached to an aluminum heat sink by means of silicone gel, epoxy adhesive or solder. In addition, an extreme case when the thickness of the attachment is zero (which is the ideal case of a bi-metallic thermostat) is examined. The assumed temperature differential is $\Delta t = 240°C$. The input and the calculated data are given in Table 5.1. In the slashed boxes (for the soldered assembly) the upper numbers are obtained by using a linear approach (i.e., ignoring nonlinearity in the behavior of the solder) and the lower numbers are calculated with consideration of the plastic effects. Some of the calculated results are plotted in Figs. 5.13, 5.14 and 5.15.

3.7 Discussion

The foregoing analysis has not only provided a set of engineering approximations for the stresses which develop in a microelectronic structure undergoing differential thermal expansion, but also serves to develop physical insight into the controlling failure mechanisms and the effect of material properties. Section 3.6 presented the numerical results of using the primary stress relation. This section will focus on the conclusions which can be drawn from both the form of the governing relations and the magnitude of the relevant stresses.

Table 5.1: Input and calculated data for example.

COMPONENT #1 (Si)		$E_1 = 17.5 \times 10^6$ psi, $\alpha_1 = 3.2 \times 10^{-6}$ 1/°C, $h_1 = 0.020$ in									
COMPONENT #2 (Aℓ)		$E_2 = 10.2 \times 10^6$ psi, $\alpha_2 = 23.6 \times 10^{-6}$ 1/°C, $h_2 = 0.080$ in									
ATTACHMENT MATERIAL		NONE	SILICONE GEL			EPOXY			SOLDER		
ATTACHMENT THICKNESS		0	10μm	20μm	30μm	0.5mil	1mil	2mils	1mil	2mils	3mils
STIFFNESS FACTOR, k, 1/in		32.31	1.876	1.328	1.084	18.40	14.21	10.58	31.98	31.66	31.35
MAX. STRESSES, psi — NORMAL IN THE COMPONENT	COMPONENT #1	41600	17730	10840	7800	41600			41600 / 19300		
	COMPONENT #2	34070	14520	8880	6390	34070			34070 / 15800		
MAX. STRESSES, psi — IN THE ATTACHMENT	SHEAR	18956	900	525	370	10795	8337	6207	18760/4330	18570/4310	18390/4290
	PEEL	-5271	-18	-9	-6	-1709	-1019	-565	-5160/-1190	-5060/-1170	-4960/-1160
MAXIMUM BOW, mils		10.92	7.42	6.65	6.32	10.92					

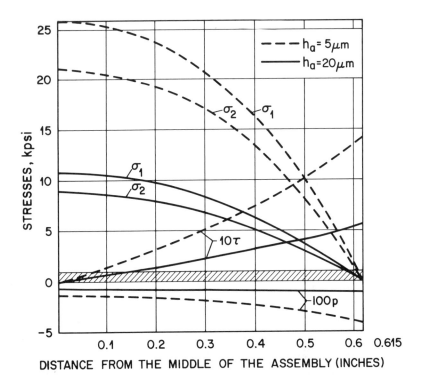

Figure 5.13: Stress distribution along the Si/Al assembly bonding material: silicon gel.

3.7.1 Stresses

Stresses arising in microelectronic assemblies, due to differential expansion between the constituents include:

- Normal stresses, acting over the cross-sections of the components
- Shearing and transverse normal (peeling) stresses in the interface.

It is to be noted, that there are also transverse shearing stresses, arising in the component cross-sections due to shearing forces $N(x)$. However, in the majority of important cases, these stresses are small and may be ignored.

Stresses of the first category are responsible for the mechanical failure of the adherends themselves, while the stresses of the second category lead to cohesive and the adhesive failure of the attachment material.

The stresses, acting over the cross-sections of the adherends are maximum in their mid-cross-sections and turn to zero at the edges. The interfacial stresses, on the contrary, reach their maximum values at the end cross-sections. The peeling stresses can be reduced by designing the assembly in such a way that the flexural rigidities of the components are proportional to the thickness of the components.

3.7.2   Effect of Assembly Size

For sufficiently long assemblies with stiff attachments ($kl > 4$ as, for instance, in the case of epoxy bonded or soldered substrate/heat-sink assemblies, the maximum stresses in the assembly components are independent of the assembly size. The maximum shearing stresses in the interface become independent from the assembly size in the region $kl > 2.5$.

In cases where extremely compliant adhesives (such as silicone gel) are used and/or in the case of small assemblies (such as chip/substrate), the maximum stresses in the adherends and the shearing stresses in the interface do depend on the assembly size and diminish with a decrease in the size of the assembly. Thus, it may be expected that if the components of an assembly having $kl > 4$, or the attachment material of an assembly with $kl > 2.5$, survived certain test conditions, an assembly of larger size will also survive, and therefore, there is no need for stress reduction measures. However, if the above assemblies failed during testing, only a substantial reduction in the assembly size and/or the stiffness of the attachment (resulting in significantly smaller kl magnitudes than the above critical values) could improve the situation. Obviously, there is an incentive to use compliant attachments for lower stresses in the components of relatively small size assemblies, while in large assemblies, application of attachments of moderate compliance may not result in a smaller stress level. If a small assembly is increased in size, this always causes greater stresses and could possibly result in a physical failure. In this case, a compliant attachment can be successfully utilized for greater reliability and the results obtained previously can be used to guide the physical design of such an attachment.

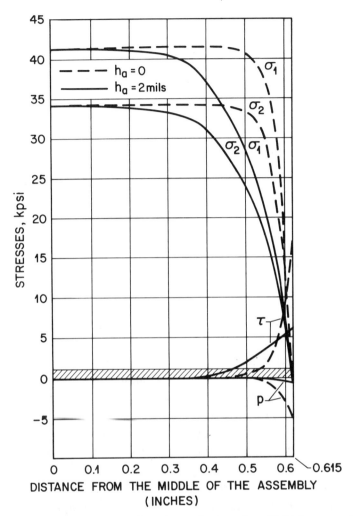

Figure 5.14: **Stress distribution along the Si/Al assembly. Bonding material: epoxy.**

The interfacial stresses in large assemblies with stiff attachments concentrate near the assembly ends and drop exponentially as the center is approached. This trend is particularly important in the interpretation of experimental data of different geometry, or when the strength of assemblies of different size is compared. In effect, for long-and-stiff attachments, neither the maximum shearing stress, nor the effective half length of an adhesively bonded joint (i.e., the half length of the area subjected to the shearing stresses exceeding, say, 5% of the maximum shearing stress value) depend on the size of the bonded area. Therefore, if the average stress value, based on the geometrical half length (instead of the effective half length) is used as a major stress criterion, attachments with greater overlaps may be mistakenly expected to reduce the thermal stresses.

3.7.3   Effect of Attachment Compliance

As has been indicated above, if the $k\ell$ value is greater than 4, the maximum values of the normal stresses acting in the mid-cross-sections of the assembly components are independent of the attachment compliance. This is not true, however, for the maximum interfacial stresses which decrease with an increase in the attachment compliance for any $k\ell$ value. In small size assemblies, both the normal stresses in the components and the stresses in the attachment material decrease with an increase in the attachment compliance. The effect of the attachment compliance increases as the $k\ell$ value decreases.

The expression for the coefficient of the attachment compliance indicates that the compliance of a sufficiently thin attachment is directly proportional to its thickness and is inversely proportional to the shear modulus of the material. The increased attachment compliance results in the redistribution of the interfacial stresses in such a way, that the maximum stresses at the edges decrease and the stresses in the inner part of the assembly increase, leading to a more uniform distribution of the interfacial stresses. The effect of the increased compliance is greater for greater compliances. For long and stiff assemblies, the above redistribution of the shearing stresses does not result in essentially smaller maximum normal stresses in the component mid-cross-sections. However, for small assemblies and/or when extremely compliant attachment materials (such as, silicone gel) are used, the increased attachment compliance could be very effective as a stress reduction measure, both for the adhesive and the adherends. It also results in reduced bowing of the assembly.

3.7.4   Effect of the Non-linear Behavior of the Attachment Material

Non-linear behavior is thought to be of the primary importance for soldered assemblies and is due to the strongly pronounced non-

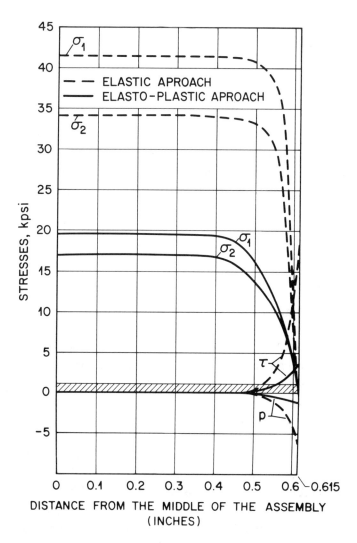

Figure 5.15: **Stress distribution along the Si/Al assembly.**
Bonding material: 2 mil thick solder.

linearity of the stress/strain curve in shear for solders. The theoretical stress model, developed in previous sections, indicates that the effect of the actual stress/strain curve for the given solder material can be accounted for by using the yield point, τ_s, of the material and the fullness coefficient, β, of this curve as governing parameters.

The elasto-plastic behavior of solders results in smaller maximum stresses and in a more uniform distribution of the interfacial

stresses along the assembly. The interfacial stresses are affected by
the attachment compliance non-linearity more strongly than the normal stresses in the components. In the executed example, this nonlinearity resulted in halving of the normal stresses in the assembly
components and a four-fold reduction in the maximum stresses in
the solder attachment itself.

The length of the plastic zone, located at the end portions of
the assembly, increases with an increase in the interfacial compliance, and in the differential thermal expansion mismatch of the constituents. This length reduces with the increase in the axial compliance, in the yield stress level and in the fullness of the stress/strain
curve. In the sample calculation, length of the plastic zone (one one
side of the assembly) was about 8% of the assembly half length.

3.8   Conclusions

The following major conclusions can be drawn from the foregoing
analysis:

- An engineering theory of thermally induced stresses arising
  in adhesively bonded and soldered assemblies due to the thermal expansion mismatch of the materials is developed. The
  suggested approach enables one to evaluate the magnitude
  and the distribution of the normal stresses acting in the assembly components, as well as the shearing and the peeling
  stresses in the interface. This approach can be utilized for
  stress/deflection evaluation and rational physical design of
  assemblies operated, shipped or stored at elevated or low temperatures. It can also be used as guidance for experimental
  planning. The relationships obtained are simple, easy-to-use,
  and clearly indicate the role of major factors affecting the mechanical behavior of the assembly. Calculations can be done
  manually and no computerization is necessary.

- The Timoshenko theory of bi-metal thermostats can be utilized for the prediction of the maximum stresses in the components and the bow for long assemblies with stiff attachments. This theory is of no help, however, as far as the
  strength or the effect of the attachment material is concerned,
  and could result in a substantial overestimation of stresses

acting over the cross-sections of small size assemblies such as, for instance, die/substrate structures.

- Although the analysis here has been performed for assemblies used in microelectronics, the obtained results are equally applicable to other areas of engineering and physics where shear lap joints, subjected to thermally induced or external shear loading, are utilized (see, for instance, Luryi and Suhir (1986)).

4 SUMMARY

1. The major thermal stress induced failure modes in microelectronic components can be due to transistor junction breakdown, fatigue, ductile rupture, brittle fracture, thermal shock, creep, excessive elastic or plastic deformation, thermal relaxation and stress corrosion.
2. Results obtained in the area of thermal stresses in various non-electronic systems can be successfully applied for the analysis of thermal stresses and thermal stress failure in microelectronic devices.
3. The majority of studies, dealing with thermal stresses and reliability of electronic components, are experimental. It is believed, however, that a theoretical and, in particular, an analytical approach must always be considered as an important part of any reliability analysis concerning thermal stress failures.

5 ACKNOWLEDGEMENT

The author acknowledges, with thanks, the comments made by C. J. Bartlett, J. M. Segelken, W. Engelmaier, and especially P. M. Hall.

6 NOMENCLATURE

a	Non-continuous attachment length (on one side of the assembly), in
C_i	Constant of integration
c	Fatigue ductility parameter
D_i	Flexural rigidity of the i-th assembly component, treated as a rectangular plate, in-lb
$D = \sum_i D_i$	Total flexural rigidity of the assembly, in-lb
E_i	Young's modulus for the i-th assembly component, psi
E_i^o	"Generalized" Young's modulus for the i- th assembly component, psi
E_f, E_f^o	Young's modulus and "generalized" Young's modulus, respectively, for the film, psi
E_s, E_s^o	Young's modulus and "generalized" Young's modulus, respectively, for the substrate, psi
G_i	Shear modulus for the i-th assembly component, psi
G_a	Shear modulus for the attachment material, psi
G_e	Shear modulus for the attachment material in the elastic range, psi
h	Solder bump height, in
h_i	Thickness of the i-th assembly component, in
h_f, h_s	Thickness of the film and the substrate, respectively, in
h_a	Thickness of the attachment material, in
I_i	Moment of inertia per unit width of the cross-sectional area for the i-th assembly component, in^3
i	Assembly component number, interface number constant of integration number
k	Stiffness parameter (eigenvalue)
ℓ	Half the assembly length, in
M_i	Bending moment (per unit width) for the i-th assembly component, lb
M_i, max	Maximum value of the above bending moment, lb
N_o	Concentrated force (per unit width) at the end of the assembly, lb/in
N_f	Number-of-cycles till failure
n	Exponent of the stress-strain curve in shear for solder mat
P	External shearing force (per unit width), lb/in
$p(x)$	Peeling stress, psi

T	Internal shearing force (per unit width), caused by the thermal expansion (contraction) mismatch of the assembly components, *lb/in*
T_i	Shearing force (per unit width) acting in the i-th interface i.e., in the interface between the i-th and the $(c + 1)$-st components
T_{max}	Maximum value of the above forces, *lb/in*
$\overline{T}_s$	Mean cycling temperature, $^{\circ}C$
t_D	Dwell time, hours
Δt	Temperature change, $^{\circ}C$
u_i^T	Longitudinal displacement of the upper extreme fiber for the i-th component, *in*
u	Shearing strain parameter
$w(x)$	Deflection function, *in*
w_o, w_{max}	Maximum deflection, *in*
x	Abscissa of the given cross-section, *in*
x_e	Abscissa of the elastic zone

Greek

α_i	Coefficient of linear thermal expansion for the i-th component, $1/^{\circ}C$
α_f, α_s	Coefficients of linear thermal expansion for the film and the substrate, respectively, $1/^{\circ}C$
$\Delta\alpha$	Difference between the coefficients of thermal expansion, $1/^{\circ}C$
β	"Fullness" coefficient for the stress-strain curve for solder material
γ	Shear strain
γ_s	Shear strain corresponding to the yield stress τ_s
γ_u	Ultimate shear strain (corresponding to the ultimate stress τ_u)
$\Delta\gamma$	Change in the shear strain during thermal cycling
δ	Free term in the equation for the eigenvalue k
ϵ	In-plane thermal expansion estimated strain for the assembly components; parameter of the allowable shear stress
ϵ_f'	Fatigue ductility parameter
ϵ_f, ϵ_s	In-plane thermal expansion strains for the film and the substrate, respectively

η, η_τ Shear stress parameters

χ_i Interfacial compliance for the i-th assembly component, in^3

$\chi_{i,i+1}$ Interfacial compliance for the
i-th interface, in^3/lb

χ Total interfacial compliance, in^3/lb

χ_a Interfacial compliance for the attachment material, in^3/lb

χ_e Interfacial compliance for the attachment material
in the elastic zone, $^3/lb$

λ_i Axial compliance of the i-th assembly component, in/lb

$\lambda_{i,i+1}$ Axial compliance due to bending, in/lb

$\lambda, \lambda^o_{i,i+1}$ total axial compliance, in/lb

μ Peeling stress parametric; stress-strain fullness parameter

ν_i Poisson's ratio for the i-th assembly component

ν_s Poisson's ratio for the substrate material

ρ Radius of curvature, in

σ_i Normal stress acting over the cross-section of the
i-th assembly component, psi

$\sigma_{i,max}$ Maximum value of the above stress, psi

σ_f, σ_s Normal stresses in the film and in the substrate,
respectively, psi

σ_∞ Normal stress in a indefinitely large die, psi

$[\sigma]$ Allowable normal stress, psi

τ_i Shear stress in the i-th interface, psi

$\tau_{max,\bar\tau}$ Maximum and mean values of the shear stress, psi

τ_s Field shear stress, psi

$[\tau]$ Allowable shear stress, psi

$\chi(x)$ Function reflecting the distribution of the normal
stress $\sigma(x)$ along the assembly

χ_{max} Maximum value of the above function

$\chi'(x)$ Function reflecting the distribution of the shear stress
$\tau(x)$ along the assembly

χ'_{max} Maximum value of the above function

ω Area under the stress-strain curve in shear for solders,

7 REFERENCES AND BIBLIOGRAPHY

Agrawala, B N (1985). Thermal Fatigue Damage in Pb-In Solder
Interconnections, 23rd Reliability Physics Symp., 198-205.

Aleck, B J (1949). Thermal stresses in a rectangular plate clamped along an edge, *J Appl Mech*, **16**, 118–122.

Allman, D J (1977). A theory of plastic stresses in adhesive bonded lap joints, *Q J Mech & Appl Math*, **30**, (4), 415–436.

Anderson, J E et al (1985). Prediction of the temperature field for an electronic device operating under unsteady electro-thermal conditions, 5th Ann Int Elec Pack Conf, 508–525.

Avdonin, N A, et al (1972). Influence of the temperature field and the thermal stress field upon the formation of the dislocation structure in Gallium-Arsenide single Ccystals grown by the Czochralski method, *Soviet Phys-Dokl*, **16**, 772–775.

Baker, E (1970). Calculation of thermally induced mechanical stresses in encapsulated assemblies, *IEEE Trans Parts, Mater & Pack*, **PMP-6**, 121–128.

Baker, E (1970). Thermal ratchet in encapsulated assemblies, Eastern Electr Pack Conf, MIT, Cambridge, MA, 23.1-23.8.

Baker, E (1972). Some effects of temperature on material properties and device reliability, *IEEE Trans Parts, Hybrids & Pack*, **PHP-8** (4), 4–14.

Baker, E (1975). Understanding cycling thermal stress in electronic assemblies: the key to improving accelerated life testing, *Insulation/Circuits*, 49–56.

Baker, E (1979). Stress relaxation in tin-lead solders, *Mat Science & Eng*, **38**, 241–247.

Balde, J W (1984). Surface attach technology - introduction and overview, *Proc ISHM*, Dallas, TX.

Balde, J W (1984). Problems in the shift to leaded chip carrier constructions - an overview, Int Elec Pack Soc, Baltimore, MD.

Bartlett, C J, Segelken, J M and Teneketdges, N (1987). Multi- chip packaging design for VLSI-based systems, 37th Elec Comp Conf, 518–525.

Baxter, G K, Anslow, J W, (1977). High temperature characteristics of microelectronic packages, *Trans on Parts, Hybrids and Packaging*, **PHP-13** (4).

Becker, G (1979). Testing and results related to the mechanical strength of solder joints, IPS Fall Meeting, San Francisco, CA.

Becker, G (1983). Creep and fatigue testing of micro-solder joints, Symp on Thermal Fatigue in Surface-Mounted Electronic Components, The Swedish Inst for Metals Research, Stockholm, Sweden.

Belton, D J (1986). The effect of high temperature storage life and thermal shock tests upon the time dependent properties of epoxy molding compounds, *Int Symp on Microelectr Proc*, 393–399.

Benson, N K (1961). The mechanics of adhesive bonding, Appl. Mech. Rev.

Benson, N K (1966) Influence of stress distribution on the strength of bonded joints, Int Conf on Adhesion, Nottingham Univ.

Berkehile, M H (1967). Investigation of solder cracking problems on printed circuit boards, Marshall Space Flight Center, Huntsville, AL, **N67-36783**.

Biot, M A (1956). Thermoelasticity and irreversible thermodynamics, *J Appl Phys*, **27** (3), 240–253.

Boah, J K and De Vore, J A (1978). Thermal fatigue failure of soft soldered silicon triacs, General Electric Co., **R78EGP10**, Syracuse, NY.

Boley, B A and Weiner, J H (1960). *Theory of thermal stresses*, John Wiley and Sons, New York, NY.

Boley, B A and Testa, R B (1969). Thermal stresses in composite beams, *Int J Solid Struct*, **5**, 1153–1169.

Bolger, J C and Mooney, C T (1983). Failure mechanism for epoxy die attach adhesives, *33rd Elec Comp Conf, Proc*, 227–231.

Brenner A and Stenderoff S (1949). Calculation of stress in electrodeposits from the curvature of a plated strip, *J Res the Nat Bureau of Stand,* Research Paper RP/954, **42**, 105–123.

Burges, J F, et al (1984). Solder fatigue problems in power packages, *IEEE CHMT-7*, 4, 405–410.

Carlson, R O, et al (1985). Thermal expansion mismatch in electronic packaging, Elec Pack Mat Science, *Symp Proc*, 177–190.

Chadwick, P (1960). Thermoelasticity, the dynamic theory, in *Progress in Solid Mechanics*, **1**, (6), Amsterdam, North-Holland Publ. Co., 263–328.

Chang Fo-Van (1983). Thermal contact stresses of bi-metal strip thermostat, *Appl Math & Mech*, **4**, (3), Tsing-hua Univ., Beijing, China, 363–376.

Chen, D, et al (1982). Thermal stress in laminated beams, *J Thermal Stresses*, 67–84.

Chen, P E, et al (1972). Time and temperature effects on the mechanical behavior of tin-lead solders, *Mech Behavior of Mat*, III, Soc Mat Science, Japan, 242–254.

Chin, W T and Nelson C W (1979). Thermal stress in bonded joints, *IBM J Res Develop*, **23**, (2), 178–188.

Cherry, B W and Harrison, N L (1970). Optimum profile for a lap joint, *J Adhesion*, **2**, p. 125.

Chiang, S S and Shukla R K, (1984). Failure mechanism of die cracking due to imperfect die attachment, 34th Electr Comp Conf, IEE CHMT, 195–202.

Clatterbaugh, G V and Charles, H K Jr (1985). Design optimization and reliability testing of surface mounted solder joints, *1985 Int Symp Microelectr Proc.*

Coffin, L F Jr (1954). A study of cyclic thermal stress in a ductile metal, *Trans ASME*, **76**, 931–950.

Coffin, L F Jr (1954). The problem of thermal stress fatigue in austenitic steels at elevated temperatures, *ASTM Spec Tech Publ* **165**.

Coffin, L F Jr (1962). Low cycle fatigue: a review, *Appl Mat Research*, **1**, (3), 129–141.

Coffin, L P (1968). Introduction to high temperature low cycle fatigue, *Experimental Mechanics*, 218–224.

Coombs, V D (1976). Fatigue properties of pure-metal solders, 1976 Int Microelectr Symp, 67–72.

Cooper, P A and Sawyer, J W (1979). A critical examination of stresses in an elastic single lap joint, *NASA Tech Paper* **1507**.

Dance, F J (1983). Low thermal expansion rate clad metals for chip carrier applications, Northcon/83, Electr Show & Convention, 14/2/1-5.

Dash, W C (1955). *Phys Rev*, **98**, p. 1536(A).

Dean, R (1983). Thermal stress in electronic systems colloquium on high temperature electronics, 1/1-11, IEE, London, England, 1–12.

Delale, F, et al (1981). Stresses in adhesively bonded joints: a closed-form solution, *J Composite Mat*, **15**, 249–271.

Der Marderosian, A, et al (1982). A rapid technique of evaluating thermally induced strains in leadless ceramic chip carriers mounted to polymeric substrates, 20th Reliability Physics Symp, 1–5.

De Vore, J A (1982). Fatigue resistance of solders, *Proc NEPCON'82*.

De Vore, J A (1982). Fatigue resistance of solders, *Nat Elect Pack Conf Proc*, 409–414.

Dillon, O W Jr (1967). Coupled thermoelasticity of bars, *Trans ASME*, Ser. E, **34**, (1), 137–145.

Dorn, I E, Ed (1961). *Mechanical behavior of materials at elevated Temperatures*, McGraw-Hill, New York, NY.

Du Chen and Shun Cheng, S C (1983). An analysis of adhesive-bonded single-lap joints, *J Appl Mech*, Trans of ASME, **50**, 110–115.

Ebel, G W, et al (1982). Wirebonding reliability and techniques and analysis, *IEEE Trans in Components, Hybrids & Manuf Tech*, **CHMT-5**, (4), 441–445.

Edwards, D R, et al (1987). Shear stress evaluation of plastic packages, 37th Elec Comp Conf, *1987 Proc*, 84–95.

Eley, D D Ed (1969). Adhesion - fundamentals and practice, *Maclaren & Sons*, London, England.

Engel, P A (1984). Thermal stress analysis of soldered pin connectors for complex electronic modules, Comput Mech Eng, **2**, (6), 59–69.

Engelmaier, W (1982). Effects of power cycling on leadless chip carrier mounting reliability and technology, *Int Elect Pack Soc Proc*.

Engelmaier, W (1983). Fatigue life of leadless chip carrier solder joints during power cycling, *IEEE Trans, Comp, Hybrids, & Manuf Tech*, **CHMT-6**, (3), 232–237.

Engelmaier, W (1983). Fatigue life of leadless chip carrier solder joints, *NEPCON'83 Proc*.

Engelmaier, W (1984). Test method considerations for SMT solder joint reliability, *Int Electr Pack Soc Conf Proc*, 360–369.

Engelmaier, W (1984). Functional cycling and surface mounting attachment reliability, Surface Mount Technology, *Tech Monograph Series 6984-02*, ISHM.

Engelmaier, W (1985). Functional cycling and surface mounting attachment reliability, *Circuit World*, 11, (3), 61–72.

Engelmaier, W (1985). Solder joint reliability and testing considerations for leaded chip carriers, Natl Elect Pack & Prod Conf, NEPCON'85.

Evans, A G and Ruhle, M (1985). On the mechanics of failure in ceramic/metal bonded systems, Elect Pack, Mat Science, *MRS Symp Proc*, 153–166.

Evans, A G and Hsueh, C H (1986). Residual stresses and damage in multilayer ceramic/metal packages, Elect Pack Mat Science II, *MRS Symp Proc*, 91–112.

Finnie, I and Heller, W R (1959). Creep of engineering materials, *McGraw-Hill Book Co*, New York, NY.

Fox, A (1971). Stress relaxation and fatigue of two electromechanical spring materials strengthened by thermomechanical processing, *IEEE Trans*, Parts, Mat & Pack, **PMP-7**, (1), 34–47.

Fox, L R, et al (1985). Investigation of solder fatigue acceleration factors, *IEEE Trans on Comp, Hybrids & Manuf Tech*, **CHMT-8**, (2), 275–281.

Gaffney, J (1968). Internal lead fatigue through thermal expansion in semiconductor devices *IEEE Trans Electron Devices*, **ED-15**, p. 617.

Gatewood, B E (1957). Thermal stresses, *McGraw-Hill Book Co*, New York, NY.

Glang, R, Holmwood, R A and Rosenfeld, R L (1965). Determination of stress in films on single crystalline silicon substrates, *The Review of Scientific Instruments*, **36**, (1), 7–10.

Glascock, H H and Webster, H J (1984). Structural copper: a pliable high conductance material for bonding to silicon power devices, *IEEE Transactions on Components, Hybrids & Manuf Tech*, **CHMT-6**, (4).

Goland, M and Reissner, E (1944). The stresses in cemented joints, *J Appl Mech, Trans ASME*, 11, 17–27.

Goldenblat, I E and Nikolaenko, N A (1964). Calculation of thermal stresses in nuclear reactors, Consultants Bureau, New York, NY.

Goldman, L S (1969). Geometric optimization of controlled collapse interconnections, *IBM J Res Develop*, 13, 251–265.

Goodier, J N (1953). Thermal stresses, in design data and methods, Am Soc Mech Eng, New York, NY, 74–77.

Grant, P J (1976). Strength and stress analysis of bonded joints, **SOR(P)** (109), *British Aerospace*, Warton.

Grimado, P B (1978). Interlaminar thermoelastic stresses in layered beams, *J Thermal Stresses*, 1, 75–86.

Groothuis, S, et al (1985). Computer aided stress modeling for optimizing plastic package reliability, 23rd Ann Reliability Physics Symp, 182-191.

Hagge, J (1982). Predicting fatigue life of leadless chip carriers using Manson-Coffin equations, *Int Elec Pack Soc Proc*.

Hall, P M (1983). Solder attachment of ceramic chip carrier, *Solid State Tech*, 103–107.

Hall, P M, et al (1983). Thermal deformation observed in leadless ceramic chip carriers mounted to printed wiring boards, 33rd Elec Comp Conf, 350–359.

Hall, P M (1984). Strain measurements during thermal chamber cycling of leadless ceramic chip carriers soldered to printed boards, 34th Elec Comp Conf, 107–116.

Hall, P M (1984). Forces, moments and displacements during thermal chamber cycling of leadless ceramic chip carriers soldered to printed boards, *IEEE Transactions on Comp, Hybrids & Manuf Tech*, **CHMT-7** (4), 314–327.

Hall, P M (1987). Creep and stress relaxation in solder joints in surface-mounted chip carriers, 37th Elec Comp Conf, 579–588.

Hawkins, S P, et al (1986). The mechanical properties of soldered joints in surface mounted devices, *Brazing & Soldering*, 10, 4–6.

Heller, P (1984). Thermal cyclic and cooling effectiveness tests of leadless chip carrier assemblies for military avionics, *Proc NEPCON'84*, Anaheim, CA.

Heywood, R B (1962). Designing against fatigue, *Chapman and Hall*, Ltd, London.

Hieker, H (1985). Thermomechanical relaxation of thin-film metallizations, Elect Pack Mat Science, *MRS Symp Proc*, 191–202.

Hill, R (1950). The mathematical theory of elasticity, *Clarendon Press*, Oxford, U.K.

Hilton, H H (1967). Thermal stresses in bodies exhibiting temperature-dependent elastic properties, *J Appl Mech*, **19**.

Hoff, N (1956). Analysis of structures, *John Wiley and Sons*, New York, N.Y.

Homa, T R (1979). Circuit package with improved fatigue life, *IBM Tech Disclosure Bull*, **22**, (3), p. 950.

Howard, R T, et al (1983). A new package-related failure mechanism for leadless ceramic chip carriers solder-attached to alumina substrates, *Solid State Tech*, 115–122.

Howes, M A H (1976). A study of thermal fatigue mechanisms, *Thermal Fatigue of Materials & Components*, **ASTM STP 612**, 86–105.

Hsueh, C H and Evans, A G (1985). Residual stresses and cracking in metal/ceramic systems for microelectronics packaging, *J Am Ceram Soc*, **68**, (3) 120–127.

Hu, S M (1979). Film-edge-induced stress in substrate, *J Appl Phys*, **50**, p. 4661.

Huang, C K, et al (1986). The effect of window edge stress on dopant diffusion in silicon, *MRS Symp Proc*, **76**.

Hund, T D and Burchett, S N (1983). Stress production and relief in the gold-silicon eutectic die attach process, Int Microelectr Symp, Int Soc Hybrid Microelect, **6**, (1), Oct 243–250.

Inayoshi, H, et al (1979). Moisture-induced aluminum corrosion and stress on the chip in plastic-encapsulated LSIS, 17th Ann Rel Physics Symp, 113–117.

Isagawa, M, et al (1980). Deformation of Al metallization in plastic encapsulated semiconductor devices caused by thermal shock, 18th Ann Rel Physics Symp, 171–177.

Isomae, S (1981). Stress distribution in silicon crystal substrates with thin films, *J Appl Phys*, **52**, p. 2782.

Isomae, S (1985). Stresses in silicon at $Si_3N_4SiO_2$ film edges and viscoelastic behavior of SiO_2 films, *J Appl Phys*, **57**, p. 216.

Iwaki, T and Kobayashi, N (1981). Residual stresses of Czochralski-grown crystal, *ASME J Appl Mech*, **48**, 866–870.

Iwaki, T and Kobayashi, N (1986). Thermal and residual stresses of Chochralski-grown semiconducting material, *Int J Solids Structures*, **22**, (3), 307–314.

Jaccodine, R J and Schlegel, W A (1966). Measurement of strains at Si-SiO$_2$ interface, *J Appl Phys*, **37**, (6), p. 2429.

Jarboe, D M (1980). Thermal fatigues of solder alloys, Bendix Corp, Kansas City, **BDX-613-2341**.

Johns, D J (1965). Thermal stress analysis, *Pergamon Press*, New York, NY.

Johnson, J E (1974). Die-bond failure modes, *12th Ann Int Rel Phys Symp Proc*, 150–154.

Johnson, W S and Mall, S (1984). Bonded inst strength; static versus fatigue, *Proc, 5th Int Congr Exper Mech*, Montreal.

Jordan, A S and Berkstresser, G W (1980). Thermal stress analysis of composite encapsulants with a spherical adhesive interface, Microelectr & Reliability (GB), **20**, (4), 495–499.

Jordan, A S, et al (1980). A thermoelastic analysis of dislocation generation in pulled GaAs crystals, *Bell System Tech J*, **59**, 593–637.

Jordan, A S, et al (1986). The theoretical and experimental fundamentals of decreasing dislocations in melt grown Ga As and InP, *J Crystal Growth*, **76**, 243–262.

Kasem, Y M and Feistein, L G (1987). Horizontal die cracking as a yield and reliability problem in integrated circuit devices, 37th Elect Comp Conf, 96–104.

Keer, L M and Chantaramunghorn, K (1972). Stress analysis for a double lap joint, *Trans ASME, Series E, J Appl Mech*, **42**, p. 353.

Kelsey, S and Benson, N K (1966). ISD Rep 10, Inst Für Statik und Dynamik der Luft-und Raumfahztkonstruktionen, Univ of Stuttgart, West Germany.

Kempner, J and Pohle, V (1952). On the non-existence of a finite critical time for linear viscoelastic columns, *J Aeron Sciences*, **10**, p. 19.

Kohara, M, et al (1984). Thermal-stress-free package for flip-chip devices, 34th Elect Comp Conf, 388–393.

Kotlowitz, R W and Engelmaier, W (1986). Impact of lead compliance on the solder attachment reliability of leaded surface mounted devices, Int Elect Pack Soc Conf, 841–865.

Kreibel, F and Lochmann, S (1981). Deformations and mechanical stresses in soft soldered solid state circuits chips (in German), Feingeraetetechnik, **30**, (10), 452–454.

Kubik, E C and Li, T P L (1982). Thermal shock and temperature cycling effects on solder joints of hermetic chip carriers mounted on copper thick films, *Int J Hybrid Microelect*, **5**, (2), 314–321.

Kumar, A H and Miller, L F (1978). Stress-reduced conductive vias, *IBM Techn Discl Bull*, **21**, (1), p. 144.

Kuraniski, M (1959). The behavior of sandwich structures involving stress, temperature and time dependent factors, in non-homogeneity in elasticity and plasticity, **26**, *Pergamon Press*, New York, NY, 323–338.

Lade, J K and Wild, R N (1983). Some factors affecting leadless chip carrier solder joint fatigue life, 28th Natl SAMPLE Symp, 1406-1414.

Lake, J K and Wild R N (1983). Some factors affecting leadless chip carrier solder joint fatigue life, *28-th Nat SAMPLE Symp Proc*, 1406–1414.

Landis, R C (1985). Load distribution analysis for plastic chip carriers under stress, 5th Ann Int Elect Pack Conf, 271–278.

Lang, G A, et al (1970). Thermal fatigue in silicon power devices, *IEEE Trans Elec Devices*, **ED-17**, 787–793.

Lau, J H and Rice, D W (1985). Solder joint fatigue in surface mount technology: State of the Art, Solid State Tech, **28**, (10), 91–104.

Lau, H H and Rice, D W (1986). Effects on interconnection geometry on mechanical responses of Surface sount component, *IEEE/CHMT Int Elect Manuf Tech Symp Proc*, 205–217.

Lau, J, et al (1987). Experimental analysis of SMT solder joints under mechanical fatigue, 37th Elect Comp Conf, 589–597.

Lawrence, J E (1966). Diffusion induced stress and lattice disorders in silicon, *J Electrochem Soc*, **113**, p. 819.

Lawrence, J E (1967). Solute diffusion in plastically deformed silicon crystals, *Brit J Appl Phys*, **18**, p. 405.

Lawson, L (1986). Thermal fatigue testing of 97Pb-3Sn solder, MS Thesis, Northwestern Univ, Evanston, IL.

Lessen, M (1956). Thermoelasticity and thermal shock, *Phys Solids*, **5**, 57–61.

Levine, E and Ordonez, J (1981). Analysis of thermal cycle fatigue damage in microsocket solder joints, *IEEE Trans, Comp, Hybrids & Manuf Tech*, **CHM7-4**, 515–519.

Lewis, T E and Adams, D L (1982). VLSI thermal management in cost driven systems, 32nd Elect Comp Conf.

Lichtenberg, L R (1985). Comparison of environmental thermal cycle tests on reflow soldered assemblies, *Test & Meas World*, **5** (1), 59–66.

Liliental-Weber, Z, et al (1987). The structure of GaAs/Si(211) heteroepitaxial layers, *MRS Symp Proc*, **91**, 91–98.

Liu, A T (1976). Linear elastic and elastoplastic stress analysis for adhesive lay joints, Ph.D Thesis, Univ of Illinois.

Loo, M C and Du, K (1985). Die attach of large dice with Ag/glass in multilayer package, 5th Ann Int Elect Pack Conf, 402–412.

LoVasco, F and Britman, D A, (1987). Personal communication to E. Suhir.

Luryi, S and Suhir, E (1986). New approach to the high quality epitaxial growth of lattice-mismatched materials, *Appl Phys Letters*, **49**, (3), 140–142.

Manko, H H (1967). Solders and soldering, *McGraw-Hill*, New York, NY.

Manson, S S (1961). Mechanical behavior of materials at elevated temperatures, *McGraw-Hill*, New York, NY, 419–476.

Manson, S S (1981). Thermal stress and low-cycle fatigue, *Robert E Krieger Publ Co*, Malabar, FL.

Mark, R (1977). Photoelastic analysis of microelectronic-component thermal stresses, *Exp Mech*, **17**, (4), 121–127.

Merrell, L J (1971). A methodology for analysis of fatigue in solder joints, *Sandia Report* **SC-RR-710326**.

Mikoshiba, H (1981). Stress-sensitive properties of silicon-gate MOS devices, **24** (3), 221–232.

Moghadam, F K (1983). Development of adhesive die attach technology in cerdip packages, Materials Issues, *Int J Hybrid Microelect*, **6**, (1), 79–87.

Moeschke B, and Wlodarsky W (1979). Analysis of the effects of mechanical stress on the properties of p-channel MOS structures, *Electron Technology*, **12** (1), 29-48, **2**, 37-58.

Mohler, J B (1971). Solder joints versus time and temperature, machine design, *3*, 84-87.

Morton, G A and Forgue, S V (1959). *Proc I R E*, **47**, p. 1607.

Natarajan, B and Bhattacharyya, B (1986). Die surface stresses in a molded plastic package, 36th Elect Comp Conf, 544-551.

Niranjan, V (1970). Bonded joints – a review for engineers, *UTIAS Rev*, **28**, Univ of Toronto.

Norris, K C and Landzberg, A H (1969). Reliability of controlled collapse interconnections, *IBM J Res Devel*, **13**, (3), 266-271.

Nowacki, W (1962). Thermoelasticity, Oxford-London-New York-Paris, *Pergamon Press*.

Nowakowski, M F and Villella, F (1971). Thermal excusion can cause bond problems, 9th Ann Reliability Conf, *IEEE Catalog* **71-C-9-PHY**, Las Vegas, NV, 172-177.

Ojalvo, I U and Eidlnoff, H L (1978). Bond thickness effects upon stresses in single lap adhesive joints, *AIAA J*, **16**, (3), p. 204.

Okikawa, S, et al (1983). Stress analysis of passivation film cracks for plastic molded LSI caused by thermal stress, Int Symp for Testing and Failure Analysis, 275-280.

Okikawa, S, et al (1984). Stress analysis of poor gold-silicon die attachment for LSI's, Int Symp Testing & Failure Analysis, 180-189.

Olsen, G H and Ettenberg, M (1977). Calculated stresses in multilayered heteroepitaxial structures, *J Appl Physics*, **48** (6), 2543-2547.

Pahoja, M H (1972). Stress analysis of an adhesive lap joint subjected to tension, shear force and bending moments, it T&AM Rep **361**, Univ of Illinois, 2543-2547.

Parkus, H (1958). Stationare warmespanungen, wien, *Springer Verlag*.

Parkus, H (1959). Unstationare warmespanungen, wien, *Springer Verlag*.

Parkus, H (1968). Thermoelasticity. *Blaisdell Publ Co*, Waltham, MA.

Penning, P (1958). Generation of imperfections in germanium crystals by thermal strain, *Philips Res Rpts*, **13**, 79–97.

Pirvics, J (1974). Two-dimensional displacement stress distributions in adhesive bonded composite structures, *J Adhesion*, **6**, (3), 207–228.

Prager, W (1959). Introduction to plasticity, *Addison-Wesley Publishing Co*, Reading, MA.

Rabotnov, Y N (1969). Creep problems in structural members, *North Holland Publishing Co*, Elsevier, New York, NY.

Rao, K B S, et al (1985). On the failure conditions in strain-controlled low cycle fatigue, *Int J Fatigue*, **7**, 141-147.

Rathore, H S, et al (1973). Fatigue behavior of solders used in flip-chip technology, *J Test & Eval*, **1**.

Ravi, K V and Philofsky, E M (1972). Reliability improvements of wire bonds subjected to fatigue stresses, 10th Ann Reliability Physics Conf, **IEEE Catalog 72** (CH0628-8-PHY), 143–148.

Raynor, D and Skelton, R P (1985). The onset of cracking and failure criteria in high temperature fatigue testing, *Elsevier Appl Science Publ*, London & New York, 143–145.

Reinhart, F K and Logan, R A (1973). Interface stress of $Al_xGa_{1-x}As$-GaAs layer structures, *J Appl Phys*, **44**, (7), 3171–3175.

Reshey, J (1985). Thick-film thermal stress induced by lid-seal epoxy, Int Symp Microelectr, 513–519.

Riemer, D E (1983). The effect of thick-film materials on substrate breakage during processing, *Int J Hybrid Microelectr*, **6**, (1), 599–602.

Riemer, D E and Saulsberry, C W (1984). Power cycling of ceramic chip carriers on ceramic substrates (An Analysis of Test Results), 1984 Int Symp Microelectr.

Riney, T D (1961). Residual thermoelastic stresses in bonded silicon wafers, *J of Appl Physics*, **32**, (3), 454–460.

Ristic, S and Cvekic, V (1978). Some effects of localized stress on silicon planar transistors, *Phys Status Solidi*, **50** (1), 153–157.

Röll, K (1976). Analysis of stress and strain distribution in thin films and subtrates, *J Appl Physics*, **47**, (7), 3224–3229.

Sawyer, J W and Cooper, P A (1980). Analysis and test of bonded single lap joints with preformed adherends, *AIAA/ASME/AS-*

CE/Ash 21st Structures, Structural Dynamics Mat Conf, Seattle, WA.

Schafft, H A (1972). Testing and fabrication of wire-bond electrical connections - a comprehensive survey, *NBS Tech Note* **726**.

Schroen, W H, et al (1981). Reliability tests and stress in plastic integrated circuits, 19th Ann Rel Physics Symp, 81–87.

Seraphim, D P and Feinberg, I (1981) Electronic packaging evolution in IBM, *IBM J Res Develop*, **25**, (5), p. 617.

Shah, H J and Kelly, J H (1970). Effect of dwell time on thermal cycling of the flip-chip joint, Int Hybrid Microelectr Symp, 341–346.

Shaw, D W (1987). Epitaxial GaAs-on-Si: progress and potential applications, MRS Spring Meeting, **91**, *MRS Symp Proc* **91**, 15–30).

Sherry, W M, et al (1985). Analytical and experimental analysis of LCCC solder joint fatigue life, 35th Elect Comp Conf, 81–90.

Sherry, W M and Hall, P M (1986). Materials, structures and mechanics of solder joints for surface-mount microelectronics, 3rd Int Conf Interconnection Tech Electr, Fellback, West Germany.

Shukla, R K and Mencinger, N P (1985). A critical review of VLSI die-attachment in high reliability applications, *Solid State Tech*, 67–74.

Sinha, A K, et al (1978). Thermal stresses and cracking resistance of dielectric films (SiN, Si_3N_4 and SiO_2) on Si substrates, *J Appl Phys*, **49** (4), 2423–2426.

Smith, W K and Robinson, A T (1958). Strength of metals undergoing rapid heating, in short-time high temperature testing, Am Soc Metals, Novelty, OH, 5–35.

Soloman, H D (1986). Creep, strain rate sensitivity and low cycle fatigue of 60/40 solder, brazing and soldering, **11**.

Soloman, H D (1986). Fatigue of 60/40 solder, 36th Elect Comp Conf, 622–629.

Spencer, J (1981). Calculating stress and mobility in silicon chips using strain gage measurements, *TI Semi Eng*, **1**, 34–37.

Spencer, J L, et al (1981). New quantitative measurement of IC stress introduced by plastic packages, 19th Ann Reliability Physics Symp, 74–80.

Stone, D, et al (1985). Kinetics of cavity growth in solder joints during thermal cycling, Electr Pack Mat Science, *MRS Symp Proc*, 117–122.

Stone, D, et al (1985). The effects of service and material variables on the fatigue behavior of solder joints during the thermal cycle, 35th Elect Comp Conf, 46–51.

Stone, D, et al (1986). Mechanics of damage accumulation in solders during thermal fatigue, 36th Elect Comp Conf, 630–635.

Strinivas, S, (1975). Analysis of bonded joints, **NASA TN-D-7855**.

Suhir, E (1986). Stresses in adhesively bonded bi-material assemblies used in electronic packaging, Elect Pack Mat Science-II, **MRS Symp Proc**, 133–138.

Suhir, E (1986). Stresses in bi-metal thermostats, *Trans, ASME, J Appl Mech*, **53**, (3), 657–660.

Suhir, E (1986). Calculated thermally induced stresses in adhesively bonded and soldered assemblies, Int Symp Microelectr, 383–392.

Suhir, E (1987). Stresses in multilayered thin films on a thick substrate, Heteroepitacy-on-Silicon II, *MRS Symp Proc*, **91**, 73–80.

Suhir, E (1987). Die attachment design and its influence on thermal stresses in the die and the attachment, 37th Elect Comp Conf, 508–517.

Taira, S (1973). Relationship between thermal fatigue and low-cycle fatigue at elevated temperatures, *ASTM STP 520*, ASTM, 80–101.

Takenaka, T, et al (1984). Reliability of flip- chip interconnections, Int Symp Microelectr.

Taylor, T C (1959). Thermally-induced cracking in the fabrication of semiconductor devices, *IRE Trans Electron Devices*, **6**, 299–310.

Taylor, T C and Yuan, F L (1962). Thermal stress and fracture in shear constrained semiconductor device structures, *IRE Trans Electron Devices*, **ED-9**, 303–308.

Taylor, J R and Pedder, D J (1982). Joint strength and thermal fatigue in chip carrier assembly, *Int J Hybrid Microelect*, **5**, (2), 209–214.

Thamm, F (1976). Stress distribution in lap joints with partially thinned adherends, *J Adhesion*, **7**, p. 301.

Thomas, R E (1985). Stress-induced deformation of aluminum metallization in plastic molded semiconductor devices, 35th Elect Comp Conf, 37–45.

Thongcharoen, V (1977). Optimization of bonded joints by finite element and photoelasticity methods, Ph.D Thesis, Iowa State Univ.

Thwaites, C J (1982). Soft-soldering handbook, Int Tin Research Institute, England.

Thwaites, C J (1986). Some metallurgical aspects of SMD technology, brazing & soldering, 38–42.

Tien, J K, et al (1985). Flow and fracture at elevated temperatures, ASM Metals Park, OH, 179–214.

Timoshenko, S (1925). Analysis of bi-metal thermostats, *J Optical Soc Am*, **11**, 233–255.

Timoshenko, S P and Goodier, J N (1970). Theory of elasticity, *McGraw-Hill*, New York, NY.

Usell, R J and Smiley, S A (1981). Experimental and mathematical determination of mechanical strains within plastic IC packages and their effect on devices during environmental tests, 19th Ann Rel Physics Symp, 65–73.

Van Kessel, C J M, et al (1983). The quality of die attachment and its relationship to stress and vertical die-cracking, 33rd Elect Comp Conf, 237–244.

Vaynman, S and Fine, M E (1987). Prediction of fatigue life of lead-based low tin solder, 37th Elect Comp Conf, 598–603.

Vidano, R P, et al (1987). Mechanical stress reliability factors for packaging GaAs, MMIC and LSIC components, 37th Elect Comp Conf, 74–83.

Vilms, J and Kerps D (1982). Simple stress formula for multilayered thin films on a thick substrate, *J Appl Phys*, **53** (3).

Volkerson, O (1938). Die nietkraftverteilung in zubeanspruchten mit kenstanten laschonquerschnitten, Luftfahrtforschung, 15, p. 41.

Wahl, A M (1944). Analysis of the Valverde thermostat, *J Appl Mech*, **A-183-A-189**.

Waine, C A, et al (1982). Thermal fatigue failure of solder joints in printed circuit assemblies, 5th Europ Conf in Electromechanics, Copenhagen, 231–235.

Waller, D L et al (1983). Mount thermal and thermal stress performance, 33rd Elect Comp Conf, 534–545.

Weller, D L et al (1983). Analysis of surface mount thermal and thermal stress performance, *IEEE Trans, Components, Hybrids and Manuf Tech*, **CHMT-6**, (3), 257–266.

Wild, R N (1972). Fatigue properties of solders, welding & research, **37**, (11), 521–526.

Wild, R N (1975). Some fatigue properties of solders and solder joints, IBM Federal Systems Dir, Oswego, NY.

Wilkinson, W C (1983). HCC-compatible substrate, NORTHCON'83, Elect Show & Convention, 14/1/1-3.

Wilson, E A (1981). An analytical investigaion into the strain distribution within IC bumps, *Int J Hybrid Microelect*, 233–239.

Wilson, E A, and Anderson, E P (1983). An analytical investigation into geometric influence on integrated circuit bump strain, Elect Comp Conf.

Witt, G R (1972). Some effects of strain and temperature on the resistance of thin gold-glass cermet films, *Thin Solid Films*, **13**, 109–115.

Wong, C P (1986). Integrated circuit encapsulants - polymers in electronics, *Encyclop Polymer Science & Eng*, **5**, *John Wiley & Sons*, New York, NY.

Yamada, Y et al (1976). Low stress design of flip-chip technology for Si-on-Si multichip modulus, *Int Symp Elect Pack Proc*.

Zeyfang, R (1971). Stresses and strains in a plate bonded to a substrate: semiconductor devices, *Solid State Elec*, **14**, 1035–1039.

Zommer, N D et al (1976). Reliability and thermal impedance studies in soft soldered power transistors, *IEEE Trans Elect Devices*, **ED-23**, 843–850.

Chapter 6

BIBLIOGRAPHY OF HEAT TRANSFER IN ELECTRONIC EQUIPMENT - 1986

R. E. Simons
IBM Corporation
Poughkeepsie, New York

1 INTRODUCTION

Electronic equipment in all forms has rapidly permeated virtually all facets of human endeavor; from applications as mundane as miniature FM receivers to sophisticated computers that support vital health, business, and defense systems. Although, in some instances failure of electronic equipment results only in minor inconvenience, in a growing number of applications such failures can result in a major disruption of vital services and can even take on life threatening dimensions. As a consequence, efforts to improve the reliability of electronic equipment have become increasingly important. A major element in assuring the reliability of any electronic component is, of course, the satisfactory control of its operating temperature.

The introduction of microelectronics in the form of LSI (Large Scale Integration) and VLSI (Very Large Scale Integration) technologies, offering respectively 100-1000 and 1000-8000 gates per chip, has generally allowed the volume occupied by electronic equipment to decrease. The demand for faster circuits and increased capacity, however, has led to both increases in the power dissipation of each circuit, and an increase in the number of circuits per unit volume. The result has been increasing power densities at the chip, module, and system levels of packaging, and a need for continuous improvements in the methods of heat removal.

This demand for improved thermal control of electronic components and equipment, coupled with the need for more efficient methods of heat removal, has resulted in increased attention to the important task of thermal management. The level of importance that thermal management or heat transfer in electronic equipment has attained, is attested to by the number of papers published each year on the various aspects of the subject. A recent bibliography prepared by Antonetti and Simons covering the period from 1970-1984, listed 237 publications related to various aspects of heat transfer in electronic equipment. It was noted that the majority of the papers listed was published in the last three years of the period covered. An annotated bibliography of 92 papers and articles published in 1986 related to heat transfer in electronic equipment is provided here. For the reader also interested in the broad area of heat transfer and its more fundamental aspects, an excellent review of the 1985 literature was provided by Eckert et al.

The material presented here was obtained from technical journals, trade magazines, and conference proceedings. The journals and magazines surveyed included the International Journal of Heat and Mass Transfer, the IEEE Transactions on Components, Hybrids, and Manufacturing Technology, the IEEE Transactions on Nuclear Science, the Applied Mechanics Review, the Microelectronics Journal, International Communications in Heat and Mass Transfer, Heat Transfer Engineering, Hybrid Circuits, EDN, Electronics, Electronic Products, Electronic Packaging and Production, Machine Design, and Mechanical Engineering. Among the conferences and symposiums offering sessions devoted to the cooling of electronic equipment were the IEEE 36th Electronic Components Conference (Seattle -

May 5-7), the AIAA/ASME 4th Thermophysics and Heat Transfer Conference (Boston - June 2-4), the ISHM International Electronics Packaging Symposium on Microelectronics (Atlanta - October 6-8), the IEPS 6th Annual International Electronics Packaging Conference (San Diego - November 17-19), and the SEMI-THERM3 Semiconductor Thermal and Temperature Measurement Symposium (Scottsdale - December 8-11).

The sections that follow provide groupings of the papers and articles in 11 topic areas covering State-Of-The-Art, Materials, Thermal Contact Resistance, Air Cooling, Liquid Cooling, Thermal Measurements and Sensing, Thermal Analysis Techniques, Cooling Techniques, Cooling Devices/Hardware, and Thermal Assembly Processes. The content or contribution of each paper is briefly highlighted.

2 STATE-OF-THE-ART

Several papers reported on the state-of-the-art of heat transfer in electronic equipment, providing an overview of both past and recent developments and applications. Among the sessions devoted to heat transfer in electronic equipment at the AIAA/ASME 4th Thermophysics and Heat Transfer Conference was a session providing a historical overview of electronics thermal control. In a keynote paper for the session, Bergles discussed selected aspects of cooling technology for electrical apparatus and electronic devices over the past 60 years, emphasizing the past 35 years. Examples of electronic cooling applications ranged from 1935 vintage air-cooled and water-cooled high power vacuum tubes, to the recent IBM Thermal Conduction Module, that was cited as an example of a synergistic approach that introduced thermal considerations at the earliest stages of design. It was noted that electronic cooling technology has evolved to meet the challenges of microminiaturization, and that heat transfer considerations are now an integral part of the design procedure for microelectronic systems. Although the other papers in this session were originally presented as early as 1942, they are nonetheless worthy of inclusion here as landmark papers in the evolution of the art and science of electronic cooling. Each of the papers was presented in its original form, along with a contemporary assessment of its relevance.

The problem of cooling modern-day computers was discussed in an article by Oktay, Hannemann, and Bar-Cohen. They noted that a two-orders-of-magnitude increase in power dissipation has occurred over the past two decades resulting in a rise in chip level heat flux to as high as 10^6 W/m^2. Considerations in the design of cooling systems for both cost-performance and high-performance electronics were discussed, and several high-performance electronic modules were described which are cooled via either air-cooled heat sinks or water-cooled cold plates.

In a keynote paper at the 8th International Heat Transfer Conference, Chu provided an overview and assessment of the most recent developments in heat transfer and thermal packaging for electronic systems. Examples were given of the use of heat sinks, turbulators, parallel air flow, impinging air flow, and water-cooled interboard heat exchangers to enhance air cooling capability. It was noted that it was necessary to develop conduction cooled packages using water-cooled cold plates at the module level, in order to support the packaging density and heat flux levels existing in some of today's machines. Examples were given of packages of this type that were developed by IBM, NEC, Honeywell, and Sperry Univac. Direct liquid cooling using either forced convection or boiling was also discussed. The phenomena of temperature overshoot prior to boiling with fluorocarbon liquids was identified as a problem yet to be resolved.

A detailed review of technology and research topics related to the thermal management of electronic equipment was provided by Nakayama. This paper examined several categories of electronic packaging attempting to define relevant research topics in each, provide a summary of the present state of knowledge, and suggest some possible future trends. Considering the current trend of increasing power, the question was raised in the concluding remarks as to how much power consumption is justified for information processing? It was concluded nonetheless, that "the ground laid down in the era of bipolar technology will continue to serve as the base of more sophisticated heat transfer engineering in the era of new technologies."

Under the sponsorship of the National Science Foundation and Purdue University, a group of 58 industry and university researchers in electronic cooling met at Andover, MA, to conduct a 3-day (June 4-6) workshop on Research Needs In Electronic Cooling. As reported by Incropera, the objectives of the workshop were to: 1) critically assess past and current activities, 2) identify areas of future research which have high potential for advancing electronic cooling technologies, and 3) strengthen university/industry interfaces. Areas covered by the presentations and discussions included air cooling, liquid cooling (single and two phase), internal resistances, computer modeling, future challenges, and other topics. Conclusions and recommendations were formulated in each of these areas and may be found in the proceedings. It is appropriate to note here, however, that cooling requirements for the early 1990s were projected to be in excess of $100 \ W/cm^2$ at the chip level, $25 \ W/cm^2$ at the module level, and $10 \ W/cm^2$ on a printed circuit board (PCB).

3 MATERIALS

The choice of materials is an important consideration in the thermal design or evaluation of an electronics package. Before heat can be removed from any external surface it must be transported to the surface by the process of thermal conduction. In many electronic modules containing integrated circuit chips, thermal conduction through the substrate is the principal path for heat flow from the chip to the external module surface. Consequently, the thermal conductivity of the substrate material is a property of principal interest in thermal design. Of course, other material characteristics such as thermal coefficient of expansion, electrical insulation capacity, low dielectric constant, and mechanical strength are equally important in terms of other design considerations.

Although, alumina (Al_2O_3) has long been the dominant substrate material, in recent years there has been growing interest in silicon carbide (SiC) and aluminum nitride (AlN) as substrate materials. During the past year, two papers discussing AlN substrates were published. Kurokawa et al reported on the development of an AlN substrate with a thermal conductivity (at room temperature) of $230 \ W/mK$ compared to $20 \ W/mK$ for Al_2O_3 and $270 \ W/mK$ for SiC. The paper discussed the process for fabricating the green sheet and substrate, and the metallization techniques used to provide electrical paths. The results of infrared temperature measurements were cited as a demonstration of the excellent heat dissipation capability of an AlN substrate in comparison with Al_2O_3 Kuramoto et al also reported on the development of an AlN ceramic substrate. The thermal conductivity of this substrate material was reported to be 140 W/mK at room temperature.

Because of its high thermal conductivity (260 W/mK), beryllia (BeO) is another ceramic that is of interest for application as a substrate material on which to mount microelectronic chips. An article by Hill discussed the use of beryllia in such applications. For one multilayer package under development it was shown that for "a BeO chip carrier containing a 1 square cm chip mounted on a glass-epoxy board, the thermal resistance between the silicon chip and a heat sink mounted to the other side of the board is four times (2.5 vs. 10.0 $^\circ C/W$) lower than that for the alumina equivalent." It was noted that, although, beryllia may typically cost 5 to 10 times as much in IC applications as an equivalent alumina part, the additional cooling hardware required with alumina to achieve the same overall thermal resistance could more than offset the higher cost of a beryllia part. The safety considerations required in handling beryllia were also discussed. It was concluded that years of experience with the material have shown that where the potential for hazard does exist, risks can be controlled with the proper safeguards.

In another paper, SinghDeo et al discussed a dual-in-line (DIP) package that utilized a copper alloy base and cap. Comparing results for the metal DIP package and a conventional package with an alumina substrate, reductions in junction-case thermal resistance of as much as 27% were shown for the metal DIP package. It was also explained, that the high thermal coefficient of expansion (TCE) of the copper alloy forming the base and lid of the package configuration, necessitated the development of matched TCE materials for the leadframe and glass seal. A soft solder die attach system was developed to keep residual stress in the die attach area within acceptable levels.

The results of a study assessing heatspreader thermal performance, for molding compounds and leadframes of different thermal conductivities, were published in a paper by Aghazadeh and Natarajan. Both a 48 lead plastic DIP with bilateral symmetry, and a 68 lead square plastic leaded chip carrier package were considered under natural convection cooling conditions. It was found that the thermal resistance for both package configurations was relatively insensitive to the heatspreader thickness and thermal conductivity. For the conditions of the study, it was concluded that the key criterion for choosing a heatspreader material is its mechanical match with the molding compound. It was noted that, although, package thermal performance is sensitive to changes in heatspreader area, increasing the heatspreader area beyond a certain size does not reduce thermal resistance significantly.

4 THERMAL CONTACT RESISTANCE

In virtually all electronic packages and equipment, there are physical, mechanical, or metallurgical joints between different materials or parts of the structure across which heat is conducted. The thermal resistance across such interfaces is often referred to as a thermal contact resistance, and is generally a complex function of the geometric, physical, and thermal properties of the contacting solids and the interstitial substance.

As noted in the previous section, in a majority of integrated circuit packages today, the principal heat flow path is from the chip into the substrate to which it is bonded. It is therefore, important that the thermal contact or "die bond" resistance between the chip and its substrate carrier be as small in magnitude as possible. Voids or cracks which impede the flow of heat can develop as a result of imperfections in the manufacturing processes, or after thermal and power cycling. This problem was considered in papers by Khory and Abuaf and Kadambi.

In the paper by Khory, a simple analytical method is proposed to estimate the effect of die bond voids on die bond thermal resistance. The increase in thermal resistance may be used to estimate the increase in junction temperature, and the effect on mean time to failure (MTTF) based upon an electromigration failure mechanism. Conversely, the analytic method may be used to "set an upper threshold on the thermal resistance of the device, or a reasonable estimate of the minimum expected device reliability."

In order to model the difference between hot and cold voids in the solder joints of a power semiconductor chip package, Abuaf and Kadambi performed a coupled finite element analysis of the electrical and thermal fields in the silicon chip. Whereas a cold void configuration will interrupt both heat flow and electrical current, a hot void will interrupt only the heat flow. Based upon a 100 W power dissipation, calculations showed that the maximum thermal resistance of the hot void is 5.5 times the resistance without a void, and the maximum thermal resistance of the cold void was 3.5 times that without a void. It was also shown that hot voids can result in greater temperature gradients across the top surface of the chip, which could affect the accuracy of the forward voltage drop technique for measuring thermal resistance.

In the case of some large multi-chip modules with many chips, the substrate can no longer be used effectively as the principal thermal conduction path. For such situations, methods must be developed to remove heat from the back of the chip. Thus, the thermal contact resistance between the chip and the alternate heat flow path is matter of some importance.

One example of heat removal from the back of the chip is provided by the thermal conduction module (TCM) used in some IBM computers. TCM modules are filled with helium gas, and heat is removed from the chips by spring-loaded metal pistons with a crowned-tip. Heat transfer from the chip occurs both through the solid contact and through the helium filled gap. Considering this cooling approach, Fisher and Yovanovich studied the piston/chip contact and developed a theoretical model to predict the effect of a surface layer on the thermal constriction resistance of the piston/chip contact. Comparisons of model results with experimental results showed good agreement for light loads in the elastic range. Eid and Antonetti also considered the thermal contact resistance between an aluminum piston and a silicon chip. In another paper Antonetti and Eid provided an overview of thermal contact resistance theory, with an illustrative example of a thermal contact resistance calculation procedure. The techniques and apparatus used to measure thermal contact resistance on test specimens were also discussed.

Paal and Pease investigated the thermal contact conductance achieved using microcapillary attachment of a chip to a substrate containing microcapillary channels filled with a fluid such as silicone oil. The paper describes the method for making the channels and presents the results of performance modeling and experiments. An average thermal contact conductance of 21.7 W/cm^2 $^{\circ}C$ was measured.

Another area where thermal contact resistance can be important is in the attachment of a heat sink to a module. This is particularly true if a non-metallurgical joint is used. A paper by Hultmark et al described the use of a silicone rubber adhesive to attach the ceramic cap to the aluminum heatsink for a multi-chip module. The process developed offered a cost savings over the previous method of soldering the heatsink to the cap, with only an increase in R_{ext} of 0.03 $^{\circ}C/W$ above the value of 0.23 $^{\circ}C/W$ achieved with a solder-joint.

5 AIR COOLING

For many years air has been the most widely used cooling medium in electronic equipment. The principal advantages of cooling with air are of course its ready availability and ease of application. For applications with low power and low packaging densities, natural convection cooling with air is often used. The advantage offered by natural convection cooling is that no fan or blower is required to move the air. The bulk of air cooling papers reported here are in the natural convection area.

Johnson provided an evaluation of heat transfer correlations pertaining to natural convection cooling of vertical electronic card arrays. The accuracy of correlations by Aung et al, Wirtz and Stutzman, Bar-Cohen and Rohsenow, Birnbrier, and Coyne, was compared using experimental data obtained at AT&T Bell Laboratories. Recommendations are made pertaining to the use of the correlations in terms of Rayleigh Number. An article by Chung discussed the use of natural convection correlations in a PC program to optimize package size and minimize temperature rise. The program input is discussed and example output shown. Although, no program listing is provided in the article, the reader is provided information on how to obtain a copy.

More recent experiments on natural convection cooling of a vertical array of simulated electronic modules were reported by Ortega and Moffat. It was shown that "under conditions in which the fluid in a buoyancy induced channel flow is well-mixed for a majority of the channel length, the heat transfer from any element in the array is driven by the globally induced channel flow and is not affected by local buoyancy effects." It was noted that, as long as the local ratio of Gr/Re^2 for the cube array is less than about 0.3, the hydrodynamics are indistinguishable from pressure-driven forced convection. A comparison of local heat transfer coefficient measured in both buoyancy induced and forced channel flow confirmed this. Further heat transfer results for buoyancy driven flows were reported in another paper by Moffat and Ortega. It was stated that, "forced convection heat transfer data can be used confidently for the design of free convection channels for electronics cooling, as long as the induced flow can be predicted."

Although natural convection cooling performance will suffer, in some applications such as desk top computers, it is sometimes necessary to mount PCBs horizontally. Heat transfer under this condition was experimentally investigated and reported by Krane and Phillips. A variety of experiments was performed to determine the effects on the maximum surface temperature by systematically varying the enclosure geometry for two values of the modified Rayleigh number. A list of recommendations/guidelines was provided for thermal designers. A paper by Grawoig discussed an empirical model to thermally characterize electronic components mounted on a horizontal card cooled by natural convection in an enclosed environment.

For those packaging situations requiring more cooling capability or a greater degree of control, forced convection heat transfer with air may be the method of choice. Forced air cooling offers higher heat transfer coefficients than natural convection, is relatively insensitive to packaging orientation, is less affected by heating of the air by upstream components, and is generally more easily controlled. The last two papers in this section address forced convection cooling with air.

Graham and Witzman reported the results of a combined analytical and experimental study to develop an analytic expression that could be used to predict external thermal resistance over a range of package sizes and air velocities. The model developed takes into account convective heat transfer from the external surfaces of the component and the printed circuit board on which it is mounted. The correlation given incorporates a multiplication constant C that must be experimentally derived for a specific package family. Once

C is determined, the equation can be used to extrapolate thermal resistance data over a range of device sizes, board thermal conductivities, board thicknesses and air velocities. Typical values of C, and a comparison between measured and calculated values of $< R_{j-a}$ were shown.

Santos and Souza-Mendes reported the results of experiments using the napthalene mass transfer technique to obtain data on heat transfer coefficients. The work reported investigated the effect of geometrical non-uniformity on heat transfer and pressure drop in an array of modules in a flat rectangular duct. A module whose height was twice that of the other modules was positioned at various locations in the array. It was concluded that "in many instances it might be advisable to locate a high-power component by the side of a taller component, since, in the present study, heat transfer enhancements of the order 50 percent were measured at this relative location."

6 LIQUID COOLING

Considerable improvements in heat transfer capability can be obtained when liquid cooling is used instead of air cooling. Liquid cooling applications can generally be categorized as indirect or direct. An indirect scheme is usually considered to be one in which the liquid does not contact the electronics or the packages housing the electronics. Since there is no electrical contact, water can be used as the coolant, taking advantage of its superior thermophysical properties. In recent years, however, there has been interest in the possibility of flowing water directly through channels within the electronic component. Several papers in this section address this possibility.

Koh and Colony considered the scheme proposed by Tuckerman and Pease in 1981. In this scheme forced convection cooling of the chip is provided by pumping water through microchannels in the back of the chip. Recognizing the similarity with transpiration cooling through a porous material, the method of analysis was modified to extend the analysis to the microchannels. The results of calculations were presented and it was stated that the analysis showed a design capable of dissipating over 1000 W/cm^2 while maintaining the structure at temperatures suitable for electronics operation. This heat flux is consistent with that claimed by Tuckerman and Pease in their original work, and more recently by Pease in a paper discussing advances in packaging for VLSI systems. Similar microchannel chip cooling techniques were considered by Reichl and Sasaki and Kishimoto. Reichl proposed the use of integral etched microcooling channels on a silicon chip attached to a silicon substrate as part of a silicon-on-silicon package. Sasaki and Kishimoto addressed the problem of the pressure drop associated with liquid flow through the microgroove channels, and found the optimum channel widths to maximize allowable power density for a fixed pressure drop.

Kishimoto and Ohsaki, on the other hand, disclosed a packaging concept utilizing a stack of chip-carrying substrates, with very fine water cooling channels in each substrate. The cooling design and package fabrication were discussed, along with the results of thermal experiments.

In direct liquid cooling the electronics or packages within which they housed are in direct and intimate contact with the cooling liquid. Direct liquid cooling offers the opportunity of removing heat directly from the chip, or surface of the heat generating device, and transferring it into the liquid with virtually no intervening thermal conduction resistance paths. Although, most of the papers in this area involved a phase-change heat transfer process, Incropera et al described experiments to obtain forced convection heat transfer data. The data was obtained for both a single heat source, and an in-line, four row array of twelve heat sources flush mounted to one side of a flow channel. Experimental results were presented for both water and the fluorochemical liquid FC-77.

Two papers discussed thermosyphon or heat pipe type cooling schemes. Kiewra and Wayner described a small scale thermosyphon that was designed to cool a disc heat source. The results of experiments using decane and hexane as the evaporative coolant were discussed. A paper by Kromann et al described what was termed an integral heat pipe concept for cooling multiple chips on a substrate. Experiments were conducted using a 16 chip test vehicle. An overall internal package thermal resistance below 1.0 $^{\circ}C/W$ was reported.

The remaining papers in this section dealt with the the boiling of a dielectric coolant as means to cool integrated circuit chips. Park and Bergles reported the results of experiments to investigate the effects of the size of simulated integrated circuit chips on boiling and critical heat flux with R-113. Three types of incipient boiling temperature overshoot were observed and discussed. Correlations were provided for estimating critical heat flux. In another paper, Park and Bergles discussed the boiling heat transfer characteristics of various heat sinks attached to simulated microelectronic chips immersed in a pool of R-113. Ma and Bergles disclosed the results of a study investigating the use of a submerged jet to enhance nucleate boiling cooling of a simulated chip using R-113. Heat fluxes in excess of 10^6 W/m^2 were reported. Although boiling offers high heat flux cooling capability, its use in commercial electronic equipment has in part been deterred by the problem of wall temperature overshoot at boiling incipience. In the final paper of this section Bar-Cohen and Simon examined the literature on incipience superheat excursions and discussed mechanisms that may be responsible. A comparison was made between observed values for pool and flow boiling, and a proposed analytical approximation for the superheat excursion was provided.

7 THERMAL MEASUREMENTS AND SENSING

Temperature measurement of prototype hardware has been a long-standing method used to confirm the adequacy of a thermal design. With the continuing trend towards higher power and packaging densities, the need for thermal design verification and accurate temperature measurements will assume even greater importance. This is attested to by the relatively recently established (1984) SEMI-THERM Semiconductor Thermal and Temperature Measurement Symposiums for the exchange and dissemination of information in this area.

Although thermocouples may be used to measure the temperature at some point on the external surface of a package, the so-called junction or chip temperature within the package is usually measured using temperature sensitive electrical parameters of the chip itself. Shanker and Lambertson considered the sources of error in package thermal resistance measurements using the voltage drop across the substrate diode at constant-current as a temperature sensitive parameter. Of the various factors considered, "measurement delay time" was found to have a particularly strong effect on the test results. This time difference between when the forward-bias heating voltage is removed, and when the reverse-bias voltage at the measurement current is sampled, leads to readings of junction temperature which are too low. Other factors which affected the measurement of thermal resistance included the choice of test socket or heat sink, and the junction temperature used. A paper by Anderman et al described the use of special thermal test chips with temperature sensing diodes, to perform evaluation of large die attach integrity.

In some instances it may be necessary to take a large number of temperature measurements off a chip or a number of chips in a short period of time. In order to quickly perform the necessary measurements and handle the large amount of data involved, computer controlled automated test systems may be used. Stazak et al described a system designed and built at the University of Arizona to thermally characterize packaged VLSI chips. Similarly, Ciminera described a PC driven system designed and built to thermally characterize IBM TCMs.

In addition to accurate measurement of temperature, it is equally important that the cooling condition under which the experiment or test of an electronic component is performed be controlled and repeatable. Tests of air-cooled components are often performed in "wind tunnels" to provide control and repeatability. Hayward and Van Andel described the design and air flow characterization of a "wind tunnel" to be used to obtain thermal resistance measurements on forced air-cooled packages.

8   THERMAL ANALYSIS TECHNIQUES

As the interval between successive generations of electronic equipment decreases, less time is available for performing prototype tests on next generation equipment. As a result increased dependence is placed on the ability to design and predict performance using analytic and numerical models. This would seem to be confirmed by the twenty papers in this section which represent almost one-quarter of all the papers presented here.

A number of papers discussed the use of finite element methods (FEM) to analyze component cooling problems. Bocchi described how the finite element method is used at the Rome Air Development Center employing a Very High Speed Integrated Circuit (VHSIC) as an example. Jennings and Rubinsky discussed the use of a finite element model to characterize the electrical and thermal performance of coplanar Josephson junctions. In another paper Agonafer and Simons reported on a thermal optimization study of the IBM TCM piston design using a CAEDS FEM model. A comparison was made with the results obtained using a one-dimensional analytic model that was also discussed in the paper. Bonnifait et al described the use of a FEM program to model integrated circuit packages, and how the results are then used to model a printed circuit board containing many components. Similarly, Pinto and Mikic described a "building blocks" strategy that merged the results of FEM computations at different package levels to provide "a methodology which does not require simultaneous handling of all details from the board to the chip." Kale and Kim described the development of a numerical thermal model based on finite element theory to provide transient analysis of junctions of a power transistor. The model was implemented as a PC program. Bullister et al proposed the use of a spectral element method for the Navier-Stokes and energy equations for solution of heat transfer problems in the cooling of electronic equipment. It was explained that the spectral element method is a high-order technique that combines the geometric flexibility of finite element schemes with the rapid convergence of spectral methods.

A number of other papers discussed finite difference and network analysis techniques for numerical solution of thermal problems. Two of the papers reported on the results of numerical heat transfer studies that utilized the control-volume, finite-difference procedure developed by Patankar. Moffat et al addressed the problem of determining the relative contributions of convection and substrate conduction to heat transfer from discrete isothermal sources (e.g. integrated circuit chips or modules) mounted to one wall of a parallel plate channel. Fully developed turbulent flow was assumed for fluids with Prandtl Numbers of 0.7, 7, and 25. Habchi and Acharya studied laminar mixed convection of air in a vertical channel with a partial blockage on one wall, such as might be caused by an electronic module on a vertical PCB. In a magazine article Eid discussed the use of a PC spreadsheet program to solve finite-difference heat transfer equations. The analysis of a water-cooled cold plate for an electronic module was used as an example to illustrate the procedure. Allen and Brace discussed the role of a preprocessor in performing numerical thermal computations, and described the use of one named DELTA T. Ellison and Patelzick described their use of a coupled pressure/air flow, temperature/heat flow model to conduct a numerical thermal analysis of a forced air-cooled power supply. A paper by Sharma described a network approach combined with a "Monte Carlo" sampling technique to model the thermal performance of multi-chip microelectronic modules. It was shown that the alternative "worst case" approach leads to overly pessimistic results.

Lejannou et al discussed the development of a computer code to predict the air flow and temperature field in electronic switching systems. A three step development program entailing the development of a numerical method, single row experiments, and a five shelf mock-up was discussed.

Each of the remaining papers in this section addressed the thermal analysis of electronic components by use of analytic techniques. The first three papers presented analytic techniques suitable for the evaluation of temperature distribution of integrated circuit chips or packages on substrates or PCBs. Simeza and Yovanovich described the use of the Boundary Integral Equation Method to estimate the case temperature of packaged ICs mounted on PCBs. Another paper by Negus and Yovanovich used "an approximate analytical solution to a fundamental basis problem for heat conduction in a convectively-cooled microelectronic circuit board," to study the effects of chip spacing. Based upon the results it was suggested that "in many practical applications, little gain in thermal performance of the board is realized for chip center-to-center spacings greater than twice the chip width." The paper by Pinto and Mikic discussed analytic methods of solution that can include the effects of multiple layers and anisotropic thermal conductivity, and in some cases transient behavior. A paper by Culham et al presented the development of an analytical method to predict the temperature distribution in a circular annular fin with distributed sources. Two examples were given to illustrate the application of the technique to electronic cooling problems. The last two papers of this section by Smith et al and Lee and Palisoc illustrated the application of analytic techniques to the thermal analysis of GaAs integrated circuit devices.

9 COOLING TECHNIQUES

Specific cooling techniques in electronic equipment applications can yield equipment sizes that vary from the very small to the very large. But, whether the application involves a single natural convection-cooled heat producing component, or a large, liquid-cooled, mainframe computer, the objective remains the same: to allow thermal energy to flow from source to sink within the constraints of the specified temperature, space, and often cost. This section list papers which describe a variety of cooling applications.

At the component level, Alli et al presented the results of a study to address the thermal characteristics of a class of plastic surface-mount packages known as small outline transistor (SOT) packages. It was noted that, at modest temperatures, about 80% of the heat from an SOT flows into the board, and that the coupling between the package and the board is much stronger than has been experienced in the past with DIPs. A paper by Limbach and Cater disclosed the results of a study to determine the maximum allowable power densities for leadless chip carriers on printed wiring boards. The study considered 5 forced air cooled designs and two conduction cooled designs.

The cooling technique at a slightly higher level of packaging is discussed in a trade article by Chin on PC cooling It is explained that PCs use a combination of active and passive cooling. Although fan noise is given as an important design consideration, it is noted that "most designers agree that fans are even more crucial as more computers pack in added memory and processing power." The application of CMOS technology with its lower power dissipation is cited as the way that manufacturers of "lap-top" portables have solved the thermal management problem.

The cooling of digital transmission equipment is discussed in a paper by Carvalho et al. The unit was described as a slim rack with a number of boxes or modules stacked vertically. Within each module is a stack of horizontal printed circuit boards. The development of a theoretical model to predict temperatures, and the experimental investigation to verify the model are discussed.

A number of papers discussed techniques for cooling FASTBUS electronics. FASTBUS is the standard in the physics community for high-speed data acquisition bus hardware/software to meet the requirements for the next generation of large-scale physics experiments. A paper by Haldeman et al described the cooling system for a large FASTBUS data manipulation system. The system consisted of thirty-five FASTBUS racks housing up to three FASTBUS crates each, with each crate dissipating up to 2.5 KW. The crates contained printed circuit boards on which were mounted plastic and ceramic dual-in-line packages. Although these components were air-cooled, water-cooled interboard heat exchangers were used between the crates to remove heat from the air circulating in the closed rack. Tanaka et al provided a similar description of the rack cooling system for the VENUS detector used to support colliding beam experiments. In this system air was distributed in parallel to FASTBUS crates stacked in racks on the first floor. The exhaust air was then used to cool power supply racks dirctly above on the second floor. In another paper Chato and Golliher discussed advanced cooling techniques for FASTBUS electronics. The techniques considered included impinging air jets and cold plate conduction cooling.

Each of the remaining papers in this section addressed some aspect of mainframe computer cooling applications. Emoto et al described the first level package developed for the HITACHI M680 mainframe. Information was provided comparing the single chip module package which utilized a SiC substrate, with earlier packages that utilized an alumina substrate. A further description of the M680 packaging and cooling technology was provided by Kobayashi et al . The first level (module) and second level (PCB) packages were described, along with some aspects of the system cooling. In another paper Timko and Plucinski traced the evolution of air flow control in IBM intermediate mainframe computers. Areas discussed include air flow network analysis, air leakage/recirculation, and air flow sensing.

Although, today a number of computers use indirect liquid cooling (e.g. IBM 3080/3090, NEC SX-2, and FUJITSU FACOM-780), only one computer uses direct liquid cooling. As described by Danielson et al the CRAY-2 is cooled by flowing FC-77 fluorocarbon coolant directly through circuit modules consisting of stacks of 8 interconnected circuit boards. According to Danielson these circuit board assemblies are arranged in 14 module columns within a sealed 155 gallon tank. It was also noted that the "auxiliary cooling equipment is actually much larger than the computer itself, with two cabinets housing pumps and chilled-water heat exchangers."

10   COOLING DEVICES/HARDWARE

Various types of devices and hardware are utilized to enhance and control cooling capability in electronic equipment. Fans and blowers are used to provide and control the flow of cooling air. Heat sinks and heat pipes are used to transport heat more effectively to a sink. Mechanical or thermoelectric refrigeration devices are used to pump heat or provide sink temperatures below ambient. The papers included in this section describe the use of some of these devices and hardware.

A fan unit designed to provide cooling air flow in racks housing FASTBUS electronics and power supplies was discussed in a paper by Frisch. The fan assembly consisted of six fans, each capable of delivering 180 scfm at a static pressure of 0.3 in of H_2O. The requirements and basis for selecting the fans were discussed. Data on air velocity distribution at various levels in the rack was provided.

A trade article by Travis discussed some of the heat sinks and thermoelectric devices that are available today to help control the temperature of semiconductor components, and provided some cost and vendor information. The use of thermoelectric coolers was also discussed in a paper by Wedal et al. This paper described the thermal analysis of a laser diode cooling design that utilized thermoelectric modules to maintain a constant temperature under varying load and sink temperatures. The alternative schemes considered were also discussed.

Several papers discussed the use of heat pipes in electronic cooling applications. Scott and Tanzer addressed the application of heat pipes to conduction cooled avionic packages employing Very High Scale Integration Chip (VHSIC) technology. The paper included descriptions of the application, along with test data showing the increase in allowable power expected using heat pipes. An article by Basiulis and Minning also discussed the use of heat pipes for cooling avionic packages. Embedded heat pipe and vapor chamber concepts for cooling printed wire boards were described and compared. The effect that the use of heat pipes can have in improving component reliability was also discussed. A paper by Peterson discussed the analysis and computer model of a bellows type heat pipe for cooling electronic components. Papers by Murase et al described heat pipe heat sink designs for cooling thyristors and other semiconductor components.

A paper by Husted and Kuzmin discussed the testing and evaluation of line replaceable modules (LRM) for use in packaging military electronics. Electronic components are mounted on the web of the LRM which also provides a thermal conduction path to a cooling sink at two edges of the web. The paper focused on determining the interrelationship of web thickness to other design parameters.

11   THERMAL ASSEMBLY PROCESSES

Although not directly related to electronic equipment cooling, the area of thermal assembly processes warrants inclusion here. Many of the steps and processes in the manufacture and assembly of semiconductor chips and packages require controlled application of heat and temperature. Failure to achieve the requisite degree of control may be reflected in poor quality and degraded reliability.

The four papers included here all addressed solder processes for surface mount technology (SMT). The two papers by Cox discussed both the vapor phase and infrared reflow processes. In the first paper, Cox explained that the thermal impact of reflow soldering on electrical components is a growing concern. Surface mount components are subjected to higher temperatures and higher heat transfer rates, because the electrical device and the solder joint are much closer together and have less thermal mass to absorb the energy. This paper included heat transfer theory and examples of temperature profiles for the two processes. Cox concluded that IR reflow soldering is effective for surface mount components, but that precise control is required to dry and reflow solder without thermal damage to the components. In another paper Hutchins and King also discussed vapor phase and infrared reflow processes. They concluded that both types of processes can be used to produce reliable SMT assemblies with a high yield. Lichtenberg and Brown discussed an experimental investigation to study the effects of conveyor speed, zonal gas flow rates, and zonal power inputs on the infrared solder reflow process. Average component temperature histories are shown for surface mount ICs, metal cans and capacitors for different lamp conditions.

12 REFERENCES

12.1 State-of-the-Art

Antonetti, V W, and Simons, R E (1985). Bibliography of heat transfer in electronic equipment, *IEEE Trans Components, Hybrids, Manuf Tech*, **CHMT-8**, (2), 289–295.

Eckert, E R, Goldstein, R J, Pfender, E, Ibele, W E, Ramsey, J W, Simon, T W, Decker, N A, Kuehn, T H, Lee, H O, and Girshick, S L, (1986). Heat transfer - a review of 1985 literature, *J Heat and Mass Transfer*, **29**, (12), 1767–1842.

Bergles, A E (1986). The evolution Of cooling technology for electrical, electronic, and microelectronic equipment, *ASME HTD*, **57**, 1–9.

Landis, F (1986). W. Elenbaas' paper on heat dissipation of parallel plates by free convection, *ASME HTD* **57**, 11–21.

Gardner, K A (1986). Efficiency of extended surface, *ASME HTD* **57**, 23–33.

Kraus, A D (1986). An appraisal of Gardner's pioneering extended surface effort, *ASME HTD* **57**, 35–39.

London, A L (1986). Air coolers for high-power vacuum tubes, *ASME HTD* **57**, 41–50.

Rohsenow, W M (1986). A method of correlating heat-transfer data for surface boiling of liquids, *ASME HTD* **57**, 51–60.

Oktay, S, Hannemann, R, and Bar-Cohen, A (1986). High heat from a small package, *Mech Eng*, 36–42.

Chu, R C (1986). Heat transfer in electronic systems, *Proc 8th Intl Heat Transfer Conf*, **1**, 293–305.

Nakayama, W (1986). Thermal management of electronic equipment: a review of technology and research topics, *Appl Mech Rev*, **39**, (12), 1847–1868.

Incropera, F P (editor) (1986). *Proc NSF/Purdue sponsored workshop on Research Needs in Electronic Cooling*, Andover, MA.

12.2   Materials

Kurokawa, Y, Hamaguchi, H, Shimada, Y, Utsumi, K, Takamizawa, H, Kamata, T, and Noguchi, S (1985). Development of highly thermal conductive AlN substrate by green sheet technology, *Proc 36th Electronics Components Conf*, 412–418, Seattle, WA.

Kuramoto, N, Taniguchi, H, and Aso, I (1986). Translucent AlN ceramic substrate, *IEEE Trans Components, Hybrids, Manuf Tech*, **CHMT-9**, (4), 386–390.

Hill, B, and Beryllia, A (1986). Packages for PC board heat dissipation, *Electrionics*, **32**, (3), 51–53.

SinghDeo, N N, Cherukuri, S C, and Butt, S H (1986). Metal DIP with superior thermal conductivity, *6th Annu Intl Electronic Packaging Society (IEPS) Conf Proc*, 615–619.

Aghazadeh, M, and Natarajan, B (1986). Parametric study of heat-spreader thermal performance in 48 lead plastic DIP's and 68 lead plastic leaded chip carriers, *IEEE Trans Components, Hybrids, Manuf Tech*, **CHMT-9**, (4).

12.3   Thermal Contact Resistance

Khory, N F (1986). The impact of die bond voids in power semiconductor devices on thermal resistance and long term reliability (an analytical approach), *Proc 1986 Intl Symp Microelectronics*, 275–280.

Abuaf, N and Kadambi, V (1986). Thermal investigation of power chip packages: Effect of voids and cracks, *6th Annu Intl Electronic Packaging Society (IEPS) Conf Proc*, 821–828.

Fisher, N J, and Yovanovich, M M (1986). Thermal constriction resistance Of sphere/layered flat contacts: Theory and experiments, *ASME HTD*, **57**, 219–229.

Eid, J C, and Antonetti, V W (1986). Small scale thermal contact resistance of aluminum against silicon, *Proc 8th Intl Heat Transfer Conf*, **2**, 659–664.

Antonetti, V W, and Eid, J C (1986). Measuring thermal contact resistance, *6th Annu Intl Electronic Packaging Society (IEPS) Conf Proc*, 502–516.

Paal, A, and Pease, R F (1986). Extending microcapillary attachment to rough surfaces, *Proc IEEE Intl Electronic Manuf Tech Symp*, 169–172, San Francisco, CA.

Hultmark, E, Horvath, J L, Trestman-Matts, A, and Park, C (1986). The use of silicone adhesives in microelectronic packaging, *6th Annu Intl Electronic Packaging Society (IEPS) Conf Proc*, 340–348.

12.4   Air Cooling

Johnson, C E (1986). Evaluation of correlations for natural convection cooling of electronic equipment, *ASME HTD*, **57**, 103–111.

Chung, J (1987). Maximizing heat transfer from PCBs, *Machine Design*, 87–92.

Ortega, A, and Moffat, R J (1986). Buoyancy induced convection in a non- uniformly heated array of cubical elements on a vertical channel wall, *ASME HTD*, **57**, 123–134.

Moffat, R J, and Ortega, A (1986). Buoyancy induced forced convection, *ASME-HTD*, **57**, 135–144.

Krane, R J, and Phillips, T J (1986). Natural convection cooling of a horizontally-oriented component board mounted in a low-aspect- ratio enclosure, *ASME HTD*, **57**, 113–122.

Grawoig, B C (1986). Thermal modeling of natural convection surface mount devices, *6th Annu Intl Electronic Packaging Society (IEPS) Conf Proc*, 809–820.

Graham, K, and Witzman, S (1986). Experimental correlations for thermal resistance in forced convective heat transfer from PCB's, *Proc SEMI-THERM3 Semiconductor Thermal and Temp Measurement Symp*.

Santos, W F N, and Souza Mendes, P R (1986). Heat transfer and pressure drop experiments in aircooled electronic component arrays, *AIAA Paper* 86-1301.

12.5   Liquid Cooling

Koh, J C, and Colony, R (1986). Heat transfer of microstructures for integrated circuits, *Intl Comm Heat Mass Transfer*, **13**, 89–98.

Pease, R F (1986). Advances in packaging for VLSI systems, *Proc Intl Electron Devices Meeting*, IEEE Cat No 86CH2381-2, 480–483.

Reichl, H (1986). Silicon substrates for chip interconnection, *Hybrid Circuits*, **11**, 5–7.

Sasaki, S, and Kishimoto, T (1986). Optimal structure for microgrooved cooling fin for high-power LSI devices, *Electronics Letters*, **22**, (25), 1332–1333.

Kishimoto, T, and Ohsaki, T (1986). VLSI packaging technique using liquid-cooled channels, *IEEE Trans Components, Hybrids, Manuf Tech*, **CHMT-9**, (4), 328–335.

Incropera, F P, Kerby, J S, Moffatt, D F, and Ramadhyani, S (1986). Convection heat transfer from discrete heat sources in a rectangular channel, *Intl J Heat and Mass Transfer*, **29**, (7), 1051–1057.

Kiewra, E W, and Wayner, P C (1986). A small scale thermosyphon for the immersion cooling of a disc heat source, *ASME HTD*, **57**, 77–82.

Kromann, G B, Hannemann, R J, and Fox, L R (1986). Two-phase internal cooling technique for electronic packages, *ASME HTD*, **57**, 61–65.

Park, K A, and Bergles, A E (1986). Effects of size of simulated microelectronic chips on boiling and critical heat flux, *ASME HTD*, **57**, 95–102.

Park, K A, and Bergles, A E (1986). Boiling heat transfer characteristics of simulated microelectronic chips with detachable heat sinks, *Proc 8th Intl Heat Transfer Conf*, 4, 2099–2104.

Ma, C F, and Bergles, A E (1986). Jet impingement nucleate boiling, *Intl J Heat and Mass Transfer*, **29**, (8), 1095–1101.

Bar-Cohen, A, and Simon, T W (1986). Wall superheat excursions in the boiling incipience of dielectric fluids, *ASME HTD*, **57**, 83–94.

12.6 Thermal Measurements and Sensing

Shanker, B J, and Lambertson, R T (1986). Sources of error in package thermal resistance measurements, *Proc 36th Electronics Components Conf*, 161–168, Seattle, WA.

Anderman, J, Tustaniwsky, J, and Usell, R (1986). Die attach evaluation using test chips containing localized temperature measurement diodes, *IEEE Trans Components, Hybrids, Manuf Tech*, **CHMT-9**, (4), 410–415.

Staszak, Z J, Cooke, B, Shope, D, Fahey, W J, Prince, J L, and Nowrozi, D (1986). Thermal characterization system for packaged VLSI chips, *Proc SEMI-THERM3 Semiconductor Thermal and Temp Measurement Symp*.

Ciminera, J R (1986). Automated thermal measurement of a high-density integrated circuit package, *Proc SEMI-THERM3 Semiconductor Thermal and Temp Measurement Symp*.

Hayward, J D, and Van Andel, R (1986). How fast is it blowing?: Design and characterization of a laboratory-scale wind tunnel, *Proc SEMI-THERM3 Semiconductor Thermal and Temp Measurement Symp*.

12.7 Thermal Analysis Techniques

Bocchi, W J (1986). Finite element engineering analyses applied to microelectronics, *Proc IEEE 1986 Natl Aerospace and Electronics Conf*, 4, 1150–1153, Dayton, OH.

Jennings, R, and Rubinsky, B (1986). A finite element study of a coplanar electrode Josephson junction with respect to electric potential and temperature, *Intl Comm Heat Mass Transfer*, **13**, 55–65.

Agonafer, D, and Simons, R E (1986). Optimization trade-off analyses on a thermal conduction module package using CAEDS, *ASME HTD*, **57**, 231–237.

Bonnifait, M, and Cadre, M (1986). Thermal simulations for electronic components using finite elements and nodal networks, *ASME HTD*, **57**, 183–188.

Pinto, E J, and Mikic, B B (1986). Methodology for evaluation of temperature and stress fields in substrates and integrated circuit chips, *ASME HTD*, **57**, 209–217.

Kale, V, and Kim, J (1986). Thermal analysis for transient power applications, *Proc 1986 Intl Symp Electronics*, 368–374.

Bullister, E T, Karniadakis, G E, Mikic, B B and Patera, A T (1986). A spectral element method applied to the cooling of electronic components, *ASME HTD*. **57**, 153–160.

Moffat, D F, Ramadhyani, S, and Incropera, F P (1986). Conjugate heat transfer from wall embedded sources in turbulent channel flow, *ASME HTD*, **57**, 177–182.

Habchi, S, and Acharya, S (1986). Laminar mixed convection in a partially blocked vertical channel, *ASME HTD*, **57**, 189–197.

Eid, J C (1986). Spreadsheets for thermal analysis, *Machine Design*, 121–125.

Allen, E, and Brace, M (1986). Delta T: A simplified approach to thermal analysis, *6th Annu Intl Electronic Packaging Society (IEPS) Conf Proc*, 791–808.

Ellison, G N, and Patelzick, D L (1986). The thermal design of a forced-air-cooled power supply, *6th Annu Intl Electronic Packaging Society (IEPS) Conf Proc*, 829–838.

Sharma, A (1986). Statistical thermal modeling of multi-chip modules, *Proc 36th Electronic Components Conf*, 138–140, Seattle, WA.

Lejannou, J P, Cadre, M, Latrobe, A, and Viault, A (1986). Thermal field prediction in electronic equipment, *Proc 8th Intl Heat Transfer Conf*, **6**, 2983–2988.

Simeza, L M, and Yovanovich, M M (1986). Application of BIEM to thermal analysis of multiple sources on PCBs, *ASME HTD*, **57**, 161–166.

Negus, K J, and Yovanovich, M M (1986). Thermal analysis and optimization of convectively-cooled microelectronic circuit boards, *ASME HTD*, **57**, 167–176.

Pinto, E J, and Mikic, B B (1986). Temperature prediction on substrates and integrated circuit chips, *ASME HTD*, **57**, 199–207.

Culham, J R, Yovanovich, M M, and Graham, K D (1986). Thermal analysis of circular annular configurations with distributed heat sources, *6th Annu Intl Electronic Packaging Society (IEPS) Conf Proc*, 517–530.

Smith, D H, Frase, A, and O'Neil, J (1986). Measurement and prediction of operating temperatures for GaAs ICs, *Proc SEMI-THERM3 Semiconductor Thermal And Temp Measurement Symp.*

Lee, C C, and Palisoc, A L (1986). Thermal analysis of GaAs integrated circuit devices, *GaAs IC Symp Tech Digest 1986*, IEEE Cat 86CH2372-1, 115–118.

12.8 Cooling Applications

Alli, M M, Mahalingham, M, and Andrews, J (1986). Thermal characteristics of plastic small outline transistor (SOT) packages, *IEEE Trans Components, Hybrids, Manuf Tech*, **CHMT-9**, (4), 353–363.

Limbach, R H, and Cater, R A (1986). Characterization of thermal systems containing leadless chip carriers, *Proc SEMI-THERM3 Semiconductor Thermal and Temp Measurement Symp*, **CHMT-9**, (4), 353–363.

Chin, S (1986). Keeping the PC cool, *Electronic Products*, 47–50.

Carvalho, R D, Goldstein, L, and Milanez, L F (1986). Heat transfer analysis of digital transmission equipment with horizontally arranged printed circuit boards, *ASME HTD*, **57**, 145–152.

Haldeman, M, Holm, S, Merkel, B (1986). FASTBUS cooling, *IEEE Trans Nucl Sci*, **33**, (1), 825–830.

Tanaka, R (1986). Venus rack cooling system, *IEEE Trans Nucl Sci*, **33**, (1), 813–834.

Chato, J C, and Golliher, E L (1986). Advanced cooling techniques for FASTBUS electronics, *IEEE Trans Nucl Sci*, **33**, (1), 841–844.

Emoto, Y, Tsuchiya, M, Ogiwara, S, Sasaki, T, Kobayashi, F, and Otsuka, K (1986). A high performance package approach on 1st level packaging for main frame, *Proc 36th Electronic Components Conf*, 564–570, Seattle, WA.

Kobayashi, F, Anzai, A, Yamada, M, Takahashi, A, Yamazaki, S, and Toda, G (1986). Packaging technologies for the ultrahigh-speed processor Hitachi M-680H/M-682H, *Proc 36th Electronic Components Conf*, 571–577, Seattle, WA.

Timko, N, and Plucinski, M D (1986). A history of thermal control in IBM intermediate mainframes, *6th Annu Intl Electronic Packaging Society (IEPS) Conf Proc*, 489–500.

Danielson, R D, Krajewski, N, and Brost, J (1986). Cooling a superfast computer, electronic packaging and production, 44–45.

12.9   Cooling Devices/Hardware

Travis, W (1986). Heat-removal devices hold semiconductors within operating ranges, *EDN*, 77–82.

Wedal, R K, Gal, G, and Horine, G A (1986). Thermal analysis of a laser diode design, *ASME HTD*, **57**, 239–243.

Scott, G W, and Tanzer, H J (1986). Evaluation of heat pipes for conduction cooled level II avionic packages, *ASME HTD*, **57**, 67–75.

Basiulis, A, and Minning, C P (1986). Improving circuit reliability with heat pipes, *Electronic Packaging and Production*, **26**, (9), 104–105.

Peterson, G P (1986). Analytical development and computer modeling of a bellows type heat pipe for the cooling of electronic components, *ASME Paper* 86-WA/HT-69.

Murase, T, Endo, S, and Koizumi, T (1986). Heat pipe cooling system POWERKICKER for power semiconductor devices, *Proc 12th Intl PCI Conf*, 366–372.

Murase, T, Koizumi, T, and Ishida, S (1986). Heat pipe heat sink HEAT KICKER for cooling of semiconductors, *Proc 12th Intl PCI Conf*, 373–382.

Husted, M D, and Kuzmin, G F (1986). Line replaceable modules (LRM): The next generation military packaging, *6th Annu Intl Electronic Packaging Society (IEPS) Conf Proc*, 759–766.

12.10   Thermal Assembly Processes

Cox, N R (1986). Keeping your IC chips cooler using infrared soldering, *Microelectronics J*, **17**, (4), 27–34.

Cox, N R, and Lamp I R (1986). Reflow soldering of surface mount devices, *6th Annu Intl Electronic Packaging Society (IEPS) Conf Proc*, 176–183.

Hutchins, C, and King, S (1986). An empirical study of the surface mount solder reflow process, *6th Annu Intl Electronic Packaging Society (IEPS) Conf Proc*, 659–671.

Lichtenberg, L R, and Brown, L L (1986). Component thermal management in infrared solder reflow, *Proc 1986 Intl Symp Microelectronics*, 895–901.

Name Index

Subject Index

condenser, 55
condenser blockage, 285
conductive castings, 82
conductivity ratio, 82
conforming rough surface models
    contact conductance correla-
        tion, 90-93
    dimensionless joint conductance,
        95-98
    example of correlation, 92-93,
        96-98
    gap conductance, 93-954
    general, 89-104
    experimental verification, 98-
        104
conservation equations, 130, 199
constriction parameter, 110, 120
constriction parameter correction
    factor, 106, 107, 110
constriction resistance, 87, 92, 109,
    121
contact coefficient of heat trans-
    fer, 87
contact
    conductance
        brazing improvement, 119
        coated joint in vacuum, 105
        dimensionless, 92
        enhancement of, 104-121
        gap correlation, 90-93
        general, 87, 90, 92
        of coated joint, 110
        parameters, 91
        solder improvement, 119
        thermal, 101, 103
    hardness, 98, 101
    pressure, 18, 29, 30, 88, 89,
        101
    spot density, 91
    spot radius, 110
    resistance
        analysis assumptions, 89
        mechanical joints, 86-89

thermal, 87
contact resistance correlations
    constriction parameter coef-
        ficients, 85
    dimensionless constriction re-
        sistance, 83
    microelectronic application, 84
convective flow interaction, 36
coolants
    air, 14
    water, 27
cooling systems
    coolant selection, 15
    environmental aspects, 9
    fabrication cost, 9
    power consumption, 9
    heat transfer paths in, 15
    research needs, 15
    types, 7
    strategy evaluation, 8
    reliability of, 8
    cross-disciplinary nature, 8
copper diaphragm, 27
copper membrane interface, 30
correlation coefficient, Vickers mi-
    crohardness, 99, 101
CPU, single board, 65
CRAY-2 computer, 46
creep
    failure, 341, 342
    stages of, 342
    thermal, 338, 339
critical dimension, 131
critical point, 49
cryogenic fluids, 288
cumulative pressure coefficient, 161

D

database
    establishment of, 12-15
    primary factors, 15
    thermal packaging, 2
deformation